Galilean Mechanics and Thermodynamics of Continua

Series Editor
Noël Challamel

Galilean Mechanics and Thermodynamics of Continua

Géry de Saxcé
Claude Vallée

iSTE

WILEY

First published 2016 in Great Britain and the United States by ISTE Ltd and John Wiley & Sons, Inc.

ISTE Ltd
27-37 St George's Road
London SW19 4EU
UK

www.iste.co.uk

John Wiley & Sons, Inc.
111 River Street
Hoboken, NJ 07030
USA

www.wiley.com

Library of Congress Control Number: 2015957652

British Library Cataloguing-in-Publication Data
A CIP record for this book is available from the British Library
ISBN 978-1-84821-642-6

Contents

Foreword

In this very well-written book, my colleagues and friends Géry de Saxcé and Claude Vallée present a general framework in which the laws of mechanics are formulated, which govern equilibria and motions of material bodies in space. They show that this framework is also convenient for the formulation of the laws of thermodynamics which involve, besides the notions of mass, velocity and acceleration already in use in mechanics, the notions of heat, entropy and temperature. To allow the readers to fully appreciate the originality and the interest of their work, I am going to briefly remind them of the evolution, from the 17th Century up to our time, of the ideas of scientists about the physical space and time and about material bodies' motions.

According to Isaac Newton's (1642–1727) ideas expressed in his famous book[1], physical time and space are two separate entities, both having an absolute character. Physical time can be mathematically represented by a straight line, spreading to infinity in both directions, endowed with a geometric structure which allows the comparison of the lengths of two time intervals, even when they are separated by several centuries of millenia. Consequently, the choice of a unit of time and an arbitrary date as an origin establishes a one-to-one correspondence between elements of the physical time and elements of the real line $\mathbb{R}$. As for the physical space, Newton identifies it with the three-dimensional space of Euclidean geometry; the choice of a unit of length allows its identification with (in modern mathematical language) a three-dimensional Euclidean affine space, in which we can measure distances and angles, and apply all theorems of Euclid's geometry. Each material body occupies, during its existence, a certain position in space, which may depend on time. It is at rest if the positions of all its material elements remain fixed along the time, and in motion in the opposite case. Its motion is described by all the curves,

1 Isaac Newton, *Philosophiae naturalis principia mathematica*, 1687.

drawn in the physical space and parametrized by the time, made by the successive positions of each of its material elements.

Newton was well aware of the fact that the position and the motion of a material body in the physical space are always appreciated relatively to the positions of other physical bodies used to determine a reference frame. He formulated the fundamental law of dynamics (which states that the acceleration of a punctual material body is equal to the quotient by its mass of the force which acts on it) for the absolute motion of a material body in the physical space. But immediately, he noticed that this fundamental law remains valid for the *relative motion* of the material body with respect to a reference frame whose absolute motion is a motion by translations at a constant velocity.

Newton's ideas about time, space and the mathematical description of motions of material bodies were soon criticized, notably by Gottfried Wilhelm Leibniz (1646–1716), who believed that the concept of an absolute space was useless, had no real existence and that the laws governing material bodies' motions should be formulated in a way involving only the relative position of each body with respect to all the other bodies. Unfortunately the mathematical concepts needed to translate Leibniz' ideas into a usable theory were not available at his time. Much later, Ernst Mach (1838–1916)[2], who thought that the inertia of a material body was due to the actions on it of all other material bodies present in the universe, criticized Newton's ideas about the absolute character of space and time, as well as the principles of inertia and equality of action and reaction. Mach's ideas could not be incorporated into a usable theory in mechanics, but they influenced Albert Einstein when he developed the theory of general relativity.

In spite of these criticisms, Newton's ideas about space, time and motion are, essentially, still in use nowadays in classical mechanics. Of course, the progress of astronomy has shown that nothing is at rest in the universe, leading many scientists to become doubtful about the existence of an absolute space. But they found a way to avoid the use of that concept: when Newton's laws of dynamics can be applied to the *relative motions* of material bodies with respect to some reference frame, that frame was said to be *inertial*, or *Galilean*. It is then easy to show that when a given reference frame R_1 is inertial, another reference frame R_2 is also inertial if and only if it moves relatively with respect to R_1 by translations at a constant velocity. Instead of assuming the existence of an absolute space, it is enough to assume the existence of *one* inertial reference frame, which implies the existence of *an infinite number* of such frames, each in relative motion by translations at a constant velocity with

2 Ernst Mach, *Die Mechanik in ihrer Entwicklung, Historisch-kritisch dargestellt*, 1883. First English translation by T. J. McCormack, under the title *The Science of Mechanics*, Chicago, 1893.

respect to each other. Mechanicians became accustomed to only considering relative motions of material bodies with respect to (preferentially approximately inertial) reference frames, and to use concepts (such as velocity and kinetic energy) which depend on the considered body and also on the reference frame with respect to which its motion is studied. The use of *fictitious forces* (centrifugal and Coriolis' forces) even made possible the use of non-inertial frames and the study, for example, of the relative motion of the *Foucault's pendulum* with respect to the Earth's reference frame (discussed in Chapter 3). By preventing any interrogation about the mathematical tools used to represent the physical space and time, the immoderate use of reference frames seems to have delayed the discovery of the special theory of relativity. I am going to recall the main steps of this discovery. Then, I will speak about that of the general theory of relativity.

At the beginning of the decade 1860–1869, James Clerk Maxwell (1831–1879) established the equations governing electromagnetic phenomena and introduced the concept of *field*. According to these equations, perturbations of an electromagnetic field propagate as waves at a finite velocity, which does not depend on the motion of the source of the perturbations and is the same in all directions of propagation. Observing that the numerical value of that velocity was close to that of light's velocity, Maxwell understood that light is an electromagnetic wave. In classical kinematics, a phenomenon can propagate at the same velocity in all directions only with respect to a particular reference frame. Physicists, who no more really believed in Newton's absolute space, then assumed the existence of a very subtle medium, filling empty space and impregnating all material bodies, in which the propagation of electromagnetic waves occurred. They called *luminiferous ether* that hypothetic medium. They thought that it was in a reference frame with respect to which the luminiferous ether is at rest that the relative velocity of light was the same in all directions. Under this assumption, careful measurements of the velocity of light in various directions, made at different dates at which the Earth's velocity on its orbit around the Sun takes different values, could detect the relative velocity of the Earth with respect to the luminiferous ether. These measurements were made around 1887 by Albert Abraham Michelson (1852–1931) and Edward William Morley (1838–1923). No perceptible velocity of the Earth with respect to the luminiferous ether could be detected. No satisfying explanation of this result was found until 1905.

Albert Einstein (1879–1955) offered[3], in 1905, a truly revolutionary explanation. He clearly understood that light's property to propagate at the same velocity in all directions, whatever the reference frame with respect to which its relative velocity is evaluated, is incompatible with the absolute character of the notion of simultaneity of

3 Albert Einstein, *Zur Electrodynamic bewegter Körper*, translated by M. Saha under the title *On the Electrodynamics of Moving Bodies*, Calcutta, 1920.

two events occurring at two different places in space. He proposed as a new principle, called the *Principle of Relativity*, the fact that all inertial reference frames are equivalent, for electromagnetic phenomena as well as for mechanical phenomena. He proposed as a second principle the fact that the light propagates at the same velocity in all directions, independently of the motion of its source and the inertial reference frame with respect to which that velocity is evaluated. On these two founding principles, giving up the absolute character of simultaneity of two events occurring at different places, therefore the concept of an absolute time, he succeed in building a coherent theory. The principle of relativity led him to give up the notions of absolute rest and absolute motion, and he clearly saw that in his new theory the concept of luminiferous ether was no more useful.

In the same year, Jules Henri Poincaré (1854–1912) published a Note in the *Comptes rendus de l'Académie* and a much longer paper[4] in which he introduced "local times" at which an event occurs, depending on that event *and on the frame in which that event is perceived by an observer* which, together with the three space coordinates of the place at which that event occurs, make a system of four coordinates of the event in space-time. He studied the transformation laws, which he called *Lorentz transformations* in honor of Hendrik Anton Lorentz (1853–1928), which give the four space-time coordinates of an event in some inertial reference frame as functions of the four space-time coordinates of *the same event* in *another inertial reference frame*. Lorentz transformations can also be seen as the transformation laws which give the four space-time coordinates of an event in some inertial reference frame as functions of the four space-time coordinates of *another event* in *the same inertial reference frame*. Poincaré proved that the set of all Lorentz transformations is a group and determined its invariants. He saw that the local times of the same event seen by two different observers are different, since he determined the formula which links these two local times. He also saw that Lorentz transformations are in agreement with the fact that light propagates at the same velocity in all directions, which does not depend on the reference frame with respect to which that velocity is evaluated, since the light velocity appears as an invariant of the group of all Lorentz transformations. But, he did not state as clearly as it was stated by Einstein the fact that the concept of an absolute time should be discarded, nor the fact that the concept of luminiferous ether is useless.

4 Henri Poincaré, *La Mécanique nouvelle*, the book brings together in a single volume the text of a conference in Lille of the Association française pour l'avancement des sciences in 1909, the note of 23 July 1905 entitled *Sur la dynamique de l'électron*, published by Rendiconti del Circolo matematico di Palermo **XXI** (1906) and a note by the Académie des Sciences, of the same title (15 June 1905, **CXL**, 1905, p. 1504); Gauthier-Villars, Paris, 1924; reprinted by Éditions Jacques Gabay, Paris, 1989.

Hermann Minkowski (1864–1909) precisely described, in 1908[5], the geometric structure of space-time in the special theory of relativity, which today bears his name: it is a four-dimensional affine space endowed with a pseudo-Euclidean scalar product of signature $(+, -, -, -)$. Lorentz transformations are linear automorphisms of the associated pseudo-Euclidean vector space which leave invariant that scalar product. Without giving its formal definition, Poincaré already considered this space-time and studied its geometric properties in his papers published in 1905. For this reason the group of affine transformations of the Minkowski space-time which preserve its structure today is called *Poincaré's group*.

As in classical mechanics, there exists in the theory of special relativity privileged reference frames: the inertial frames. These reference frames are global: each of them totally includes space and time. Einstein wanted to build a theory only using *local* reference frames, none of them being privileged. He also noticed that the special theory of relativity does not explain a disturbing fact: the equality of inertial and gravitational masses. This equality implies (and is equivalent to) the identity of nature between acceleration fields and gravitational fields: in a reference frame suitably accelerated, it is possible either to annihilate, or to create a gravitational field in a limited part of space-time. Einstein proposed to consider this fact as a principle, and called it the *equivalence principle*. To account for this principle, he had the brilliant idea of including gravitational fields into the geometric properties of space-time. The mass–energy equivalence, which he discovered while developing the special theory of relativity, led him to think that not only mass, but all forms of energy (for example, electromagnetic energy) must contribute to the creation of a gravitational field. With these ideas, he built a coherent mathematical theory, which he called the *general theory of relativity*, and published it in four successive papers in 1915[6].

In the general theory of relativity, space-time is no longer an affine space, as it is in classical mechanics and in the special theory of relativity: it is a four-dimensional differential manifold endowed, once a unit of time (or, equivalently, of length) is chosen, with a pseudo-Riemannian metric of signature $(+, -, -, -)$. It is no longer a frame with fixed geometric properties in which physical phenomena occur. In

5 H. Minkowski, Talk presented in Cologne on the 21th september 1908, published in the book by H. A. Lorentz, A. Einstein and H. Minkowski *Das Relativitätsprinzip; eine Sammlung von Abhandlungen*, B. G. Teubner, Leipzig, Berlin 1922. Analyzed in the book by René Dugas *Histoire de la Mécanique*, Éditions du Griffon, Neuchâtel, 1950, reprinted by Éditions Jacques Gabay, Paris, 1996, pp. 468–473.

6 Albert Einstein, *Fundamental Ideas of the General Theory of Relativity and the Application of this Theory in Astronomy*, *On the General Theory of Relativity*, *Explanation of the Perihelion Motion of Mercury from the General Theory of Relativity*, *The Field Equations of Gravitation*, Preussische Akademie der Wissenschaften, Sitzungsberichte, 1915 part 1 p. 315, part 2 pp. 778–786, 799–801, 831–839, 844–847.

Maxwell's theory of electromagnetism, an electromagnetic field in which an electrically charged particle is moving acts on the motion of that particle. Conversely, the motion of that electrically charged particle creates an electromagnetic field, therefore modifies the field in which its motion takes place, and acts on the motions of other electrically charged particles. A similar reciprocity exists in the theory of general relativity: the mass of a material body, and more generally any kind of energy, is acted upon by the gravitational field (included in the geometry of space-time) and that action affects its motion; conversely, that mass or energy participates in the creation of the gravitational field, therefore acts on the geometry of space-time. The notion of a straight line on which a particle moves at a constant velocity is no longer valid in general relativity: the world line of a material particle acted upon by gravitation only is a time-like geodesic of the Levi-Civita connection associated with the pseudo-Riemannian metric of space-time. There is no perfectly rigid body, nor instantaneous action of a body on another non-coincident body in general relativity space-time. Actions at a distance only occur by fields, which propagate at a velocity not exceeding the velocity of light, whose evolution must be determined together with the motions of material bodies on which they act. It makes rather cumbersome the practical use of general relativity theory, which explains why classical, non-relativistic mechanics keeps its usefulness when the mechanical phenomena under consideration only involve material bodies each of whose relative velocity with respect to the other bodies is very small compared to the velocity of light.

The great French geometer Élie Cartan (1869–1951) very soon understood that the mathematical tools and the ideas of general relativity, especially the concept of space-time, could be used advantageously in classical mechanics. In a two-part paper published in 1923 and 1924[7], he introduced and investigated the notion of an *affine connection* on a differential manifold and explained how that notion could be used to include gravitational forces in the geometry of space-time in the framework of classical mechanics. The notion of an affine connection has its sources in the works of Tullio Levi-Civita (1873–1941), Hermann Weyl (1885–1955) and Élie Cartan himself. It was generalized and made clearer by Charles Ehresmann (1905–1979). A smooth path in a differential manifold being given, an affine connection allows us to define the parallel transport of an affine frame of the tangent space at a point of that path toward the tangent spaces at all other points of the path. It offers, therefore, a way to identify all the spaces tangent at various points of the path to a single affine space, which can be used as a local model of the manifold in a neighborhood of the path. From 1929 until 1932, Élie Cartan had a regular correspondence with Albert

7 Élie Cartan, *Les variétés à connexion affine et la théorie de la relativité généralisée*, I et II, Ann. Ec. Norm. 40, 1923, pp. 325–342 and 41, 1924, pp. 1–25. Ces articles se trouvent aussi dans ses *œuvres complètes*, partie III 1, pp. 659–746 and 799–823. Éditions du CNRS, Paris, 1984.

Einstein[8] about the concept of parallel transport, which the latter wanted to use in a new theory in which electromagnetic fields would be included in the geometry of space-time in a way similar to that in which he included gravitational fields in that geometry.

In the four-dimensional space-time of classical mechanics considered by Élie Cartan, the time keeps an absolute character: to each element (called "event") in space-time corresponds a well-defined element of the time, the instant when that event occurs. The set at all events which occur at a given instant is a three-dimensional submanifold of space-time, called the *space at that instant*. By assuming that the space at each given instant is (once chosen an unit of length) an affine Euclidean three-dimensional space, we make valid the notion of a perfectly rigid solid body and usable all the theorems of Euclid's geometry, exactly as they are in usual classical mechanics. Instantaneous actions at a distance can also be considered in that space-time, which for these reasons is better suited for the treatment of problems usually encountered in classical mechanics than the special or the general theories of relativity. A reference frame in that space-time is determined by a three-dimensional body R (which may be material or conceptual, as for example the set of three straight lines which join the Sun's center to three distant stars) which remains approximately rigid during some time interval I. For each pair (t_1, t_2) of instants in I, there exists a unique isometry of the space at t_1 onto the space at t_2 which maps the position of the body R at t_1 onto its position at t_2. By using these isometries to identify between themselves all these affine Euclidean spaces, we obtain an "abstract" three-dimensional affine Euclidean space in which the body R is at rest. The part of space-time made by all events which occur at an instant in I can, therefore, be identified with the product of that abstract three-dimensional Euclidean space with the time interval I. To study the relative motion of a mechanical system with respect to the reference frame determined by the body R amounts to use that identification. The reference frame determined by the body R is inertial if, with that identification, the motion of a material point which is not submitted to any force occurs on a straight line at a constant velocity. This is the *principle of inertia* discovered by Galileo Galilei (1564–1642), later included by Newton in his fundamental laws of mechanics.

With the exception of Jean-Marie Souriau and his coworkers, not many scientists working in classical mechanics granted much interest to the ideas of Élie Cartan about the use of space-time in their field of research. However, inconsistencies appearing in some formulations of constitutive laws governing large deformations of material bodies, Walter Noll (born in 1925) formulated his *principle of material objectivity*, which he renamed later, in agreement with his former scientific advisor Clifford Ambrose Truesdell (1919–2000) *principle of material frame indifference*.

8 Élie Cartan and Albert Einstein, *Letters on absolute parallelism 1929–1932*, Princeton University Press and Académie Royale de Belgique, Princeton, 1979.

On the webpage[9] at Carnegie Mellon University presenting his recent, still unpublished works, he gives the following formulation of this principle, which applies to any physical system: *the constitutive laws governing the internal interactions between the parts of the system should not depend on whatever external frame of reference is used to describe them.* He then indicates several examples of application of this principle and writes: *"It is possible to make the principle of material frame indifference vacuously satisfied by using an intrinsic mathematical frame-work that does not use a frame-space at all when describing the internal interactions of a physical system.* The use of space-time, even in classical mechanics, is in my opinion a very good way to put this idea into practice. It is the approach used in this book by my colleague and friend Géry de Saxcé.

Charles-Michel MARLE
November 2015

9 http://www.math.cmu.edu/wn0g/

Introduction

General Relativity is not solely
a theory of gravitation which is reduced to the
prediction of tiny effects such as bending of light
or corrections to Mercurys orbital precession
but may be above all it is a consistent framework
for mechanics and physics of continua...

I.1. A geometrical viewpoint

"*Αγεωμετρητος μηδεις εισιτω*" ("Let none but geometers enter here"). According to the tradition, this phrase was inscribed above the entrance to Plato's academy. Because of the simplicity and beauty of its concepts, geometry was considered by Plato as essential preamble in training to acquire rigor. It is in this spirit that this book was written, setting the geometrical methods into the heart of the mechanics. This is precisely the philosophy of general relativity that is adopted here but restricted to the Galilean frame to describe phenomena for which the velocity of the light is so huge as it may be considered as infinite. This general point of view does not prevent allowing us occasionally short incursions into standard general relativity.

Mechanics is an experimental and theoretical science. Both of these aspects are indispensable. Even if this book is devoted to the modeling, we have to keep in mind that a mechanical theory makes sense only if its predictions agree with the experimental observations. Among the physical sciences, mechanics is certainly the oldest one and, precisely for this reason, it is the most mathematical one. It might also be said it is the most physical science among the mathematical ones. At the hinge between physics and mathematics, this book presents a new mathematical frame for continuum mechanics. In this sense, it may be considered as a part of applied mathematics but it also turns out to be what J.-J. Moreau called "Applied Mechanics to the Mathematics" in the sense that we revisit some pages of mathematics.

But why is mathematics needed to do mechanics? Of course, it is possible to do mechanics "with the hands" but mathematics is a language allowing us to describe the reality in a more accurate way. As J.-M. Souriau says in the "Grammaire de la Nature" [SOU 07]: "*Les chaussures sont un outil pour marcher; les mathématiques, un outil pour penser. On peut marcher sans chaussures, mais on va moins loin*" ("The shoes are a tool to walk; the mathematics, a tool to think. One can walk without shoes, but one goes less far").

I.2. Overview

Our aim is to present a unified approach of continuum mechanics not only for undergraduate, postgraduate and PhD students, but also for researchers and colleagues, without, however, being exhaustive. The sound ideas structuring mechanics are systematically emphasized and many topics are skimmed over, referring to technical works for more detailed developments. The presentation is progressive, inductive and bottom-up, from the basic subjects, at the Bachelor and Master degree levels, up to the most advanced topics and open questions, at the PhD degree level. Each degree level corresponds to part of the book, the latter providing a canvas for revisting the former two parts in which special comments and cross-references to the third part are indicated as "comments for experts". Useful mathematical definitions are recalled in the final chapter of each part.

I.2.1. *Part 1: particles and rigid bodies*

Except for Chapter 6, the first part corresponds to subjects taught at Bachelor degree level, needing only elementary mathematical tools of linear algebra, differential and integral calculus recalled in Chapter 7 at the end of the first part.

Chapter 1 is devoted to the modeling of the space-time of 4 dimensions and the principle of Galilean relativity. It is essential and must not be skipped. The Galilean transformations are coordinate changes preserving uniform straight motion, durations, distances and angles, and oriented volumes. The statements of the physical laws are postulated to be the same in all the coordinate systems deduced from each other by a Galilean transformation. This principle will be an Ariadne's thread all the way through this book.

The method used in the following four chapters is founded on a key object called a torsor which will be given for the continuous media of 1 and 5 dimensions. Chapter 2 deals with the statics of bodies. Introducing the force torsor, an object equipped with a force and a moment, we deduce the transport law of the moment in a natural way. Usual tools to study the equilibrium are the free body diagram, internal and external forces.

Chapter 3 is devoted to the dynamics of particles and gravitation. Tackling the dynamics is simply a matter of recovering an extra dimension, the time, leading to the dynamical torsor. The boost method reveals its components, the mass, the linear momentum, the passage and the angular momentum. After representing the rigid motions due to the Galilean coordinate systems, we model the Galilean gravitation, an object with two components, gravity and spinning, and we deduce the equation of motion. We state Newton's law of gravitation and solve the 2-body problem. We define the minimal properties expected from the other forces. As for application, we discuss Foucault's pendulum and model rocket thrust.

Chapter 4 applies the concepts of Chapter 2 to arches, slender bodies which, if they are seen from a long way off, can be considered as geometrically reduced to their mean line. Generalizing the methods developed previously, we obtain the local equilibrium equations of the arches and, using a frame moving along this line, a generalized corotational form of these equations. The concepts are illustrated by applications, a helical coil spring, a suspension bridge, a drilling riser and a cantilever beam.

Chapter 5 extends the tools developed in the previous chapters to study the dynamics of rigid bodies. The Lagrangian or material description is opposed to the Eulerian or spatial one. The body motion can be characterized by the co-torsor, an object equipped with a velocity and a spin. After introducing the mass-center, we construct the dynamical torsor and the kinetic energy of a body as extensive quantities. Next, we generalize the equation of motion to study the motion of the body around it. As for application, we present Poinsot's geometrical construction for free bodies and we deduce three integrals of the motion for a body with a contact point, i.e. Lagrange's top.

Chapter 6 is devoted to the calculus of variation which allows us to deduce from the minimum of a function, called the action, the equations of motion in a more abstract way than in Chapter 3. The principle of least action has over all a mnemonic value which allows deducing these laws in a consistent and systematic way. Such a principle presupposes that the Galilean gravitation is generated by a set of 4 potentials, not unique but defined *modulo* an arbitrary gauge function. We also introduce the Hamiltonian formalism and the canonical equations.

I.2.2. *Part 2: continuous media*

The second part corresponds to subjects taught at Master degree level, requiring more advanced mathematical tools of linear algebra and analysis such as partial derivative equations and tensorial calculus. In particular, if you are not familiar with the affine tensors which is of outstanding importance all throughuot the part, this would be a good time to consult Chapter 14 before tackling the present part.

Chapter 8 lays the foundations of the statics of continuous media of 3 dimensions by making our first move in the tensorial calculus and elasticity. Modeling the internal forces leads to the concept of the stress tensor based on Cauchy's tetrahedron theorem and obeying local equilibrium equations. Next, we generalize the concept of the torsor to a continuum. Usual three-dimensional (3D) bodies of which the behavior is represented by a stress torsor are called Cauchy's continua.

Chapter 9 tackles the elasticity and elementary theory of beams. To describe the kinematics of elastic bodies, we introduce the displacement vector and the strain tensor obeying Saint–Venant compatibility conditions. Next, we state Hooke's law for 3D bodies and study in particular the structure of the elasticity tensor for isotropic materials. The elastic beams are analyzed, merging displacement and stress methods and introducing the concept of transversely rigid body.

Chapter 10 is devoted to the dynamics of continuous media of 3 dimensions. After modeling their motion, we shed a new light on the equations of motion of particles and rigid bodies introduced in Chapters 3 and 5 due to the covariant derivative and the affine tensor calculus. Next, we introduce the stress-mass tensor, reveal its structure and show that it is governed by Euler's equations of motion, the cornerstone of elementary mechanics of fluids. Finally, we lay the foundations of constitutive equations with illustrations to hyperelastic materials and barotropic fluids.

Chapter 11 allows us to model all the intermediate continua between the particle trajectory of 1 dimension and the bulky body of 3 dimensions. Although general balance equations are proposed for continua of arbitrary dimensions perceived as Cosserat media, we focus our attention on the dynamics of one-dimensional (1D) material bodies (arch if solid, flow in a pipe or jet if fluid).

Chapter 12 returns to the variational methods introduced in Chapter 6, proposing an action principle for the dynamics of continua. In order to recover the balance equations, we use a special form of the calculus of variation consisting of performing variations not only on the value of the field but also on the variable.

Chapter 13 is devoted to the thermodynamics of reversible and dissipative continua. The cornerstone idea is to add to the space-time an extra dimension linked, roughly speaking, to the energy. The status of the temperature is a vector. The cornerstone tensors are its gradient called friction and the corresponding momentum tensor. For reversible processes, introducing Planck's potential reveals its structure and allows us to deduce classical potentials, internal energy, free energy and the specific entropy. The modeling of the dissipative continua is based on an additive decomposition of the momentum tensor into reversible and irreversible parts. The first principle of thermodynamics claims that it is covariant divergence free. The second principle is based on a tensorial expression of the local production of entropy. The constitutive laws are briefly discussed in the context of thermodynamics and illustrated by Navier–Stokes equations.

I.2.3. *Part 3: advanced topics*

The third part is devoted to research topics. The readers are asked whether they know the classical tools of differential geometry, some of them being recalled in Chapter 18.

In Chapter 15, the tangent space to a manifold is equipped with a differential affine structure by enhancing the concept of chart, due to a set of one parameter smooth families of charts, called a film library. In particular, we show how the fields of points of the affine tangent space can be viewed as differential operators on the scalar fields. So, we recover the concept of particle derivative, usual in the mechanics of continua.

In Chapter 16, the affine structure is enriched by Galilean, Bargmannian and Poincarean structures allowing us to derive the equation of motion in a covariant form compatible with the classical mechanics. Besides the torsors widely used in the former two parts, we introduce a new affine tensor relevant for mechanics called momentum tensor. We determine the most general transformation law of Galilean momenta. We deduce the Galilean coordinate systems from the study of the corresponding G-structure and we calculate the Galilean curvature tensor. The end of the chapter is devoted to torsor and momentum affine tensors for Bargmanian and Poincarean structures and to the underlined geometric structure of Lie group statistical mechanics.

In Chapter 17, the affine mechanics is discussed with respect to the symplectic structure on the manifold. In the framework of the coadjoint orbit method, the main concepts are the symplectic action of a group and the momentum map, allowing us to give a modern version of Noether's theorem. Bargmann's group, introduced in Chapter 13 by heuristic arguments, is now constructed as a link to the symplectic cohomology. Finally, we construct a symplectic form based on the factorization of the connection 1-form and the differential of the momentum tensor.

I.3. Historical background and key concepts

Before starting, let us give some words to briefly explain the key concepts underlying the structure of the book. The present section is addressed to experts and can be bypassed, in an initial reading, by undergraduate and postgraduate students.

General relativity is not solely a theory of gravitation which is reduced to the prediction of tiny effects such as bending of light and corrections to mercury's orbital precession but – maybe above all – it is a consistent framework for mechanics and physics of continua. It is organized around some key-ideas:

– the space-time, equipped with a metrics which makes it a Riemannian manifold;

– a symmetry group, Poincaré's one;

– associated with this group, a connection which is identified to the gravitation and of which the potentials are the 10 components of the metrics;

– a stress-energy tensor, representing the matter and divergence free;

– its identification to a tensor linked to the curvature of the manifold provides the equations allowing us to determine the 10 potentials.

More details can be found in Souriau's book "Géométrie et relativité" [SOU 08] or in the survey "Gravitation" by Misner, Thorne and Wheeler [MIS 73].

Is this scheme transposable to classical mechanics? The idea is not new and many researchers tried their hand at doing it, among them, for instance Souriau [SOU 07, SOU 97], Küntzle [KUN 72], Duval and Horváthy [DUV 85, DUV 91]. Let us draft the rough outline of this approach:

– working in the space-time but with another symmetry group, Galileo's one;

– it preserves no metrics, then tensorial indices may be neither lowered nor raised;

– the associated connection, structured into gravity and spinning, leads to a covariant form of the equation of motion and derives from 4 potential;

– Galileo's and Poincaré's groups are both subgroups of the affine group, from which follows the idea of identifying the common elements of classical and relativistic theories: affine mechanics [SOU 97];

– it hinges on torsor, a divergence free skew-symmetric 2-contravariant affine tensor [DES 03].

The moment of a force, due to Archimedes, is a fundamental concept of mechanical science. Its modeling by means of standard mathematical tools is well known. In the modern literature, it sometimes appears under the axiomatic form of the concept of a torsor [PER 53], an object composed of a vector and a moment, endowed with the property of equiprojectivity and obeying a specific transport law. Although the latter invokes a translation of the origin, very little interest has been taken in wondering about the affine nature of this object. These elementary notions can be presented with a minimal background of vector calculus. At a higher mathematical level, another, no-lesser overlooked keystone of the mechanics is the concept of a continuous medium, especially organized around the tensorial calculus which arises from Cauchy's works about the stresses [CAU 23, CAU 27]. The general rules of this calculus were introduced by Ricci-Curbastro and Levi-Civita [RIC 01]. They are concerned by the tensors that we will call "linear tensors" insofar as their components are modified by means of linear frame changes, then of regular linear transformations, elements of the linear group. The use of moving frames allows determining these objects in a covariant way due to a connection, known by its Christoffel's symbols.

It was É. Cartan who pointed out the fact that the linear moving frames could be replaced by affine moving frames, introducing so the affine connections [CAR 23], [CAR 24]. His successors only remember the concepts of principal bundle and connection associated with some groups: the linear group, the affine group, the projective group and so on. It is the application to the orthogonal group which above all will hold the attention on account of the Riemannian geometry and the Euclidean tensors. The interest that Cartan originally took in the connections of the affine group became of secondary importance. Perhaps only the name of "affine connection" remains, while oddly used for any group, even if it is not affine. Certainly, we can find in the continuous medium approach a unifying tool of the mechanics, even if the dynamics of the material particles and the rigid bodies remain on the fringe and if the torsor – so essential to the mechanics – seems to escape from any attempt of getting it into the mold of the tensorial calculus. The contemporaries will instead find responses to this concern of unifying and structuring the mechanics in the method of virtual powers or works, initiated by Lagrange, and the variational techniques [SAL 00]. Without denying the power of these tools, their abstract character and the traps of the calculus of variations must not be underestimated yet.

More recently, Souriau proposed revisiting mechanics emphasizing its affine nature [SOU 97]. It is this viewpoint that we will adopt here, starting from a generalization of the concept of torsor under the form of an affine object [DES 03, DES 11]. It allows structuring the mechanics due to a unique principle which, by declining it for each kind of continuous medium, provides the classical equations of the statics and dynamics. Our starting point is closely related to Souriau's approach on the ground of two key ideas: a new definition of torsors and the crucial part played by the affine group of $\mathbb{R}^n$. This group forwards on a manifold an intentionally poor geometrical structure. Indeed, this choice is guided by the fact that it contains both Galileo and Poincaré groups [SOU 97], which allows involving the Galilean and relativistic mechanics at one go. This viewpoint implies that we do not use the trick of the Riemannian structure. In particular, the linear tangent space cannot be identified to its dual one and tensorial indices may be neither lowered nor raised.

A class of tensors corresponds to each group. The components of these tensors are transformed according to the action of the considered group. The standard tensors discussed in the literature are those of the linear group of $\mathbb{R}^n$. We will call them linear tensors. A fruitful standpoint consists of considering the class of the affine tensors, corresponding to the affine group [DES 03, DES 11]. To each group is associated a family of connections allowing us to define covariant derivatives for the corresponding classes of tensors. The connections of the linear group are known through Christoffel's coefficients. They represent, as usual, infinitesimal motions of the local basis. From a physical viewpoint, these coefficients are force fields such as gravity and Coriolis' force. To construct the connection of the affine group, we need Christoffel's coefficients arising from the linear group and additional ones describing infinitesimal motions of the origin of the affine space associated with the linear tangent space. On this basis, we construct the affine covariant divergence of torsors.

The concept of a torsor was successfully applied to the dynamics of 3D bodies and shells [DES 03] and to the dynamics of material particles and rigid bodies [DES 11]. We claim that the torsor field representing the behavior of these continua is affine divergence free, which allows recovering the equations of motion. The structure of mechanics is revealed by the analysis of a unique object, the torsor, perceived as an affine tensor and which can be given with respect to the surrounding space, the submanifold and the symmetry group. Although the affine geometry could appear as a poverty-stricken mathematical frame, we think it is sufficient to describe the fundamental tools of the continuum mechanics.

To conclude with this quick survey, let us point out that, as well as the torsors, there are two other types of affine tensors useful for mechanics. The co-torsors, in duality with the torsors, lead to revisit the notion of kinetic torsor of a rigid body and could be a new starting point to develop Lagrange's virtual power method. On the other hand, the momentum tensors, in duality with the affine connections, lead to a factorization of the symplectic form and to revisiting Kirillov–Kostant–Souriau's theorem [SUO 70, SOU97].

PART 1

Particles and Rigid Bodies

1

Galileo's Principle of Relativity

1.1. Events and space–time

DEFINITION 1.1.– An *event* $\mathbf{X}$ is just an occurrence at a specific moment and at a specific place. The *space-time* (or *universe*) is the set $\mathcal{U}$ of all the events.

Lightning striking a tree, a crash, the battle of Fontenoy, a birthday, the reception of an e-mail by a computer are some examples of events. Most events are relatively blurred, without either beginning or end or precisely defined localization. The events which, within the limits imposed by our measuring instruments, seem instantaneous and pointwise are called *punctual events*. In the following, when talking about events, readers are referred only to punctual events.

DEFINITION 1.2.– A *particle* is an object appearing as a pointwise phenomenon endowed with some time persistence.

We can see it as a sequence of events. A trace can be kept, for instance, due to a film consisting of frames recorded by a camera. Of course, this kind of observation has a discontinuous feature. If a high-speed camera is used, the observed events are closer. If we imagine that the time resolution can be arbitrarily reduced, a continuous sequence of events is obtained.

DEFINITION 1.3.– A *trajectory* is the continuous sequence of events revealing the persistence of a particle and represented by a continuous map $t \mapsto \mathbf{X}(t)$.

1.2. Event coordinates

1.2.1. *When?*

The *clock* is an instrument allowing us to measure the *durations*.

DEFINITION 1.4.– By the choice of a reference event $\mathbf{X}_0$ to which the time $t_0 = 0$ is assigned, an observer can assign to any event $\mathbf{X}$ a number t called the *date*, equal to the duration between $\mathbf{X}_0$ and $\mathbf{X}$, if $\mathbf{X}$ succeeds to $\mathbf{X}_0$, and to its opposite, if $\mathbf{X}$ precedes $\mathbf{X}_0$.

Conversely, the duration elapsed between two events $\mathbf{X}_1$ and $\mathbf{X}_2$ is calculated as the date difference $\Delta t = t_2 - t_1$. We assume that all the clocks are synchronized, i.e. they measure the same duration between any events:

$$\Delta t = \Delta t'. \qquad [1.1]$$

This means each clock measures the durations with the same unit (for instance, the second). This also entails that if a clock assigns a date t' to some event, the other one assigns to the same event a date $t = t' + \tau_0$ where τ_0 depends only on both clocks.

DEFINITION 1.5.– Two events are *simultaneous* if, measured with the same clock, their dates are identical.

Clearly, if two events are simultaneous for a clock, it is so for any other one.

1.2.2. *Where?*

The most common measuring instrument for a *distance* is the *graduated ruler*. Of course, there exist less accurate instruments (the land-surveyor's string or measuring tape), while others are much more accurate (especially due to the lasers) but, for the simplicity of the presentation, the readers are only referred to the rulers as distance measuring instruments.

Whatever, we have just to know that the ruler allows us to measure the distance Δs between two *simultaneous events* $\mathbf{X}_1$ and $\mathbf{X}_2$. We assume that all the rulers are standardized in the sense that they measure the same distance between events:

$$\Delta s = \Delta s'. \qquad [1.2]$$

This means each ruler measures the distances with the same unit (for instance, the meter). Let us have a break now to explain the meaning of the simultaneity between events. When they fit the ruler graduations, the observer is informed by light signals. The essential point is – as mentioned before – these signals arrive at the observer with an infinite velocity, and then instantaneously.

As we assigned to each event a date, we would like to assign it a position. Without entering into the details of the measurement method, which is not useful to our discussion, let us say only that – in addition to the rulers – instruments are required to measure the angles, for instance *set squares* and *protractors.* We admit that the measurement method allows an observer to assign to any event **X** three coordinates x^1, x^2, x^3. The column vector gathering them:

$$x = \begin{pmatrix} x^1 \\ x^2 \\ x^3 \end{pmatrix},$$

is called *position* of **X**.

DEFINITION 1.6.– To each event **X**, an observer can assign a time t – in the sense prescribed by definition 1.4 – and a column $x \in \mathbb{R}^3$, called the *position*, by the choice of a reference event $\mathbf{X}_0$ with position $x_0 = 0$ and in such a way that for any distinct but *simultaneous* events $\mathbf{X}_1$, $\mathbf{X}_2$ and $\mathbf{X}_3$ of respective positions x_1, x_2 and x_3:

– if $\Delta x = x_2 - x_1$, we can calculate the distance between the first two by:

$$\Delta s = \| \Delta x \| ;$$

– and the angle θ between the segments Δx and $\Delta' x = x_3 - x_1$ by:

$$\cos\theta = \Delta x \cdot \Delta' x \; / \; \| \Delta x \| \, \| \Delta' x \| \, .$$

In short, any observer has available instruments measuring durations, distances and angles. This allows him or her to assign to each event **X** a date t and a position x. In the following, we adopt the following convention:

CONVENTION 1.1.– Coordinate labels:

– Latin indices i, j, k and so on run over the special coordinate labels, usually, $1, 2, 3$ or x, y, z.

– Greek indices α, β, γ and so on run over the four space-time coordinate labels $0, 1, 2, 3$ or t, x, y, z.

DEFINITION 1.7.– To each event **X**, a column $X \in \mathbb{R}^4$:

$$X = \begin{pmatrix} t \\ x \end{pmatrix} = \begin{pmatrix} t \\ x^1 \\ x^2 \\ x^3 \end{pmatrix},$$

is assigned by an observer. Their components $X^0 = t$, $X^i = x^i$ are called *coordinates* of the event. The assignment is one-to-one. Each observer creates her or his own *coordinate system*.

Hence, an observer can record the trajectory of a particle $t \mapsto \mathbf{X}(t)$ due to an assignment $t \mapsto X(t)$ in her or his own coordinate system.

Additionally, the length and angle measures allow us to calculate the areas and volumes, at least for simple geometrical objects.

DEFINITION 1.8.– The positions being determined by an observer for simultaneous events:

– the positions of three of its vertices being x_1, x_2, x_3, the *area* of a parallelogram is calculated by:

$$S = \| \Delta x \times \Delta' x \|,$$

with $\Delta x = x_2 - x_1$ and $\Delta' x = x_3 - x_1$;

– three adjoining faces of it being defined by four of its vertices x_1, x_2, x_3, x_4, the *oriented volume* of a parallelepiped is calculated by:

$$V = (\Delta x \times \Delta' x) \cdot \Delta'' x,$$

with $\Delta'' x = x_4 - x_1$.

1.3. Galilean transformations

1.3.1. *Uniform straight motion*

Newton's first law claims the velocity of a particle or a body remains constant unless the body is acted upon by an external force. This assumes we know what a force is, at least intuitively. We prefer to take it as starting point to define the forces.

DEFINITION 1.9.– A *force* is a phenomenon modifying the velocity of a particle. Hence, a *free particle* force moves in a straight line at uniform velocity. This is the *uniform straight motion* (*USM*). If the velocity is null, the particle is said to be *at rest* in the considered coordinate system.

The problem is that gravity is a large-scale force affecting all matter equally, so there are no completely free particles, even in deep space. On the Earth, experiences of USM can be carried out only in reduced regions of the space-time, for instance during a small enough duration or with objects moving without friction on a horizontal plane. The motion of a free particle is given by:

$$x = x_0 + v\,t,$$

where the initial position $x_0 \in \mathbb{R}^3$ at $t = 0$ and the uniform velocity $v \in \mathbb{R}^3$ are constant. The event "the particle is passing through x_0 at $t = 0$" is represented in the considered coordinate system by:

$$X_0 = \begin{pmatrix} 0 \\ x_0 \end{pmatrix}.$$

Introducing the 4-column:

$$U = \begin{pmatrix} 1 \\ v \end{pmatrix},$$

the event "the particle is in x at t" is represented by:

$$X = X_0 + Ut. \qquad [1.3]$$

DEFINITION 1.10.– With respect to a given family of coordinate systems, a characteristic of an object or a quantity is *invariant* if its representation in all the systems of the family is identical. We also talk about the *invariance* of the characteristic or the quantity and say that the coordinates changes of the family preserve the considered characteristic or quantity.

For instance, let us consider the family of the coordinate systems of observers for which the motion of the same particle is straight and uniform. We would like to ask the following question: what are the coordinate changes $X' \mapsto X$ of this family?

THEOREM 1.1.– The coordinate changes preserving:

– the USMs;

– the durations;

– the distances and angles;

– the oriented volumes;

are regular affine maps of the following form:

$$X = PX' + C, \quad C = \begin{pmatrix} \tau_0 \\ k \end{pmatrix}, \quad P = \begin{pmatrix} 1 & 0 \\ u & R \end{pmatrix}, \qquad [1.4]$$

where $\tau_0 \in \mathbb{R}$, $k \in \mathbb{R}^3$, $u \in \mathbb{R}^3$ and $R \in \mathbb{SO}(3)$ (see Comment 1, section 1.4).

PROOF.– Parametrization [1.3] of the trajectory being affine, the coordinate change in $\mathbb{R}^4$ preserves straight lines and the middle of segments. As a parallelogram is a quadrilateral whose the diagonals meet in their middle, the coordinate change preserves parallelograms and, reasoning by recurrence, parallelepipeds and parallelotopes. So, the coordinate change is affine:

$$X = PX' + C, \qquad [1.5]$$

where $C \in \mathbb{R}^4$ and the 4×4 matrix P are constant. As the coordinate systems define one-to-one assignments from X into the event $\mathbf{X}$, the coordinate change is also one-to-one. Considering the difference of the columns representing two events $\mathbf{X}_1$ and $\mathbf{X}_2$ in the considered coordinate systems:

$$\Delta X = X_2 - X_1 = \begin{pmatrix} \Delta t \\ \Delta x \end{pmatrix}, \quad \Delta X' = X'_2 - X'_1 = \begin{pmatrix} \Delta t' \\ \Delta x' \end{pmatrix},$$

we obtain a linear relation:

$$\Delta X = P \Delta X'. \qquad [1.6]$$

Next, we put:

$$C = \begin{pmatrix} \tau_0 \\ k \end{pmatrix}, \quad P = \begin{pmatrix} \alpha & w^T \\ u & F \end{pmatrix},$$

where $\alpha, \tau_0 \in \mathbb{R}$, $u, w, k \in \mathbb{R}^3$ and F is a 3×3 matrix. Hence, [1.6] gives:

$$\Delta t = \alpha \Delta t' + w^T \Delta x'.$$

Identifying it with condition [1.1] ensuring the invariance of the duration gives:

$$\alpha = 1, \quad w = 0.$$

Hence, we have:

$$P = \begin{pmatrix} 1 & 0 \\ u & F \end{pmatrix}. \qquad [1.7]$$

As P is regular, F must be so. Hence, [1.6] gives for simultaneous events:

$$\Delta x = F \Delta x' \quad .$$

Invariance [1.2] of the distance reads:

$$(\Delta x')^T F^T F \Delta x' = (\Delta x')^T \Delta x.$$

The column $\Delta\, x'$ being arbitrary, we obtain:

$$F^T\, F = 1_{\mathbb{R}^3}.$$

The matrix F is orthogonal. Taking into account that oriented volumes [1.8] are transformed as:

$$V' = \det(F) V,$$

their invariance entails that F is a rotation that we denote by R afterward. As $\det(P) = \det(R) = 1$, P is regular and so is the affine map $X' \mapsto X$. ∎

DEFINITION 1.11.– The coordinate changes [1.4] are called *Galilean transformations*. Any of them can be obtained composing elementary ones from amongst:

– *clock change* τ_0 (with $k = u = 0,\ R = 1_{\mathbb{R}^3}$): $t = t' + \tau_0, x = x'$;

– *spatial translation* k: $t = t', x = x' + k$;

– *rotation* R: $t = t', x = R\, x'$;

– *Galilean boost* or *velocity of transport* u: $t = t', x = x' + ut$.

A general Galilean transformation reads:

$$x = R\, x' + u\, t' + k, \quad t = t' + \tau_0, \qquad [1.8]$$

or in matrix form:

$$C = \begin{pmatrix} \tau_0 \\ k \end{pmatrix}, \quad P = \begin{pmatrix} 1 & 0 \\ u & R \end{pmatrix}, \qquad [1.9]$$

1.3.2. *Principle of relativity*

If a particle is in USM for an observer, it is a also so for any other observer. Hence, all the coordinate systems in the sense defined by definition 1.6 are equivalent, including the ones in which the particle is at rest. In other words, we admit in particular the equivalence between the motion and rest. Galileo Galilei proposed in his famous "Dialogue concerning the two chief world systems" (1632) this point of view according to which the observations of physical phenomena do not allow us to know whether we are in motion or at rest, provided that the motion is straight and uniform. *Galileo's principle of relativity* turns this from a negative to a positive statement:

PRINCIPLE 1.1.– The statement of the physical laws of the classical mechanics is the same in all the coordinate systems in the sense of definition 1.6.

For the moment, this principle is formulated in rather general words but we will soon make it clearer in applications. By classical mechanics, let us recall that we consider phenomena for which the velocity of the light is so huge that it may be considered as infinite.

1.3.3. *Space–time structure and velocity addition*

Up to now, the space-time was a set of which the elements – the events – were parametrized by four coordinates. Considering only USMs, we need only affine transformations [1.5] for the coordinate changes. In other words, the space-time $\mathcal{U}$ may be perceived as an *affine space* of 4 dimensions and the coordinates of an event **X** change according to the transformation law for the component of one of its points. Hence, the structure of the space-time must not be imposed *a priori* but is deduced from the physical observations (the USM).

Have a look now at our starting point, the USM. In the old coordinate system, it reads:

$$X' = X'_0 + U't'.$$

Combining it with the Galilean transformation [1.4] gives:

$$X = P(X'_0 + U't') + C.$$

Taking into account [1.8], we recover [1.3], provided that:

$$X_0 = P(X'_0 - U'\tau_0) + C, \qquad [1.10]$$

$$U = PU'. \qquad [1.11]$$

What do these relations teach us?

– Without clock change, the first one reads:

$$X_0 = PX'_0 + C,$$

which is nothing other than the transformations law for the components of a point of $\mathcal{U}$. For more general transformations, the additional term in [1.10] takes into account the clock change.

– The second relation, [1.11], is the transformation law for the components of a vector $\vec{\mathrm{U}}$ of the vector space attached to $\mathcal{U}$. It will be called the 4-*velocity*.

Let us consider, for instance, a particle of velocity v' in the coordinate system X'. In another one X obtained from X' by a Galilean transformation [1.9], the 4-velocity is given by [1.11]:

$$U = \frac{dX}{dt} = \begin{pmatrix} 1 \\ \dot{x} \end{pmatrix} = \begin{pmatrix} 1 \\ v \end{pmatrix} = \begin{pmatrix} 1 & 0 \\ u & R \end{pmatrix} \begin{pmatrix} 1 \\ v' \end{pmatrix}. \qquad [1.12]$$

Thus, the velocity in the new coordinate system is:

$$\boxed{v = u + R\,v'.} \qquad [1.13]$$

In particular, for a Galilean boost u, we have:

$$v = u + v'.$$

This is the *velocity addition formula*. Also, combining two Galilean boosts u_1 and u_2, we verify that the resulting velocity of transport is:

$$u = u_1 + u_2.$$

1.3.4. *Organizing the calculus*

For convenience, an affine transformation $X' \mapsto X = PX' + C$ can be denoted by $a = (C, P)$. Applying successively a_1 and a_2 gives a new affine transformation a_3:

$$a(X) = a_2(a_1(X)) = a_2(C_1 + P_1 X) = C_2 + P_2(C_1 + P_1 X),$$

hence:

$$a_3 = a_2 a_1 = (C_2, P_2)(C_1, P_1) = (C_2 + P_2 C_1, P_2 P_1).$$

This product is associative and has an identity transformation $e = (0, 1_{\mathbb{R}^4})$ such that $e\,a = a\,e = a$. Each affine transformation $a = (C, P)$ has an inverse transformation $a^{-1} = (-P^{-1}C, P^{-1})$ such that $a^{-1}a = a\,a^{-1} = e$. It is straightforward to verify that the combination of two Galilean transformations a_2 and a_1 is also a Galilean transformation a given by:

$$u = u_2 + R_2 u_1, \quad R = R_2 R_1, \quad \tau_0 = \tau_2 + \tau_1, \quad k = k_2 + R_2 k_1 + u_2 \tau_1. \qquad [1.14]$$

It is easy to verify that the inverse transformation $X \mapsto X' = P^{-1}X + C'$ is a Galilean transformation represented by (see Comment 2 section 1.4):

$$C' = \begin{pmatrix} \tau_0' \\ k' \end{pmatrix}, \quad P^{-1} = \begin{pmatrix} 1 & 0 \\ -R^T u & R^T \end{pmatrix}, \qquad [1.15]$$

putting:

$$\tau_0' = -\tau_0, \qquad k' = -R^T(k - u\tau_0).$$

It is often convenient to organize the matrix calculation by working rather in $\mathbb{R}^5$, representing the column X and the affine transformation $a = (C, P)$, respectively, by:

$$\tilde{X} = \begin{pmatrix} 1 \\ X \end{pmatrix} \in \mathbb{R}^5 \qquad \tilde{P} = \begin{pmatrix} 1 & 0 \\ C & P \end{pmatrix}, \qquad [1.16]$$

so affine transformation [1.4] looks like a simple regular linear transformation:

$$\tilde{X} = \tilde{P}\tilde{X}', \qquad [1.17]$$

where, taking into account [1.9], the Galilean transformation a is represented by the 5×5 matrix decomposed by blocks:

$$\tilde{P} = \begin{pmatrix} 1 & 0 & 0 \\ \tau_0 & 1 & 0 \\ k & u & R \end{pmatrix}. \qquad [1.18]$$

In a similar way, owing to [1.15], the inverse transformation is represented by:

$$\tilde{P}^{-1} = \begin{pmatrix} 1 & 0 & 0 \\ \tau_0' & 1 & 0 \\ k' & -R^T u & R^T \end{pmatrix}. \qquad [1.19]$$

1.3.5. *About the units of measurement*

There is still a long way to go to cover the mechanics of continua but let us stop for a moment to have a look at the conversion of units. Let the event $\mathbf{X}$ be represented in a coordinate system by $\bar{X}$ where durations and times are measured with new units. Let us say that the time and length units in the old coordinate system are equal, respectively, to T and L in the new one. The conversion of units is given by the scaling:

$$\bar{t} = T\,t, \qquad \bar{x} = Lx,$$

or, in matrix form:

$$\bar{X} = P_u X \tag{1.20}$$

with:

$$P_u = \begin{pmatrix} T & 0 \\ 0 & L1_{\mathbb{R}^3} \end{pmatrix}. \tag{1.21}$$

Similarly, let us apply the scaling:

$$\bar{X}' = P_u X' \tag{1.22}$$

Combining Galilean transformation [1.4] and scalings [1.20] and [1.22] leads to:

$$\bar{X} = \bar{P}\bar{X}' + \bar{C},$$

with:

$$\bar{P} = P_u P P_u^{-1}, \qquad \bar{C} = P_u C.$$

Using [1.9] and [1.21] shows that

$$\bar{C} = \begin{pmatrix} \bar{\tau}_0 \\ \bar{k} \end{pmatrix}, \quad \bar{P} = \begin{pmatrix} 1 & 0 \\ \bar{u} & \bar{R} \end{pmatrix}, \tag{1.23}$$

with $\bar{C}$ being a simple scaling of C:

$$\bar{\tau}_0 = T\,\tau_0, \qquad \bar{k} = Lk,$$

and:

$$\bar{u} = (L/T)\,u, \qquad \bar{R} = R,$$

As result of the conversion of units, the rotation is invariant while the boost u is scaled as a velocity. It is worth observing that, in a conversion of units, a Galilean transformation $a = (C, P)$ turns into a Galilean transformation $\bar{a} = (\bar{C}, \bar{P})$ (see Comment 3, section 1.4). The conversion does not affect the Galilean feature of an affine transformation. Of course, calculations can be organized with 5×5 matrices:

$$\tilde{\bar{P}} = \tilde{P}_u \tilde{P} \tilde{P}_u^{-1} \qquad \text{where} \qquad \tilde{P}_u = \begin{pmatrix} 1 & 0 \\ 0 & P_u \end{pmatrix}.$$

1.4. Comments for experts

COMMENT 1.– This theorem is related to the Toupinian structure of the space-time which gives a theoretical framework to the universal or absolute time and space (see section 16.1).

COMMENT 2.– In fact, the set of all the Galilean transformations is a Lie group of 10 dimensions called Galileo's group.

COMMENT 3.– Conversely, the normalizer of Galileo's group in the affine group is composed of the Galilean transformations themselves and the conversions of units [1.20] (see section 16.2).

2

Statics

2.1. Introduction

In this chapter, the bodies occupy a volume or are pointwise (particles). The bodies can be subjected to various kinds of forces, for instance gravity, electromagnetic forces and contact forces. Our aim is to determine at which conditions a body subjected to several forces remains in uniform straight motion (USM). If the body occupies a volume, that means all its particles are in USM with the same velocity. Thus, the body is at rest in some particular coordinate systems. Before proposing general conditions, we hope to develop some intuition by discussing simple situations in which – to make the understanding easier – the body is initially at rest for the observer. Of course, if the body remains at rest in the observer coordinate system, it is in USM in any other coordinate system resulting from a Galilean transformation.

Let us consider, for instance, a body immersed in a fluid. Under the effect of gravity only, it moves downward. On the other hand, according to Archimedes' principle, buoyancy opposes it by moving the body upward. The body remains at rest when both opposite forces, gravity and buoyancy, are of the same magnitude. Hence, the balance of force is a necessary condition of static equilibrium but is not sufficient, as shown in the next example.

Let us consider weighing scales composed of two pans hung at the ends of a beam pivoting on the fulcrum at any given position (Figure 2.1). The standard weight on the left-hand pan and the unknown weight on the right-hand pan are equilibrated by the reaction of the fulcrum but, according to its position, the beam can nevertheless rotate. This is due to the moment of each force, i.e. its tendency to rotate the object. The beam is at rest when the moments of both weights are opposite and of the same magnitude.

2.2. Statical torsor

2.2.1. *Two-dimensional model*

The first step of our modeling is to consider two-dimensional systems as our weighing scales. Let R_f be the fulcrum reaction, positive upward, g be the gravity, m_u be the unknown mass, at the distance d_u of the fulcrum and m_s be the standard mass, at the distance d_s of the fulcrum. The resultant of vertical forces is:

$$F = R_f - (m_u + m_s)g.$$

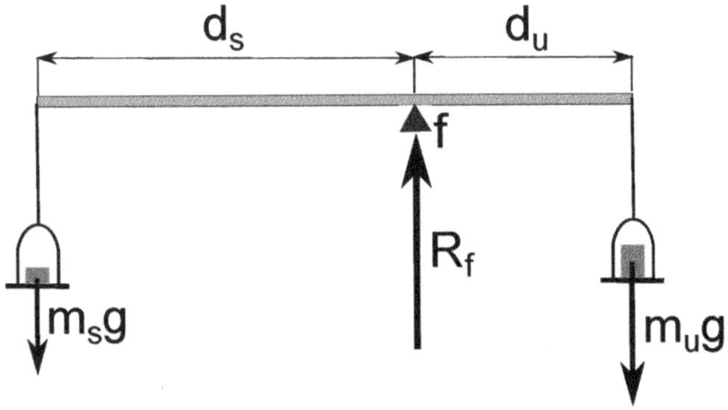

Figure 2.1. *Weighting scale*

The moment about a point is a scalar, its magnitude being the product of the force and the perpendicular distance between the point and the force axis, and its sign being positive if the force rotates the body counterclockwise and negative if clockwise. The moment resultant about the fulcrum f is:

$$M = d_s m_s g - d_u m_u g.$$

The scales are at rest provided that both forces and moments are balanced:

$$F = 0, \qquad M = 0. \tag{2.1}$$

We have calculated the moment about a particular point, the fulcrum. What happens if we calculate it about any given point on the beam, at the distance x_0 of the fulcrum (positive or negative according to this point being, respectively, at the left or the right of f):

$$M_{x_0} = (d_s - x_0)m_s g - (d_u + x_0)m_u g + x_0 R_f \ ?$$

it is worth observing that:

$$M_{x_0} = M + F x_0.$$

This is the *transport law of the moment*. If the scales are at rest, the resultants F and M vanish, then so does M_{x_0}. The balance of moments occurs, whatever the position of the reference point. The balance principle can read:

$$F = 0, \qquad M_{x_0} = 0. \qquad [2.2]$$

2.2.2. *Three-dimensional model*

As a second step, we generalize these simple ideas for three-dimensional situations. Let a body be subjected to a system of forces represented by 3-columns $F_1, F_2, \cdots, F_N$, acting upon the body, respectively, at positions $x_1, x_2, \cdots, x_N$. The force resultant and the moment resultant about the origin of the coordinate system are:

$$F = \sum_{i=1}^{N} F_i, \qquad M = \sum_{i=1}^{N} x_i \times F_i.$$

The moment resultant about any other point of position x_0, taken as new origin, is:

$$M_{x_0} = \sum_{i=1}^{N} (x_i - x_0) \times F_i,$$

hence:

$$M_{x_0} = \sum_{i=1}^{N} x_i \times F_i - x_0 \times \sum_{i=1}^{N} F_i = M - x_0 \times F,$$

and, owing to the anticommutativity of the cross-product, the transport law of the moment reads:

$$M_{x_0} = M + F \times x_0.$$

In three dimensions, the balance principle equally reads [2.1] or [2.2], excepted that the scalars are replaced by 3-columns. Considering a coordinate system for which the weighing balance is in the plan of the x^1 and x^2 axes, it is easy to verify that the moment is directed along the x^3 axis and to recover the previous expressions of M and M_{x_0}.

2.2.3. *Statical torsor and transport law of the moment*

The third step of our modeling consists of constructing an object structured into force and moment components. As time is not concerned with the equilibrium of bodies, we work temporarily with the 4-column:

$$\breve{X} = \begin{pmatrix} 1 \\ x \end{pmatrix},$$

obtained by bringing up the second component of $\tilde{X}$, and the 4×4 matrix:

$$\breve{P} = \begin{pmatrix} 1 & 0 \\ k & R \end{pmatrix},$$

obtained by bringing up the second row and column of Galilean transformation [1.18], so the *(special) Euclidean transformation* $x = Rx' + k$ looks like a simple regular linear transformation:

$$\breve{X} = \breve{P}\breve{X}'. \qquad [2.3]$$

DEFINITION 2.1.– A *statical torsor* $\breve{\tau}$ is an object represented in a coordinate system by a skew-symmetric 4×4 matrix:

$$\boxed{\breve{\tau} = \begin{pmatrix} 0 & F^T \\ -F & -j(M) \end{pmatrix},} \qquad [2.4]$$

where $F \in \mathbb{R}^3$ is its *force*, $M \in \mathbb{R}^3$ is its *moment* and of which the components, under Euclidean transformation [2.3], are modified according to the transformation law:

$$\breve{\tau} = \breve{P}\breve{\tau}'\breve{P}^T. \qquad [2.5]$$

Applying the rules of the matrix calculus, it is worth noting that if $\breve{\tau}'$ is skew-symmetric, so is $\breve{\tau}$ given by definition 2.1. Under an Euclidean transformation (and more generally under an affine transformation), the skew-symmetry property is preserved, which ensures the consistency of the definition (see Comment 1, section 2.4). To justify this from a physical point of view, we show first that it allows us to recover the transport law of the moment. By inversion of [2.5], we have:

$$\breve{\tau}' = \breve{P}^{-1}\breve{\tau}\breve{P}^{-T}, \qquad [2.6]$$

To shift the origin at x_0 in the new coordinate system, let us consider a spatial translation $x' = x - x_0$. The coordinate change is represented by:

$$\breve{P}^{-1} = \begin{pmatrix} 1 & 0 \\ -x_0 & 1_{\mathbb{R}^3} \end{pmatrix}. \qquad [2.7]$$

Introducing [2.4] and [2.7] into [2.6], the torsor is represented in the new coordinate system by:

$$\breve{\tau}' = \begin{pmatrix} 0 & F^T \\ -F & F x_0^T - x_0 F^T - j(M) \end{pmatrix}.$$

Taking into account [7.11] and the linearity of the map j, we have:

$$\breve{\tau}' = \begin{pmatrix} 0 & F^T \\ -F & -j(M + F \times x_0) \end{pmatrix}.$$

Thus, the force is not affected by the translation while the moment is transformed according to the *transport law of the moment*:

$$\boxed{F' = F, \qquad M' = M + F \times x_0.} \qquad [2.8]$$

In a similar manner, let us now consider a rotation $x' = R^T x$. The calculations are left to the readers and show that the force and moment rotate as the position:

$$F' = R^T F, \qquad M' = R^T M.$$

The translation and rotation are particular cases of a general Euclidean transformation $x' = R^T(x - k)$. The coordinate change is represented by:

$$\breve{P}^{-1} = \begin{pmatrix} 1 & 0 \\ -R^T k & R^T \end{pmatrix}, \qquad [2.9]$$

Matrix calculation [2.6] leads to the general transformation law:

$$\boxed{F' = R^T F, \qquad M' = R^T(M + F \times k),} \qquad [2.10]$$

combining the transport and rotation. It is easy to find two invariants under Euclidean transformations:

– the norm of the force: $\| F \|$ because of [7.20];

– the dot product of the force and moment, owing to [7.20] and [7.21]:

$$F' \cdot M' = (R^T F)^T R^T (M + F \times k) = F^T R R^T (M + F \times k)$$
$$= F \cdot (M + F \times k),$$
$$F' \cdot M' = F \cdot M + F \cdot (F \times k) = F \cdot M + k \cdot (F \times F) = F \cdot M.$$

The linear space $\mathbb{M}_{44}^{skew}$ of the 4×4 skew-symmetric matrices is of 6 dimensions. Let $\mathbf{T}_s$ be the set of statical torsors $\boldsymbol{\tau}$, in one-to-one correspondence with the skew-symmetric 4×4 matrices [2.4]. Due to this map, $\mathbf{T}_s$ is a linear space of 6 dimensions if we define by structure transport the addition of torsors and the multiplication of a torsor by a scalar.

2.3. Statics equilibrium

2.3.1. *Resultant torsor*

In a given coordinate system, let a body $\mathcal{B}$ be subjected to a system of forces represented by $F_1, F_2, \cdots, F_N$, acting upon the body, respectively, at positions $x_1, x_2, \cdots, x_N$.

DEFINITION 2.2.– The torsor of the force F_i about the origin of the coordinate system (or *force torsor*) is:

$$\breve{\tau}_i = \begin{pmatrix} 0 & F_i^T \\ -F_i & -j(x_i \times F_i) \end{pmatrix}, \qquad [2.11]$$

DEFINITION 2.3.– The *resultant torsor* of a body $\mathcal{B}$ is the sum of the torsors of the forces acting upon it:

$$\breve{\tau}(\mathcal{B}) = \sum_{i=1}^{N} \breve{\tau}_i.$$

If the resultant is null in a coordinate system, then it is so in any other coordinate system resulting from a Galilean transformation, because of [2.10].

2.3.2. *Free body diagram and balance equation*

DEFINITION 2.4.– To identify the forces acting upon a body, it is convenient to draw a *free body diagram* , which is a sketch of the body and all *efforts* (forces or moments) acting upon it, by performing the following three steps:

– isolate the body;

– identify the forces and, in particular, when removing all supports and connections, identify the corresponding *reactions*;

– make a sketch of the body, showing all forces acting on it.

To illustrate step 2, let us consider usual kinds of supports:

– removing a *simple support*, we draw a force perpendicular to the surface on which the roller could roll;

– removing a frictionless *hinge*, we draw a force acting at the hinge center;

– removing a *clamped* or *built-in support,* we draw a force acting at an unknown position near the support.

To solve a *statics problem*, we perform the following steps:

– draw a free body diagram;

– choose a convenient coordinate system to calculate the resultant moment;

– apply the *balance equation*:

$$\check{\tau}(\mathcal{B}) = 0; \qquad [2.12]$$

– solve it for the unknowns.

The resultant torsor has 6 scalar components, they are 6 balance equations. Let m be the number of scalar unknowns (generally components of the support reactions). If $m = 6$, the body is said to be *isostatic*. Otherwise, the number of missing equations to solve the problem is $h = m - 6$ and is called the *redundancy degree.*

As an example, let us consider a homogeneous rigid truss $\boldsymbol{PQ}$, of mass m and length L, hinged at $\boldsymbol{P}$ on a horizontal soil and supported at $\boldsymbol{Q}$ on a rough vertical wall, at the distance a of $\boldsymbol{P}$ (Figure 2.2). The friction reaction at $\boldsymbol{Q}$ is tangential to the circle of center $\boldsymbol{O}$ passing through $\boldsymbol{Q}$. To solve this statics problem, we draw the free body diagram to identify the forces acting on the truss (Figure 2.3):

– its weight acting at the mass center $\boldsymbol{G}$, middle of the truss:

$$W = \begin{pmatrix} 0 \\ 0 \\ -mg \end{pmatrix};$$

– the reaction acting at the hinge $\boldsymbol{P}$:

$$R_P = \begin{pmatrix} R_1 \\ R_2 \\ R_3 \end{pmatrix};$$

– the reaction acting at $\boldsymbol{Q}$:

$$R_Q = \begin{pmatrix} R'_n \\ R'_t \cos\vartheta \\ R'_t \sin\vartheta \end{pmatrix}.$$

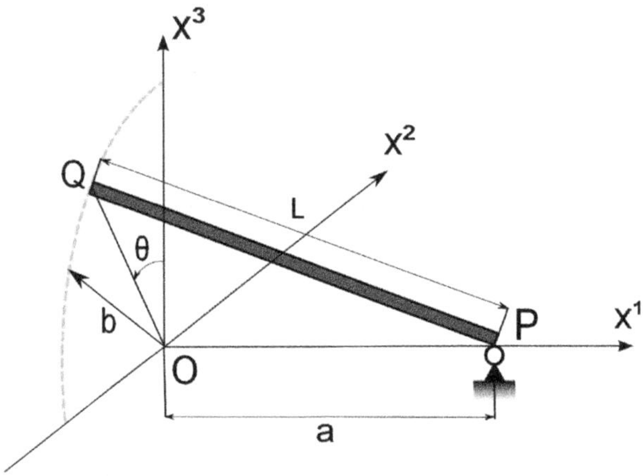

Figure 2.2. *Rigid truss hinged on a horizontal soil and supported on a rough vertical wall*

There are 6 unknowns: $R_1, R_2, R_3, R'_n, R'_t, \vartheta$. The problem is isostatic. The resultant torsor is null if its components are. The force balance leads to:

$$R_1 = -R'_n, \qquad R_2 = -R'_t \cos\vartheta, \qquad R_3 = mg - R'_t \sin\vartheta,$$

which allows us to know the components R_1, R_2, R_3 of the reaction at $\boldsymbol{P}$, after the other unknowns have been determined. As the balance of moments occurs whatever the position of the reference point, it is wise to choose $\boldsymbol{P}$, which leads to:

$$M = \begin{pmatrix} -a \\ -b\sin\vartheta \\ b\cos\vartheta \end{pmatrix} \times \begin{pmatrix} R'_n \\ R'_t \cos\vartheta \\ R'_t \sin\vartheta \end{pmatrix} + \frac{1}{2} \begin{pmatrix} -a \\ -b\sin\vartheta \\ b\cos\vartheta \end{pmatrix} \times \begin{pmatrix} 0 \\ 0 \\ -mg \end{pmatrix} = \begin{pmatrix} 0 \\ 0 \\ 0 \end{pmatrix},$$

thus:

$$-bR'_t + \frac{1}{2}mgb\sin\vartheta = 0, \qquad bR'_n\cos\vartheta + aR'_t\sin\vartheta - \frac{1}{2}mga = 0,$$

$$-aR'_t\cos\vartheta + bR'_n\sin\vartheta = 0.$$

where $b = \sqrt{l^2 - a^2}$.

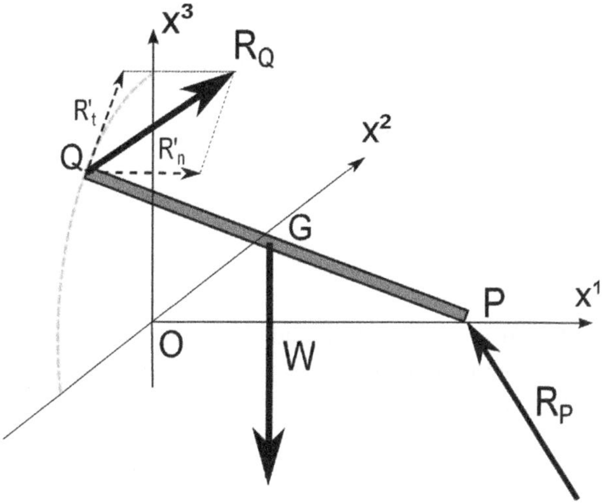

Figure 2.3. *Free body diagram of the rigid truss*

2.3.3. *External and internal forces*

If the body is in USM, so are all its parts.

DEFINITION 2.5.– We say that a body is in *static equilibrium* if the resultant torsor of each of its parts is null.

This leads us to enforce the balance equations to any of its parts, provided we correctly identify the efforts acting upon it. For that, we make a cut throught the body to isolate one part. Next, we draw a free body diagram of the considered part in which – as we identify reactions when removing a support – we identified **internal forces** when cutting the body. For instance:

– cutting a *truss*, we draw a (tension or compression) force along the truss, away from the part,

– cutting a *cable*, we draw a tension force along the cable, away from the part,

– cutting a frictionless *pulley*, we draw a tension force along the cable on both sides of the polley.

The other forces are called *external forces*.

THEOREM 2.1.– The following two statements are equivalent:

i) the mutual forces of *action and reaction* between complementary parts of a body are equal and opposite as well as their moments;

ii) the map $\mathcal{B} \rightarrow \check{\tau}(\mathcal{B})$ is an *extensive quantity*, in the sense that:

$$\check{\tau}(\mathcal{B}_1) + \check{\tau}(\mathcal{B}_2) = \check{\tau}(\mathcal{B}_1 \cup \mathcal{B}_2),$$

for any disjoint bodies $\mathcal{B}_1$ and $\mathcal{B}_2$.

PROOF.–

Indeed, let $\mathcal{A}$ be a part of the body $\mathcal{B}$ and $\mathcal{A}'$ its complementary part. Let us split the exterior forces acting upon $\mathcal{B}$ into the ones acting on $\mathcal{A}$, of resultant torsor $\check{\tau}_A^{ext}$ and the ones acting on $\mathcal{A}'$, of resultant torsor $\check{\tau}_{A'}^{ext}$. Then, one has:

$$\check{\tau}(\mathcal{A} \cup \mathcal{A}') = \check{\tau}(\mathcal{B}) = \check{\tau}_A^{ext} + \check{\tau}_{A'}^{ext}.$$

On the other hand, $\check{\tau}_A^{int}$ being the resultant torsor of internal forces acting upon $\mathcal{A}$ and $\check{\tau}_{A'}^{int}$ being the one of internal forces acting upon $\mathcal{A}'$, it holds:

$$\check{\tau}(\mathcal{A}) = \check{\tau}_A^{ext} + \check{\tau}_A^{int},$$
$$\check{\tau}(\mathcal{A}') = \check{\tau}_{A'}^{ext} + \check{\tau}_{A'}^{int}.$$

From the three previous relations, it results:

$$\check{\tau}(\mathcal{A}) + \check{\tau}(\mathcal{A}') - \check{\tau}(\mathcal{A} \cup \mathcal{A}') = \check{\tau}_A^{int} + \check{\tau}_{A'}^{int}$$

if *(i)* holds, the right hand member is null, that proves *(ii)*. Conversely, if *(ii)* is true, the left-hand member vanishes, that entails *(i)*. ■

In the following, we shall always implicitly assume *Newton's third law*:

LAW 2.1.– For any body in static equilibrium, the two equivalent statements are satisfied:

– the mutual forces of action and reaction between complementary parts of a body are equal and opposite as well as their moments;

– the resultant torsor is an extensive quantity.

It is worth observing that what we have done when we removed supports in the previous section is nothing other than make a cut (a mere semantic question). Indeed, let us call a *foundation* the joining of all the other bodies in the universe (although in fact only the vicinity of the support is relevant for us and we do not feel really concerned by what happens far away). Next we make the cut at the supports between the studied body and the foundation.

2.4. Comments for experts

COMMENT 1.– The torsor is in fact a skew-symmetric contravariant affine tensor of rank 2 and the corresponding transformation law for the components is [2.5].

3

Dynamics of Particles

3.1. Dynamical torsor

3.1.1. *Transformation law and invariants*

DEFINITION 3.1.– In this chapter, a *particle* is some matter which is pointwise (for instance, an elementary particle as an electron) or can be thought of as pointwise (for instance, if it is seen from a long way off).

In this chapter, we hope to model the motion of the particles. The torsor – introduced for the purpose of the statics – is nothing other than a "prototype" of what we shall do throughout this book. Tackling the dynamics is simply a matter of recovering an extra dimension, the time that we had provisionally arisen. Imitating the statics model, we extend the notion of torsor in a space–time framework (for the moment, it is a simple game but it will take a strong meaning later on).

DEFINITION 3.2.– The *dynamical torsor* $\boldsymbol{\tau}$ of a particle is an object represented in a coordinate system by a skew-symmetric 5×5 matrix:

$$\boxed{\tilde{\tau} = \begin{pmatrix} 0 & T^T \\ -T & J \end{pmatrix},} \qquad [3.1]$$

where $T \in \mathbb{R}^4$, $J \in \mathbb{M}_{44}^{skew}$ and the components of which, under the Galilean transformation [1.17], are modified according to the transformation law:

$$\tilde{\tau} = \tilde{P}\tilde{\tau}'\tilde{P}^T. \qquad [3.2]$$

The linear space $\mathbb{M}_{55}^{skew}$ of the 5×5 skew-symmetric matrices is of 10 dimensions. Let $\mathbf{T}_d$ be the set of dynamical torsors $\boldsymbol{\tau}$, in one-to-one correspondence with the skew-symmetric 5×5 matrices [3.1]. Due to this map, $\mathbf{T}_d$ is a linear space of 10 dimensions if we define by structure transport the addition of torsors and the multiplication of a torsor by a scalar.

Taking into account the structure of the space–time, T and J are decomposed by blocks:

$$T=\begin{pmatrix} m \\ p \end{pmatrix}, \quad J=\begin{pmatrix} 0 & -q^T \\ q & -j\,(l) \end{pmatrix}, \qquad [3.3]$$

where m is scalar and $p, q, l \in \mathbb{R}^3$. What is the physical meaning of these components? For this aim, we apply the transformation law of the torsor [3.2] or, equivalently, its inverse one:

$$\tilde{\tau}' = \tilde{P}^{-1}\tilde{\tau}\tilde{P}^{-T}. \qquad [3.4]$$

Taking into account [1.19] and [3.3], transformation law [3.4] itemizes as:

$$\boxed{m' = m,} \qquad [3.5]$$

$$\boxed{p' = R^T\,(p - m\,u),} \qquad [3.6]$$

$$\boxed{q' = R^T\,(q - \tau_0'\,(p - m\,u)) + m'\,k',} \qquad [3.7]$$

$$\boxed{l' = R^T\,(l + u\times q) + k'\times (R^T(p - m\,u)).} \qquad [3.8]$$

It is easy to spot expression [3.6] of p' in [3.8] and [3.7], these last two relations then alternatively read:

$$q' = R^T q + mk' - \tau_0' p'\,, \qquad [3.9]$$

$$l' = R^T\,(l + u\times q) + k'\times p'. \qquad [3.10]$$

To begin with, we observe that component m is invariant under any Galilean transformation, then fully characteristic of the particle. That aside, we admit that these intricate expressions of the torsor components are rather puzzling. To see things clearly, we intend annihilating some of them by suitable Galilean transformations. For our aim, we discuss only the case that the invariant component m is not null. Starting in any coordinate system X, we choose the Galilean boost:

$$u = \frac{p}{m}, \tag{3.11}$$

which annihilates p' and reduces [3.9] to:

$$q' = R^T q + m k'.$$

Next, we pick the spatial translation:

$$k' = -\frac{1}{m} R^T q,$$

which annihilates q'. As p' is null and with the boost [3.11], [3.10] is reduced to:

$$l' = R^T (l + u \times q) = R^T \left(l + \frac{1}{m} p \times q \right) = R^T \left(l - \frac{1}{m} q \times p) \right). \tag{3.12}$$

There is nothing more to do because the change clock τ'_0 occurs only in [3.9] but is multiplied by zero while the rotation R obviously cannot annihilate l'. To sum up, what is the result of this massacre? We killed components p and q. There remains m, which is pleasantly invariant, and l.

Incidentally, cast a glance at the 3-column occurring in the last relation:

$$l_0 = l - \frac{1}{m} q \times p. \tag{3.13}$$

Owing to [3.5], [3.10] and [3.9] and after obvious simplifications, this quantity becomes in any another coordinate system X':

$$l'_0 = l' - \frac{1}{m'} q' \times p' = R^T (l + u \times q) - \frac{1}{m} (R^T q) \times p'.$$

Taking into account [3.6] and [7.22], it holds:

$$l'_0 = R^T \left(l + u \times q - \frac{1}{m} q \times p + q \times u \right),$$

that leads to the transformation law:

$$\boxed{l'_0 = R^T l_0.} \tag{3.14}$$

A straightforward consequence is that the norm of l_0 is invariant. We can consider that a particle is characterized by two invariant quantities, m and $\| l_0 \|$.

3.1.2. *Boost method*

Conversely, let us consider a coordinate system X' in which the torsor has a *reduced form* (we have just finished proving the existence of such a coordinate system):

$$\tilde{\tau}' = \begin{pmatrix} 0 & T^T \\ -T & J \end{pmatrix} = \begin{pmatrix} 0 & m & 0 \\ -m & 0 & 0 \\ 0 & 0 & -j(l_0) \end{pmatrix}, \tag{3.15}$$

where there is no trouble in putting m instead of m' because we know this component is invariant. We now claim the particle is at rest in this coordinate system, for convenience at the position $x' = 0$ at time $t' = 0$. We call the *proper coordinate system of a particle* a coordinate system in which this particle is at rest at $X' = 0$. Of course, this proper coordinate system is not unique. Let $\bar{X}'$ be another proper coordinate system of the particle. As $X' = 0$ must be transformed into $\bar{X}' = 0$, the changes $X' \mapsto \bar{X}'$ of proper coordinate systems are linear transformations.

What does the dynamical torsor tell us about a free particle in uniform straight motion (USM)

To know this, let us consider another coordinate system $X = PX' + C$ with a Galilean boost v (see definition 1.11) and a translation of the origin at $k = x_0$ (hence, $\tau_0 = 0$ and $R = 1_{\mathbb{R}^3}$), providing the trajectory equation:

$$x = x_0 + v\,t, \tag{3.16}$$

of the particle moving in USM at velocity v. We can determine the new components of the torsor in X by performing matrix product [3.2] applied to [3.15] or, alternatively, using formulae [3.5]–[3.7], providing:

$$p = m\,v, \qquad q = m\,x_0, \qquad l = l_0 + q \times v,$$

or, taking into account trajectory equation [3.16]:

$$p = m\,v, \qquad q = m\,(x - v\,t), \qquad l = l_0 + x \times m\,v. \tag{3.17}$$

The last relation of [3.17] is called the *transport law of the angular momentum*. In fact, it is a particular case of general transformation laws [3.8] and [3.10] when considering only a Galilean boost.

Due to our boost method, we obtained the torsor components in a consistent way revealing their physical meaning:

– We know by experience that the quantity of matter or *mass* – measured with weighing scales – is independent of the choice of a coordinate system in the sense of definition 1.6, then we naturally identify it with m.

– The quantity p, proportional to the mass and velocity, is called the quantity of motion or *linear momentum*.

– The quantity q, proportional to the mass and initial position, provides the trajectory equation. It will be called *passage* because it indicates the particle is passing through x_0 at time $t = 0$, although it could also be called the quantity of position.

– The quantity l splits into two terms. The second one, $q \times v = x \times m v = x \times p$, is called *orbital angular momentum* to remind us of the cross-product traducing a small rotation (see section 3.2.1). The first one, $l_0 = l - q \times p / m$, is called the *spin angular momentum* and its meaning will be discussed further. Their sum, l, is called the *angular momentum*, although it could also be called the quantity of rotation.

The dynamic torsor, which was at the beginning a mere intellectual speculation, has taken now a physical sense, leading us to name its components.

DEFINITION 3.3.– The dynamical torsor is structured into two components:

– the *linear 4-momentum* T, itself substructured into:

– the *mass* m;

– the *linear momentum* p.

– and the *angular 4-momentum* J, itself substructured into:

– the *passage* q;

– the *angular momentum* l.

The invariants of the dynamical torsor are:

– the *mass* m;

– the *spin* $\| l_0 \|$.

In matrix form, the dynamical torsor reads:

$$\boxed{\tilde{\tau} = \begin{pmatrix} 0 & m & p^T \\ -m & 0 & -q^T \\ -p & q & -j(l) \end{pmatrix}.} \qquad [3.18]$$

What can we say about these components along the trajectory? We know by experience that the mass at rest is time independent. Moreover, the spin angular momentum l_0 being a characteristic at rest of the particle, it is natural to suppose that it is also time independent (this intuition will be confirmed later on). In short, $\tilde{\tau}'$ is constant along the trajectory. Using derivative, it reads:

$$\dot{\tilde{\tau}}' = 0.$$

It is worth observing that, according to transformation law [3.2] and because a Galilean transformation is time independent, it is true in any coordinate system, even if the particle is not at rest in it. On these grounds, we claim that:

LAW 3.1.– For a particle in USM, the 10 components of the dynamical torsor are constant along the trajectory, then *integrals of the motion*:

$$\dot{\tilde{\tau}} = 0. \qquad [3.19]$$

We could also find it by remarking that, as m, l_0, v and x_0, the linear momentum $p = mv$, the passage $q = mx_0$ are time independent and so is the angular momentum $l = l_0 + q \times v$. Let us observe that above all we have a first example of a physical law obeying Galileo's principle of relativity 1.1.

3.2. Rigid body motions

3.2.1. *Rotations*

Before going further, we need to acquire skills in managing rotations. Let $x' = R^T x$ be a coordinate change associated with a given rotation R. It is a complex transformation and, to grasp it better, we break it into three rotations (Figure 3.1):

– a rotation R_φ of angle φ about z's axis, bringing y's axis to $\bar{y}$'s axis, called *line of nodes*;

– a rotation R_ϑ of angle ϑ about the line of nodes, bringing the axis of $z = \bar{z}$ to that $\tilde{z}$;

– a rotation R_ψ of angle ψ about $\tilde{z}$'s axis.

The angles φ, ϑ, ψ are called *Euler's angles* and they allow us to describe any rotation by composing the three rotations:

$$x' = R_\psi^T \tilde{x} = R_\psi^T R_\vartheta^T \bar{x} = R_\psi^T R_\vartheta^T R_\varphi^T x,$$

thus:

$$R = R_\varphi R_\vartheta R_\psi. \qquad [3.20]$$

In detail, we have:

$$R = \begin{pmatrix} \cos\varphi & -\sin\varphi & 0 \\ \sin\varphi & \cos\varphi & 0 \\ 0 & 0 & 1 \end{pmatrix} \begin{pmatrix} \cos\vartheta & 0 & \sin\vartheta \\ 0 & 1 & 0 \\ -\sin\vartheta & 0 & \cos\vartheta \end{pmatrix} \begin{pmatrix} \cos\psi & -\sin\psi & 0 \\ \sin\psi & \cos\psi & 0 \\ 0 & 0 & 1 \end{pmatrix}. \qquad [3.21]$$

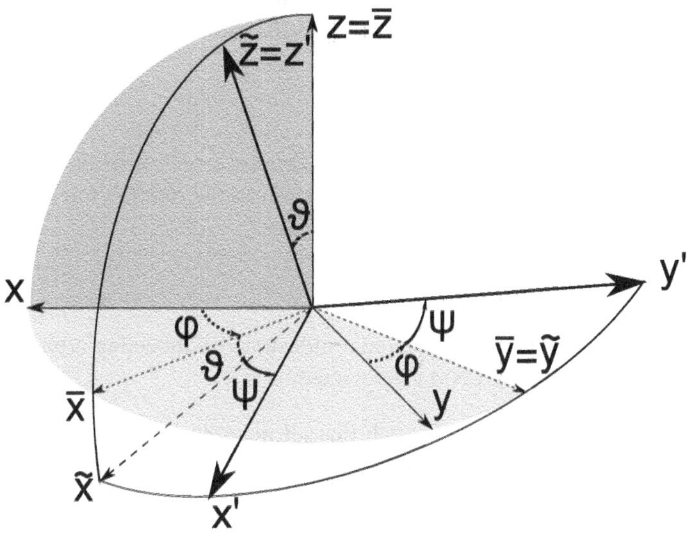

Figure 3.1. *Euler's angles*

Now, we would like to study the infinitesimal rotations. Differentiating [7.21], it holds:

$$dR\, R^T = -R\,(dR)^T = -(dR\, R^T)^T, \qquad [3.22]$$

hence, this matrix is skew-symmetric. There exists $d\psi \in \mathbb{R}^3$ such that:

$$dR\, R^T = j(d\psi),$$

then:

$$dR = j(d\psi)\, R. \tag{3.23}$$

In particular, an infinitesimal rotation around the identity is:

$$dR = j(d\psi). \tag{3.24}$$

Applying it to a vector $v \in \mathbb{R}^3$, we have:

$$dR\, v = d\psi \times v.$$

An infinitesimal rotation is represented by a cross-product and the axial vector $d\psi$ is called the *infinitesimal rotation vector*. Its direction provides the rotation axis and its norm measures the infinitesimal rotation angle. Alternatively, we can adopt the language of time derivatives instead of that of differentials, dividing by *dt* in [3.23]:

$$\dot{R} = j(\varpi)\, R, \tag{3.25}$$

where the axial vector $\varpi(t) = \dot{\psi}(t)$ is called *Poisson's vector*. Its direction provides the *instantaneous rotation axis* and its norm measures the *rotation rate*.

3.2.2. *Rigid motions*

We will reach an important conceptual milestone by modeling arbitrary motions of rigid bodies (and not only USM as previously).

DEFINITION 3.4.– A *rigid body* is such that all material lengths and angles remain unchanged by its motion. The motion of a rigid body is called a *rigid motion*.

The transformations preserving the lengths and angles are Euclidean. Let X and X' be two coordinate systems in the sense of definition 1.6, X being arbitrary given while the particle is at rest in X'. Thus, at a given time t, the position x in X of any material particle of the particle with position x' in the system X' is given by a Euclidean transformation:

$$x = R(t)\, x' + x_0(t). \tag{3.26}$$

In other words, the trajectory of this particle is modeled by the assignment:

$$t \mapsto x = R(t)\, x' + x_0(t),$$

describing the rigid motion. Hence, this brings us to consider the changes of coordinate systems of the space–time $X' \mapsto X$ composed of a rigid motion and a clock change:

$$x = R(t' + \tau_0)\, x' + x_0(t' + \tau_0), \qquad t = t' + \tau_0. \qquad [3.27]$$

A Galilean transformation [1.8] is of the previous form with $x_0(t) = u\, t + k$ and a time independent rotation R but the changes of coordinate systems [3.27] are not in general Galilean transformations. In the following, we suppose that the maps $t \mapsto x_0(t)$ and $t \mapsto R(t)$ are smooth (continuously differentiable as far as needed by the calculations).

Now, we would like to discuss what an infinitesimal rigid motion is. Differentiating [3.27] and taking into account [3.25], we have:

$$dx = (\dot{x}_0 + \varpi \times (Rx'))\, dt' + R\, dx', \qquad dt = dt'.$$

Eliminating x' due to [3.26], we have:

$$dx = (\dot{x}_0 + \varpi \times (x - x_0))\, dt' + R\, dx', \qquad dt = dt',$$

that can be recast in matrix form:

$$dX = P\, dX', \qquad [3.28]$$

where P is a linear Galilean transformation with the *velocity of transport*:

$$\boxed{u = \dot{x}_0(t) + \varpi(t) \times (x - x_0(t)).} \qquad [3.29]$$

If the changes of coordinate systems [3.27] are not in general Galilean transformations, such infinitesimal changes are linear Galilean transformations. For this reason, we introduce the following definition (see Comment 1, section 3.6).

DEFINITION 3.5.– The coordinate systems, in the sense of definition 1.6, which are deduced one from the other by changes [3.27], are called *Galilean coordinate systems*.

In a practical point of view, the Galilean transformations can be used as an approximations of [3.27], depending on the considered time and length scales at which we are working. Let us consider a particle of initial position $x_0 = 0$ and initial velocity v, moving with a uniform acceleration a. It is well known that the trajectory equation is:

$$x = v\,t + \frac{1}{2}\,a\,t^2.$$

On the right-hand member, the last term is negligeable with respect to the first one, provided that:

$$\mid t \mid \ll 2\,\frac{\parallel v \parallel}{\parallel a \parallel}.$$

During this time, the particle is almost acceleration free and we see, neglecting the last term, it covers a distance:

$$\parallel x \parallel \approx \parallel v \parallel \mid t \mid \ll 2\,\frac{\parallel v \parallel^2}{\parallel a \parallel}.$$

This leads us to consider a "space–time window" arround the initial event $X = 0$, of very small dimensions with respect to the previous time and distance thresholds, in which a particle is almost gravitation free and the changes of Galilean coordinates can be approximated by Galilean transformations (see Comment 2, section 3.6).

3.3. Galilean gravitation

3.3.1. *How to model the gravitational forces?*

Gravitation is certainly not easy to describe in a consistent way but, as it is impossible to evade it, we have decided to face it forthwith. We would like to generalize the 3.1 of the dynamical torsor in the new context of the coordinate changes [3.27], according to Galileo's principle of relativity 1.1. Even if the torsor $\tilde{\tau}'$ is constant in some of them, it is not so in others, according to transformation law [3.2] where the components of P – the rotation R and velocity of transport [3.29] – depend on time. Thus, law 3.1 is only valid for the USM and cannot be generalized on its own at the Galilean coordinate systems.

Taking into account [1.16] and [3.1], transformation law [3.2] itemizes as:

$$T = P\,T', \qquad J = P\,J'\,P^T + C(P\,T')^T - (P\,T')\,C^T. \qquad [3.30]$$

In definition 1.9, we presented the force as a phenomenon modifying the velocity of a particle. We would now like to give a more precise formulation in the context of rigid motions. As the linear momentum is the product of the mass by the velocity and the mass is constant, we claim the time derivative of the linear momentum is equal to the resultant force:

$$\dot{p} = F.$$

To develop some intuition of the suitable generalization, we consider a particle of mass m, at rest in some Galilean coordinate system X', hence $v' = 0$. For an observer turning at constant rotation velocity ϖ around an axis perpendicular to the plane containing both the observer and the particle, this last one moves in this plane along a circle, then is deflected from the straight line, revealing the presence of a force. For the observer rotating at $x_0 = 0$ and working with a Galilean coordinate system X, the velocity of the particle v is given by the velocity addition formula [1.13] and the velocity of transport [3.29]:

$$\dot{x} = v = u + v' = \varpi \times x. \qquad [3.31]$$

Poisson's vector ϖ being constant, we have:

$$\dot{p} = m\dot{v} = m\varpi \times v = m\varpi \times (\varpi \times x). \qquad [3.32]$$

For convenience, let the plane be x^1x^2, then $\varpi = \omega e_3$ and, for the observer, the particle is subjected to a force directed toward the axis:

$$\dot{p} = -m\omega^2 x. \qquad [3.33]$$

This is the *centripetal force*, responsible for the observed deflection.

In terms of torsor component T, what have we done? We derivated the first relation of [3.30] with respect to time, taking into account that T' is constant:

$$\dot{T} = \dot{P}T',$$

It can be easily checked that we recover [3.33]. This particular example suggests that forces can be generated by infinitesimal Galilean transformations. On this ground, we invent a new way to derivate, taking into account the infinitesimal variation of P:

– first, we calculate the time derivative of $T = PT'$;

– next, we consider its limit as X' approaches X.

The result is denoted by $\mathring{T}$ to distinguish it from the classical time derivative $\dot{T}$. The first step reads:

$$\frac{d}{dt}(PT') = P\dot{T}' + \dot{P}T'.$$

Next, when X' approaches X, T' approaches T and P approaches the identity:

$$\mathring{T} = \dot{T} + \dot{P}T.$$

Now, we generalize the law 3.1 by claiming that $\mathring{T}$ vanishes:

$$\dot{T} = -\dot{P}T.$$

The key idea is to consider the right-hand side models of the gravitational forces.

3.3.2. *Gravitation*

Now, we hope to provide a more precise formulation of this sketch. Provisionally, we swap the language of time derivatives for that of differentials. If in each Galilean coordinate system some assignment $X \mapsto P(X)$ is given, by differentiating it with respect to X, we obtain a map:

$$dX \mapsto dP = \Gamma(dX),$$

which is linear because of linking infinitesimal quantities. Conversely, if such a map is given, does there exist a map $X \mapsto P(X)$ of which Γ is the differential? In fact, there is a hidden trap to avoid. Indeed, let $\mathbf{X}_i$ and $\mathbf{X}_f$ be two distinct events, represented in a Galilean coordinate system by X_i and X_f. Considering a path $\mathcal{C}_1$ from $\mathbf{X}_i$ to $\mathbf{X}_f$, we have:

$$X_i - X_f = \int_{\mathcal{C}_1} dX.$$

independently of the choice of the path. $\mathcal{C}_2$ being another path from $\mathbf{X}_i$ to $\mathbf{X}_f$, let us consider the loop $\mathcal{C}$ obtained by concatenation of $\mathcal{C}_1$ and $\mathcal{C}_2$. If the sense of $\mathcal{C}$ is the same as $\mathcal{C}_1$ and opposite that of $\mathcal{C}_2$, we have:

$$\oint_{\mathcal{C}} dX = \int_{\mathcal{C}_1} dX - \int_{\mathcal{C}_2} dX = 0.$$

Considering another Galilean coordinate system X', we have to satisfy:

$$\oint_{\mathcal{C}'} PdX' = 0. \qquad [3.34]$$

$\mathcal{C}$ being the image of $\mathcal{C}'$ by the change of coordinate system $X' \mapsto X$.

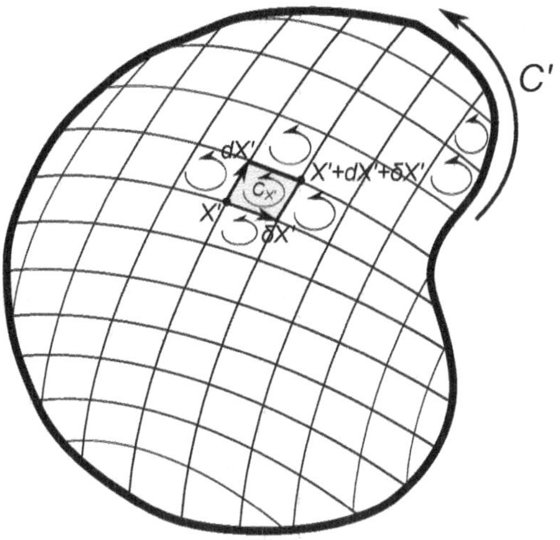

Figure 3.2. *Mesh of elementary parallelograms*

For practical purposes, let us consider a parallelogram $\mathcal{C}_{X'}$ of which the size approaches zero (Figure 3.2). We can think in infinitesimal terms. Considering two distinct infinitesimal perturbations dX' and $\delta X'$ of X', the vertices of the elementary parallelogram are $X', X' + dX', X' + \delta X', X' + dX' + \delta X'$. Along the edge from X' to $X' + dX'$ of infinitesimal length, the transformation P is constant and equal to its value at X'. Along the edge from $X' + dX'$ to $X' + dX' + \delta X'$, the transformation is constant and equal to its value $P + dP$ at $X' + dX'$. Reasoning in a similar way for the two other edges, condition [3.34] reads:

$$PdX' + (P + dP)\,\delta X' - (P\delta X' + (P + \delta P)\,dX') = 0,$$

hence, after simplification:

$$dP\,\delta X' - \delta P\,dX' = 0. \qquad [3.35]$$

Conversely, if this last condition is satisfied at any X and for any perturbations dX' and $\delta X'$, condition [3.34] is true for any loop $\mathcal{C}'$. Indeed, let us consider a surface of which the boundary is $\mathcal{C}'$ and let us mesh it with elementary parallelograms (Figure 3.2). Adding up over all the meshes, the contributions over adjoining edges annihilate, the edges being run along twice in opposite sense. There only remains the contributions along the boundary $\mathcal{C}'$. Conditions [3.34] and [3.35] are equivalent but – from a practical viewpoint – the latter is useful. Hence, we claim that the map Γ must satisfy (see Comment 3, section 3.6):

$$\forall dX', \delta X', \qquad \Gamma(dX')\,\delta X' - \Gamma(\delta X')\,dX' = 0. \qquad [3.36]$$

3.3.3. *Galilean gravitation and equation of motion*

We define the *covariant differential* of T with respect to Γ as:

$$\mathrm{d}T = d(P\,T')|_{X'=X} = (P\,dT' + dP\,T')\,|_{X'=X} = (P\,dT' + \Gamma(dX)\,T')\,|_{X'=X}\,.$$

When X' approaches X, T' approaches T and P approaches the identity:

$$\mathrm{d}T = dT + \Gamma(dX)\,T. \qquad [3.37]$$

THEOREM 3.1.– The maps $dX \mapsto dP = \Gamma(dX)$ of which the values are infinitesimal Galilean transformations and satisfying condition [3.36] are of the following form:

$$\boxed{\Gamma(dX) = \begin{pmatrix} 0 & 0 \\ j(\Omega)\,dx - g\,dt & j(\Omega)\,dt \end{pmatrix},} \qquad [3.38]$$

where $g \in \mathbb{R}^3$ is called the *gravity* and $\Omega \in \mathbb{R}^3$ is called the *spinning*.

PROOF.–

Differentiating the expression of P in [1.9] and taking into account [3.24], an infinitesimal Galilean transformation around the identity reads:

$$dP = \begin{pmatrix} 0 & 0 \\ du & j(d\varpi) \end{pmatrix}. \qquad [3.39]$$

where the 3-columns du and $d\varpi$ linearly depend on dx and dt. Thus, there exist 3×3 matrices A, B and 3-columns Ω, g such that:

$$d\varpi = A\,dx + \Omega\,dt, \qquad du = B\,dx - g\,dt. \qquad [3.40]$$

Leaving out the primes, condition [3.36] reads:

$$\begin{pmatrix} 0 & 0 \\ du & j(d\varpi) \end{pmatrix} \begin{pmatrix} \delta t \\ \delta x \end{pmatrix} = \begin{pmatrix} 0 & 0 \\ \delta u & j(\delta\varpi) \end{pmatrix} \begin{pmatrix} dt \\ dx \end{pmatrix},$$

and is reduced to:

$$du\,\delta t - \delta u\,dt + j(d\varpi)\,\delta x - j(\delta\varpi)\,dx = 0.$$

Introducing expressions [3.40] into this last equation, we obtain after simplification and some simple algebraic manipulations:

$$(Tr(A)1_{\mathbb{R}^3} - A)^T dx \times \delta x + (B - j(\Omega))\,(\delta x\, dt - dx\, \delta t) = 0.$$

The infinitesimal perturbations $dX, \delta X$ being arbitrary, it is satisfied if and only if:

$$Tr(A)1_{\mathbb{R}^3} = A, \qquad B = j(\Omega). \tag{3.41}$$

The former equation is satisfied if and only if the matrix A is null. Introducing [3.40] into [3.39] and taking into account [3.41]:

$$du = j(\Omega)\, dx - g\, dt, \qquad d\varpi = \Omega\, dt, \tag{3.42}$$

that achieves the proof. ■

Dividing both members of relation [3.37]) by dt, owing to the linearity of the map Γ and definition [1.12] of the 4-velocity U, we define the *covariant derivative*:

$$\mathring{T} = \dot{T} + \Gamma(U)\, T, \tag{3.43}$$

that allows us to claim:

LAW 3.2.– For any particle only subjected to a *Galilean gravitation* [3.38], the trajectory is governed by the equation:

$$\mathring{T} = 0.$$

Taking into account [1.12] and [3.38], we have:

$$\Gamma(U) = \begin{pmatrix} 0 & 0 \\ \Omega \times v - g & j(\Omega) \end{pmatrix}, \tag{3.44}$$

and law 3.2 reads:

$$\begin{pmatrix} \dot{m} \\ \dot{p} \end{pmatrix} + \begin{pmatrix} 0 & 0 \\ \Omega \times v - g & j(\Omega) \end{pmatrix} \begin{pmatrix} m \\ p \end{pmatrix} = \begin{pmatrix} 0 \\ 0 \end{pmatrix},$$

providing:

$$\dot{m} = 0, \qquad \dot{p} = m\,(g - 2\,\Omega \times v). \tag{3.45}$$

The first equation means the mass is time independent along the trajectory. The second equation gives the time rate of linear momentum in terms of the gravitation. Introducing the expression of p given by [3.17] into this equation and because the mass does not depend on time, we have:

$$m\dot{v} = m\,(g - 2\,\Omega \times v), \qquad [3.46]$$

which leads to *Souriau's equation of motion*:

$$\boxed{m\ddot{x} = m\,(g - 2\,\Omega \times v),} \qquad [3.47]$$

allowing us to determine the trajectory of the particle (see Comment 4, section 3.6).

3.3.4. *Transformation laws of the gravitation and acceleration*

Before applying this law, we still have a few essential details to be settled. Indeed, we must not lose track of Galileo's principle of relativity 1.1. Guided by it, we claim this law is the same in all the Galilean coordinate systems. Thus, in another Galilean coordinate system X', we must have:

$$\mathrm{d}T' = dT' + \Gamma'(dX')\,T'.$$

Introducing the first relation of [3.30] into [3.37], differentiating the products and taking into account [3.28] gives:

$$\mathrm{d}T = d(P\,T') + \Gamma(dX)\,P\,T' = P\,(dT' + (P^{-1}\Gamma(P\,dX')\,P + P^{-1}dP)\,T'),$$

that is:

$$\mathrm{d}T = P\,\mathrm{d}T', \qquad [3.48]$$

provided that the following *transformation law of the gravitation* is satisfied:

$$\Gamma'(dX') = P^{-1}(\Gamma(P\,dX')\,P + dP). \qquad [3.49]$$

Dividing both members of [3.48] by $dt = dt'$ leads to:

$$\mathring{T} = P\mathring{T}'. \qquad [3.50]$$

As the matrix P is regular, law 3.2 is valid in any Galilean coordinate system, then consistent with Galileo's principle of relativity. Dividing [3.49] by dt provides:

$$\Gamma'(U') = P^{-1}(\Gamma(U)\, P + \dot{P}).$$

THEOREM 3.2.– In a Galilean coordinate change $X' \mapsto X$, a Galilean gravitation is modified according to the transformation laws:

$$\boxed{\Omega = R\,\Omega' - \varpi,} \qquad [3.51]$$

$$\boxed{g - 2\,\Omega \times v = a_t + R\,(g' - 2\,\Omega' \times v').} \qquad [3.52]$$

where:

$$\boxed{a_t = \dot{u} + \varpi \times (v - u),} \qquad [3.53]$$

is called the *acceleration of transport*.

PROOF.–

Indeed, owing to [1.9], [1.12], [1.15], [3.25] and [3.38], we have:

$$\Gamma' = \begin{pmatrix} 1 & 0 \\ -R^T u & R^T \end{pmatrix} \left(\begin{pmatrix} 0 & 0 \\ \Omega \times v - g & j(\Omega) \end{pmatrix} \begin{pmatrix} 1 & 0 \\ u & R \end{pmatrix} + \begin{pmatrix} 0 & 0 \\ \dot{u} & j(\varpi)\, R \end{pmatrix} \right).$$

Identifying the result of the calculation to the standard form [3.44] in the Galilean coordinate system X':

$$\Gamma'(U') = \begin{pmatrix} 0 & 0 \\ \Omega' \times v' - g' & j(\Omega') \end{pmatrix},$$

owing to the linearity of j and [7.23], we have:

$$j(\Omega') = j(R^T(\Omega + \varpi)), \qquad \Omega' \times v' - g' = R^T(\Omega \times v - g + \dot{u} + \Omega \times u). \quad [3.54]$$

As the map j is one-to-one, the first relation leads to transformation law [3.51] for the spinning Ω. By obvious manipulations, the second relation of [3.54] reads:

$$g - \Omega \times v = \dot{u} + \Omega \times u + R\,(g' - \Omega' \times v').$$

We are now interested in the right-hand side of the second equation of [3.45]. Owing to the following expression, it holds:

$$g - 2\Omega \times v = \dot{u} + \Omega \times (u - v) + R\,(g' - \Omega' \times v').$$

According to [3.51], we have:

$$g - 2\Omega \times v = \dot{u} - \varpi \times (u - v) + (R\,\Omega') \times (u - v) + R\,(g' - \Omega' \times v').$$

Taking into account [1.13] and [7.22], we transform the third term of the right-hand side:

$$g - 2\Omega \times v = \dot{u} + \varpi \times (v - u) - R\,(\Omega' \times v') + R\,(g' - \Omega' \times v').$$

We obtain transformation law [3.52] for the right-hand side of the second equation of [3.45]. ■

Concerning the previous theorem, we would like to make two comments about the terminology:

– the reason Ω is called the spinning is that the corresponding transformation law [3.51], taking into account ϖ, represents a time rate of rotation;

– a_t is called acceleration of transport because, substituting u into x_0 and v into x in [3.29], it is the analogous of the velocity of transport.

Before going further, we wish to check that both members of the second equation [3.45] are identically transformed under any change of the Galilean coordinate system $X \mapsto X'$. Taking into account [3.46], we have:

$$\dot{v} = g - 2\Omega \times v.$$

Time derivating both members of [1.13] and taking into account [3.25] provides:

$$\dot{v} = \dot{u} + j(\varpi)\,R\,v' + R\,\dot{v}' = \dot{u} + \varpi \times (R\,v') + R\,\dot{v}'.$$

Taking into account [1.13], we transform the second term, which leads to the *transformation law of the acceleration*:

$$\boxed{\dot{v} = a_t + R\,\dot{v}',} \qquad [3.55]$$

which fits [3.52].

Next, we wish to discuss the structure of the acceleration of transport. Time derivating expression [3.29] of the velocity of transport and introducing it into [3.53] provides:

$$a_t = \ddot{x}_0 + \dot{\varpi} \times (x - x_0) + \varpi \times (\dot{x} - \dot{x}_0) + \varpi \times (v - u), \qquad [3.56]$$

in which we eliminate $\dot{x}_0$ due to [3.29]:

$$\dot{x} - \dot{x}_0 = v - u + \varpi \times (x - x_0),$$

which leads to the *decomposition of the acceleration of transport* into four terms:

$$\boxed{a_t = \ddot{x}_0 + \dot{\varpi} \times (x - x_0) + \varpi \times (\varpi \times (x - x_0)) + 2\varpi \times (R\,v'),} \qquad [3.57]$$

also, owing to [7.16], equal to:

$$a_t = \ddot{x}_0 + \dot{\varpi} \times (x - x_0) + (\varpi \cdot (x - x_0))\varpi - \parallel \varpi \parallel^2 (x - x_0) + 2\varpi \times (R\,v').$$

These expressions are general and allow us, for instance, to recover the centripetal force of section 3.3.1 by taking $v' = x_0 = 0$ and considering ϖ is time independent. Throughout this book, we conform to the standard terminology with a few exceptions. This is one of them, the acceleration of transport referring in the literature only to the three first terms of [3.57].

It is worth noting that theorem 3.2 gives the transformation law of Ω and $g - 2\,\Omega \times v$ but not directly of the gravity g. Let us provide a transformation law depending only on the event (through x and t) and not on the velocity v, i.e. depending also on the neighbor events on the trajectory. Taking into account [1.13], [3.51] and [7.22], transformation law [3.52] becomes:

$$g - 2\,\Omega \times v = a_t + R\,g' - 2\,(R\Omega') \times (Rv') = a_t + R\,g' - 2\,(\Omega + \varpi) \times (v - u), \qquad [3.58]$$

Let us note that the acceleration of transport [3.56] is an affine function of v:

$$a_t = a_t^* - \varpi \times u + 2\,\varpi \times v, \qquad [3.59]$$

where ϖ, u and:

$$a_t^* = \ddot{x}_0 + \dot{\varpi} \times (x - x_0) - \varpi \times \dot{x}_0, \qquad [3.60]$$

depend on x and t but not on v. Introducing [3.59] into [3.58] leads to:

$$g = a_t^* + \varpi \times u + 2\Omega \times u + R g'. \quad [3.61]$$

Eliminating $\dot{x}_0$ in [3.60] due to [3.29] and putting the expression of a_t^* into the previous relation provides the *transformation law of the gravity*:

$$\boxed{g = \ddot{x}_0 + \dot{\varpi} \times (x - x_0) + \varpi \times (\varpi \times (x - x_0)) + 2\Omega \times u + R g',} \quad [3.62]$$

where the spinning Ω is given by [3.51]. Hence, if g' and Ω' depend only on the event through x' and t', g and Ω are explicit functions of x and t due to [3.27], [3.51] and [3.62].

Before going further, let us have a look once again at the example of section 3.3.1. Let us consider a Galilean coordinate system X' such that $g' = \Omega' = 0$. In the absence of other forces, a particle initially at rest remains so later on. For an observer turning at constant rotation velocity and working with the coordinate system X, we have:

$$x = R(t)x', \qquad x_0 = 0, \qquad \dot{\varpi} = 0, \quad [3.63]$$

and [3.31]. Hence, [3.51 and [3.62] give:

$$\Omega = -\varpi, \qquad g = -\varpi \times (\varpi \times x), \quad [3.64]$$

and equation [3.46] of motion allows recovering [3.32]:

$$m\dot{v} = m\,(g - 2\Omega \times v) = m\,(-\varpi \times (\varpi \times x) + 2\,\varpi \times (\varpi \times x)) = m\,\varpi \times (\varpi \times x).$$

In this example, it is worth to noting that the second term involving Ω is the double of the gravity g and, in general, it cannot be considered small or negligeable.

3.4. Newtonian gravitation

Among all the Galilean gravitations, there exists only one corresponding to our physical world. In classical mechanics where the velocity of the light is infinite, we can state *Newton's law of gravitation*:

LAW 3.3.– There exist particular Galilean coordinate systems, called *inertial* or *Newtonian coordinate systems*, for which the gravitation resulting from a particle of mass m' passing through x' at time t is given by:

$$g = -\frac{k_g\, m'}{\| x - x' \|^2} \frac{x - x'}{\| x - x' \|}, \qquad \Omega = 0, \qquad [3.65]$$

where k_g is the *gravitational constant*, equal to $6,674 \cdot 10^{-11} Nm^2kg^{-2}$. We call this a *Newtonian gravitation* (see Comment 5, section 3.6).

Using transformation law [3.51] and [3.52] allows us to determine the expression of the gravitation in any other Galilean coordinate system, where – it is worth noting – the spinning Ω is not generally null.

We would like to determine the motion of a spinless particle of mass m around the mass m' passing, for convenience, through $x' = 0$ at every t in a Newtonian coordinate system, governed by law 3.2 then the equation of motion [3.47]:

$$m\ddot{x} = m\, g = -m\mu \frac{x}{\| x \|^3}. \qquad [3.66]$$

where $\mu = k_g m'$. The particle being spinless and taking into account [3.65], the time rate of the angular momentum given by [3.17] is:

$$\dot{l} = \frac{d}{dt}(x \times mv) = v \times mv + x \times mg = 0, \qquad [3.67]$$

because g is collinear to x. We discovered an integral of the motion. The information is invaluable. Indeed, we have:

$$x \cdot l = x \cdot (x \times mv) = 0,$$

hence, x lies in the plane orthogonal to the constant angular momentum. Picking z's axis along l and working in polar coordinates (ϱ, ϑ, z), we have:

$$x = \begin{pmatrix} \varrho \cos\vartheta \\ \varrho \sin\vartheta \\ 0 \end{pmatrix}, \qquad v = \dot{x} = \begin{pmatrix} \dot{\varrho}\cos\vartheta - \varrho\dot{\vartheta}\sin\vartheta \\ \dot{\varrho}\sin\vartheta + \varrho\dot{\vartheta}\cos\vartheta \\ 0 \end{pmatrix},$$

$$\ddot{x} = \begin{pmatrix} \ddot{\varrho}\cos\vartheta - 2\dot{\varrho}\dot{\vartheta}\sin\vartheta - \varrho\ddot{\vartheta}\sin\vartheta - \varrho\dot{\vartheta}^2\cos\vartheta \\ \ddot{\varrho}\sin\vartheta + 2\dot{\varrho}\dot{\vartheta}\cos\vartheta + \varrho\ddot{\vartheta}\cos\vartheta - \varrho\dot{\vartheta}^2\sin\vartheta \\ 0 \end{pmatrix}.$$

Performing the dot product of both members of equation [3.66] by $e_\varrho = x \;/\; \parallel x \parallel$ leads to:

$$\ddot{\varrho} - \varrho\dot{\vartheta}^2 = -\frac{\mu}{\varrho^2}. \qquad [3.68]$$

Also, the angular momentum:

$$l = m\, x \times v = \begin{pmatrix} 0 \\ 0 \\ m\varrho^2\dot{\vartheta} \end{pmatrix},$$

is time independent, thus:

$$\dot{\vartheta} = \frac{h}{\varrho^2},$$

where h is a constant. Introducing the previous expression into [3.68] leads to:

$$\ddot{\varrho} - \frac{h^2}{\varrho^3} = -\frac{\mu}{\varrho^2}. \qquad [3.69]$$

With the shrewd choice of the new variable $u = 1 \;/\; \varrho$, we obtain the linear differential equation:

$$\frac{d^2u}{d\vartheta^2} + u = \frac{\mu}{h^2},$$

or, introducing $y = u - \mu/h^2$:

$$\frac{d^2y}{d\vartheta^2} + y = 0.$$

Multiplying by $dy/d\vartheta$ leads to:

$$\frac{dy}{d\vartheta}\frac{d^2y}{d\vartheta^2} + y\frac{dy}{d\vartheta} = \frac{1}{2}\frac{d}{d\vartheta}\left(\left(\frac{dy}{d\vartheta}\right)^2 + y^2\right) = 0.$$

Hence, there exists a constant C such that:

$$\left(\frac{dy}{d\vartheta}\right)^2 + y^2 = C^2.$$

By integration, we obtain:

$$\vartheta = \int \frac{dy}{\sqrt{C^2 - y^2}} + \vartheta_0 = \arcsin\left(\frac{y}{C}\right) + \vartheta_0,$$

and by inversion:

$$y = C \sin(\vartheta - \vartheta_0).$$

We can make $\vartheta_0 = \pi/2$ by choice of the line $\vartheta = 0$. Then, returning to the variable u, we have:

$$u = \frac{\mu}{h^2} - C \cos\vartheta,$$

and the equation of the trajectory reads:

$$u = \frac{1}{\varrho} = \frac{\mu}{h^2}\,(1 + \varepsilon \cos\vartheta). \qquad [3.70]$$

The trajectory is a conic section of eccentricity ε, the mass m' being situated at a focus of the conic. It is an ellipse, parabola or hyperbola, according as $\varepsilon < 1$, $\varepsilon = 1$ or $\varepsilon > 1$. The motion is completely determined and fits the empirical laws discovered by Kepler (between 1605 and 1618).

It is worth noting the decisive part played by the angular momentum, one of the torsor components, in solving the problem. In the problem concerned, the angular momentum remains an integral of the motion because the Newtonian gravitation [3.65] generates a *central force*. Unfortunately, conversely to what happens in USM, the torsor components are no longer in general integrals of the motion, except by chance as occurred for the angular momentum. Nevertheless, here *integrals of the motion* other than the torsor components can be found:

– *The energy*. The Newtonian gravitation is such that:

$$g \cdot v = -\dot{\phi}, \qquad [3.71]$$

where:

$$\phi = -\frac{\mu}{\| x - x' \|}, \qquad [3.72]$$

is called the *gravitation potential*. Introducing the *kinetic energy*:

$$e = \frac{1}{2} m \| v \|^2, \qquad [3.73]$$

we verify the *total energy*:

$$e_T = e + m\phi, \qquad [3.74]$$

is an integral of the motion because of equation of motion [3.66]:

$$\dot{e}_T = m\dot{v} \cdot v - mg \cdot v = 0.$$

– *Laplace–Runge–Lenz vector.* This is defined as:

$$w_L = v \times l + m\,\phi\,x. \qquad [3.75]$$

Because of the angular momentum conservation [3.67], we have:

$$\dot{w}_L = \dot{v} \times l + m\,\dot{\phi}\,x + m\,\phi\,v.$$

For a spinless particle, expressions [3.17] of the momenta and [3.66] give:

$$\dot{w}_L = g \times (x \times mv) + m\,\dot{\phi}\,x + \phi\,p.$$

using vector triple product [7.16], we obtain:

$$\dot{w}_L = (g \cdot v + \dot{\phi})\,m\,x + (\phi - g \cdot x)\,p.$$

Owing to [3.71] and verifying from [3.65] and [3.72] that the second parenthesis is null, we prove the Laplace–Runge–Lenz vector is an integral of the motion. In general, the integrals of the motion are powerful tools to determine the trajectory. For instance, in Kepler's problem, using both the integrals of the motion l and w_L allows us to show geometrically the trajectory is a conic section without integrating differential equation [3.69]. Indeed, we know from the conservation of the angular momentum that the trajectory lies in the plane orthogonal to l. Next, let us observe that:

$$w_L \cdot x = (x \times v) \cdot l + m\,\phi\,\varrho^2,$$

hence, we have:

$$\| w_L \| \,\varrho\, \cos\vartheta = \frac{1}{m} \,\| l \|^2 - m\mu\varrho,$$

$$(m\,\mu + \| w_L \| \,\cos\vartheta)\,\varrho = m\,h^2,$$

which leads to equation [3.70] of a conic section with the eccentricity:

$$\varepsilon = \frac{\| w_L \|}{m\,\mu}.$$

3.5. Other forces

3.5.1. *General equation of motion*

The gravitation forces are odd, which is the reason why we have set aside a particular presentation for them. The others are very miscellaneous. We already know the reaction forces at a simple support and the internal forces identified when drawing a free body diagram, but there are many others, for instance of electromagnetic or chemical origin. As far as we are concerned here, we only define the minimal properties expected from these forces. Introducing a 4-column:

$$H = \begin{pmatrix} 0 \\ F \end{pmatrix}, \qquad [3.76]$$

where the component $F \in \mathbb{R}^3$ represents the *resultant of the other forces*, we can generalize law 3.2 as follows (*Newton's second law*):

LAW 3.4.– For any particle subjected to a Galilean gravitation and other forces, the trajectory is governed by the equation:

$$\overset{\circ}{T} = H.$$

To be consistent with Galileo's principle of relativity 1.1, taking into account [3.50], H must be transformed under a change of Galilean coordinate systems $X' \mapsto X$ according to the transformation law of a 4-vector:

$$H = P\,H', \qquad [3.77]$$

which, owing to [1.9], leaves the first component null while we recover transformation law [2.10] of the force component F. In detail, equation [3.45] is generalized in the presence of other forces to give *Newton's equation of motion*:

$$\boxed{\dot{m} = 0, \qquad \dot{p} = m\,(g - 2\,\Omega \times v) + F,} \qquad [3.78]$$

or in short:

$$\dot{p} = F_\Gamma + F,$$

where the *gravitation force*:

$$F_\Gamma = m\,(g - 2\,\Omega \times v) = F_g \; - F_C,$$

is decomposed as the difference of:

– the *gravity force* $F_g = m\,g$;

– and the *Coriolis force* $F_C = 2\,m\,\Omega \times v$.

Nevertheless, we should be careful about these notations because the nature and transformation laws of F_Γ and F are completely different.

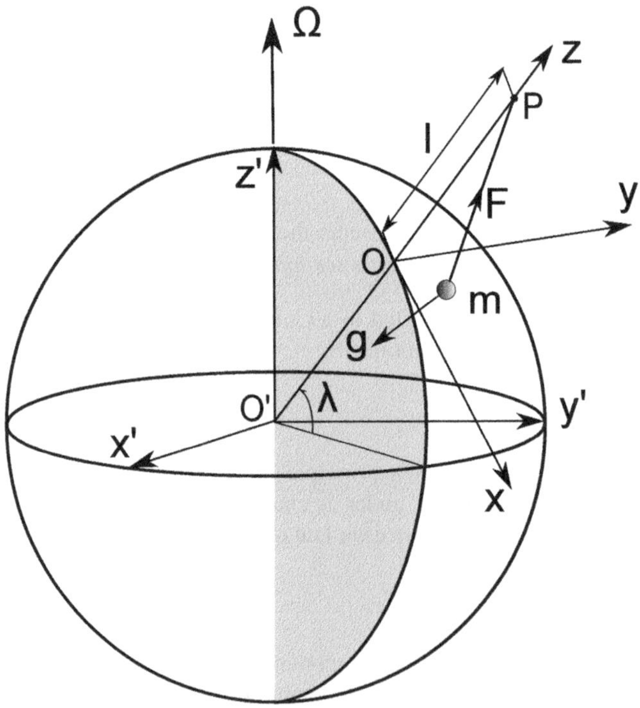

Figure 3.3. *Foucault's pendulum*

3.5.2. *Foucault's pendulum*

We would like to determine the motion of a pendulum at the latitude λ in the northern hemisphere (Figure 3.3). The bob, suspended from a point P by a light thread of length l, can be considered as a particle of mass m. To identify the forces, we draw a free body diagram of the bob. Cutting the thread, we consider the tension force F of intensity S along the thread, away from the bob. At this scale, it is reasonable to consider a constant gravity g, directed toward the Earth's center. Let us pick a Galilean coordinate system with $\boldsymbol{O}z$ directed vertically upward (as determined by a plumb line) passing through the suspension $\boldsymbol{P}$ at $z = l$, and $\boldsymbol{O}x$ pointing south. In the absence of more accurate information, the spinning Ω is assumed to be uniform with intensity $\Omega_\oplus$ and directed along the Earth's axis:

$$\Omega = \Omega_\oplus \begin{pmatrix} -\cos\lambda \\ 0 \\ \sin\lambda \end{pmatrix}. \qquad [3.79]$$

Equation [3.78] reads:

$$m\begin{pmatrix} \ddot{x} \\ \ddot{y} \\ \ddot{z} \end{pmatrix} = m \parallel g \parallel \begin{pmatrix} 0 \\ 0 \\ -1 \end{pmatrix} - 2m\Omega_\oplus \begin{pmatrix} -\cos\lambda \\ 0 \\ \sin\lambda \end{pmatrix} \times \begin{pmatrix} \dot{x} \\ \dot{y} \\ \dot{z} \end{pmatrix} + \frac{S}{l}\begin{pmatrix} -x \\ -y \\ l-z \end{pmatrix},$$

leading to:

$$m\ddot{x} = 2m\Omega_\oplus \dot{y}\sin\lambda - \frac{S}{l}x, \qquad [3.80]$$

$$m\ddot{y} = -2m\Omega_\oplus(\dot{x}\sin\lambda + \dot{z}\cos\lambda) - \frac{S}{l}y, \qquad [3.81]$$

$$m\ddot{z} = -m \parallel g \parallel + 2m\Omega_\oplus \dot{y}\cos\lambda + \frac{S}{l}(l-z). \qquad [3.82]$$

The bob moves on the sphere of center $\boldsymbol{P}$, radius l and equation:

$$x^2 + y^2 + (z-l)^2 - l^2 = 0$$

For small disturbances x, y and z, neglecting z with respect to l:

$$z = \frac{1}{2l}(x^2 + y^2),$$

showing that if x, y are small of the first order, z is small of the second order. Neglecting the terms containing z and $\ddot{z}$, equation [3.82] degenerates into:

$$S = m \parallel g \parallel - 2m\Omega_\oplus \dot{y}\cos\lambda.$$

Introducing this expression into former equations [3.80], [3.81] and neglecting terms containing small quantities of second order, $\dot{y}x$, $\dot{y}y$ and $\dot{z}$, we obtain:

$$\ddot{x} - 2\Omega_\oplus \dot{y} \sin\lambda + \omega^2 x = 0,$$

$$\ddot{y} - 2\Omega_\oplus \dot{x} \sin\lambda + \omega^2 y = 0,$$

where $\omega = \sqrt{\parallel g \parallel \; / \, l}$. Introducing $\zeta = x + i\, y$, the pair of equations read:

$$\ddot{\zeta} + 2i\Omega_\oplus \dot{\zeta} \sin\lambda + \omega^2 \zeta = 0,$$

and, neglecting $\Omega_\oplus^2$ with respect to ω^2, the general solution is:

$$\zeta = (A\, e^{i\omega t} + B\, e^{-i\omega t})\, e^{-i\Omega_\oplus t \sin\lambda}.$$

The first factor on the right represents an elliptic motion. The effect of the second factor is to make this ellipse rotate with angular velocity $-\Omega_\oplus \sin\lambda$, which is clockwise in the northern hemisphere and counterclockwise in the southern hemisphere.

The first observation of this slow shift of the ellipse was made by the physicist Léon Foucault in 1851. The usual interpretation of this experiment is that it allows us to observe "in a laboratory" the Earth's rotation about its axis and to measure the corresponding circular frequency $\omega_\oplus$ (see Comment 6, section 3.6). Indeed, let X' be a coordinate system with the space origin at the Earth's center $\boldsymbol{O}'$ and $\boldsymbol{O}'z'$ chosen as the Earth's rotation axis. Assuming this coordinate system X' is Newtonian, we have: $\Omega' = 0$. Swapping X' for X and using the spinning transformation law [3.51], we have:

$$\Omega = R^T \varpi, \qquad [3.83]$$

The rotation being described by Euler's angles $\varphi = \omega_\oplus t$, $\vartheta = \pi/2 - \lambda$ and $\psi = 0$, formula [3.21] provides:

$$R = \begin{pmatrix} \sin\lambda\cos(\omega_\oplus t) & -\sin(\omega_\oplus t) & \cos\lambda\cos(\omega_\oplus t) \\ \sin\lambda\sin(\omega_\oplus t) & \cos(\omega_\oplus t) & \cos\lambda\sin(\omega_\oplus t) \\ -\cos\lambda & 0 & \sin\lambda \end{pmatrix},$$

which entails:

$$\dot{R}R^T = \omega_\oplus \begin{pmatrix} 0 & -1 & 0 \\ 1 & 0 & 0 \\ 0 & 0 & 0 \end{pmatrix}.$$

Using [3.25] – which also reads $\dot{R}R^T = j(\varpi)$ – and [7.10], we deduce Poisson's vector:

$$\varpi = \begin{pmatrix} 0 \\ 0 \\ \omega_\oplus \end{pmatrix}$$

that allows us to recover the spinning [3.79] of the "laboratory" by [3.83], provided that the spinning frequency $\Omega_\oplus$ is equal to the Earth's rotation $\omega_\oplus$.

The experiment was conducted at the Pantheon in Paris due to a $67\,m$ pendulum. With an oscillation amplitude of $3\,m$, the thread deviates by $2°33'$ from the plumb line and the bob rises up $z = 6,7\,cm$, that is 10^{-3} of the thread length l, justifying the above approximation of small perturbations. The proper circular frequency of the pendulum is $\omega = \sqrt{9,81/67} = 0,382\,rd/s$ and its period is $T_p = 2\,\pi/\omega = 16,4\,s$. The circular frequency of the ellipse rotation at Paris's latitude $\lambda = 48°51'24''$ is $\Omega_\oplus \sin\lambda = 5,49 \cdot 10^{-5}\,rad/s$ and the period is $31h\,46'50''$. The rotation velocity of the ellipse is $11,32°$. Thus, Earth's rotation was observable but the accuracy was not better than a few percent. The difference between the sidereal day of $23h\,56'04''$ and the normal $24h$ day is so small (namely $0,27\%$) that the distinction was (and is still now) beyond observation.

3.5.3. *Thrust*

So far, we have assumed that the mass was constant but this hypothesis is not very necessary and we can model within the present framework objects with variable mass. When a body such as a rocket expels mass in a direction, this mass will cause a force of equal magnitude but opposite direction called a *thrust*. Let w be the velocity of the exhaust gases with respect to a Galilean coordinate system X' in which the rocket of mass m is at rest. The thrust is modeled by a 4-column:

$$H' = \dot{m} \begin{pmatrix} 1 \\ w \end{pmatrix}. \qquad [3.84]$$

In a coordinate system X obtained from X' by a boost v, the rocket has a velocity v and the thrust is given by its transformation law [3.77]:

$$H = \begin{pmatrix} 1 & 0 \\ v & 1_{\mathbb{R}^3} \end{pmatrix} \dot{m} \begin{pmatrix} 1 \\ w \end{pmatrix} = \dot{m} \begin{pmatrix} 1 \\ v + w \end{pmatrix}. \qquad [3.85]$$

Owing to [3.44], law [3.4] reads:

$$\dot{m} = \dot{m}, \qquad \dot{p} = m\,(g - 2\,\Omega \times v) + \dot{m}\,(v + w).$$

The former equality is satisfied for any value of the mass change rate $\dot{m}$. Taking into account $p = mv$, the latter one provides the equation of motion:

$$m\dot{v} = m\,(g - 2\,\Omega \times v) + \dot{m}\,w,$$

that must be completed by a phenomenological law governing the mass change rate, depending on the kind of thrust.

3.6. Comments for experts

COMMENT 1.– Conversely, the changes of coordinate systems of the space–time $X' \mapsto X$ such that:

$$P = \frac{\partial X}{\partial X'}$$

is a linear Galilean transformation, are the rigid motions [3.27], as it will be proved further using Frobenius method (theorem 16.6). The compatibility conditions of the system are $R = R(t)$ and [3.29].

COMMENT 2.– This "space–time windows" can be seen as an intuitive interpretation of the tangent space to the space–time manifold.

COMMENT 3.– The gravitation Γ is in fact a symmetric connection on the space–time manifold.

COMMENT 4.– This equation of motion was introduced in this form by Souriau (formula [12.47] together with [12.44], page 133, [SOU 70], English translation [SOU 97]) from symplectic mechanics arguments.

COMMENT 5.– A reason to put $\Omega = 0$ is that in Schwarzschild's solution of Einstein's equations of the general relativity, the Christoffel's symbols corresponding to Ω vanish (even if the velocity of light is finite).

COMMENT 6.– This is only a hypothesis. Rather than measuring the Earth's rotation frequency, Foucault's pendulum experiment allows us to measure the Ω component of the Newtonian gravitation directly, as in the same way observing the free falling particles allows us to measure the g component.

4

Statics of Arches, Cables and Beams

4.1. Statics of arches

4.1.1. *Modeling of slender bodies*

The aim of this chapter is to study the static equilibrium of slender three-dimensional (3D) bodies commonly called *arches*, idealized by one-dimensional (1D) material bodies from a geometrical and mechanical point of view. First, we model the geometry of a slender body by performing the following steps (Figure 4.1):

– define a *mean line* given by the piecewise smooth map $s \mapsto \boldsymbol{Q}(s)$ where s is the arclength with respect to a given reference point of the line (there is an arbitrary in this choice which is part of the modeling);

– to each point $\boldsymbol{Q}$ of the mean line, assign a *cross-section* $\mathcal{S}_Q$ locally orthogonal to the mean line.

$x(s)$ being the position in a Galilean coordinate system of a regular point of the mean line $\boldsymbol{Q}(s)$, we can define the tangent unit vector $\vec{\text{U}}(s)$ represented in the basis of the considered coordinate system by the column:

$$U(s) = \frac{dx}{ds}(s).$$

An arch is slender in the sense that the dimensions of the cross-sections are small with respect to the one of the line. If it is seen from a long way off, it can be considered in a first approximation as geometrically reduced to its mean line. In this spirit, we hope to model the internal and external efforts by an idealized sketch.

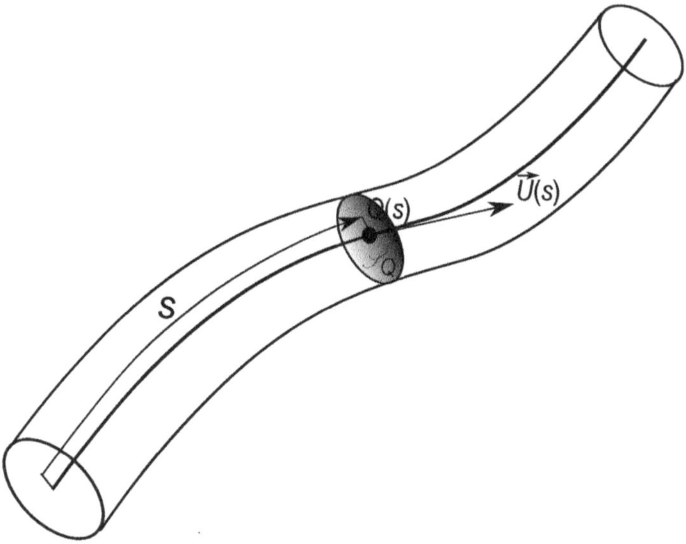

Figure 4.1. *Geometric model of the arch*

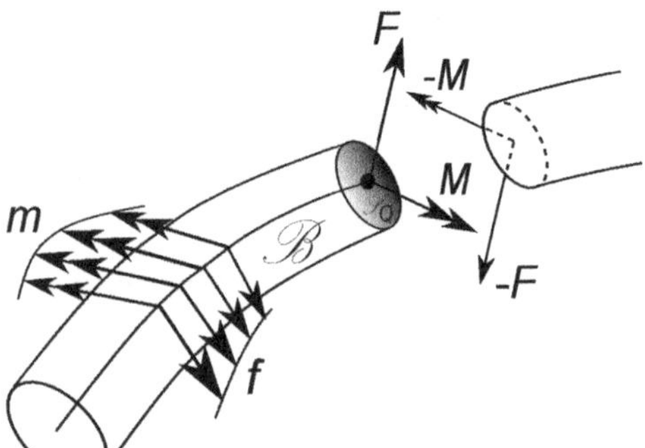

Figure 4.2. *Free body diagram of the part $\mathcal{B}$ of the arch*

We cut through the arch along the cross-section $\mathcal{S}_Q$ and we draw the free body diagram of the upstream part $\mathcal{B}$ of the arch when running along the mean line with increasing s (Figure 4.2). Let us consider the resultant torsor of the internal forces acting through the cross-section $\mathcal{S}_Q$ upon the part $\mathcal{B}$:

$$\check{\mathcal{T}}^{int} = \begin{pmatrix} 0 & F^T \\ -F & -j(M) \end{pmatrix},$$

of which the force F and the moment M with respect to Q taken as origin are represented in the figure, respectively, by a vector and a double vector, according to a usual convention in mechanics. Owing to Newton's third law 2.1, the resultant torsor of the internal forces acting through the cross-section upon the downstream part is equal and opposite to $\breve{\tau}^{int}$ as represented in the figure. The force F is decomposed into:

– the *normal force* $N = F \cdot U$, positive in tension and negative in compression;

– and the *shear force* $T = F - (F \cdot U)U$, tangent to the cross-section.

while the moment M is decomposed into:

– the *torque* $M_t = M \cdot U$, responsible for the torsion of the arch around its mean line;

– and the *bending moment* $M_b = M - (M \cdot U)U$, which tends to modify the curvature of the mean line and to rotate the cross-section.

In the same spirit, we consider that the external efforts can be modeled by a piecewise smooth distribution of exterior forces and moments of which the torsor by length unit is:

$$\frac{d\breve{\tau}^{ext}}{ds} = \begin{pmatrix} 0 & f^T \\ -f & -j(m) \end{pmatrix}, \qquad [4.1]$$

Thus, the modeling defines piecewise smooth assignments $s \mapsto \breve{\tau}^{int}(s)$ and $s \mapsto \frac{d\breve{\tau}}{ds}^{ext}(s)$.

4.1.2. *Local equilibrium equations of arches*

To know if an arch is in equilibrium, we should verify that the resultant torsor of each of its parts is null, which is a difficult task because the number of its parts is infinite. To avoid this pitfall, we would like to establish local equations that are easy to test. The key idea of classical modeling is to consider an arch slice $d\mathcal{B}$ of infinitesimal length ds, leading to differential equations. Let Q be the point on the mean line corresponding to the upstream extremity cross-section and Q' the point corresponding to the downstream extremity one that will be taken as reference origin when adding the moments. If the beam is in static equilibrium, the resultant torsor of the slice vanishes.

To write the equilibrium equation of the slide, we perform the following steps:

– we draw the free body diagram of the slide (Figure 4.3):

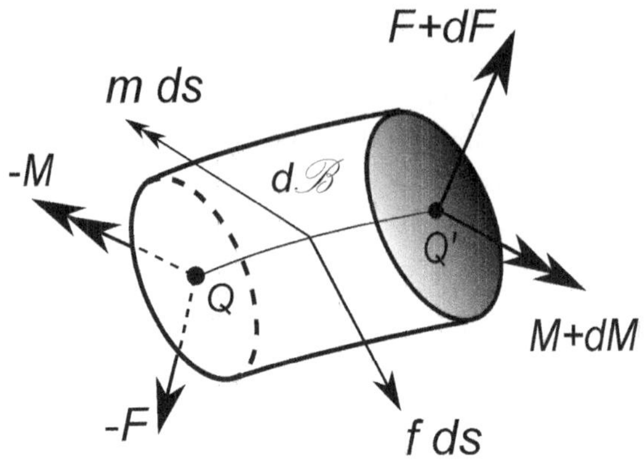

Figure 4.3. *Free body diagram of the slide $d\mathcal{B}$*

- with the external efforts $-F$ and $-M$ acting upon the slice through the upstream extremity cross-section (according to Newton's third law 2.1),

- the external efforts $F + dF$ and $M + dM$ acting upon the slice through the downstream extremity cross-section (taking into account their infinitesimal variation when running from $\boldsymbol{Q}$ to $\boldsymbol{Q}'$),

- and the resultants $f\,ds$ and $m\,ds$ of the external efforts distributed on a length ds with respect to the slide barycenter;

– we construct the resultant torsor of the slide. According to section 2.3.1, the torsors of the forces must be given with respect to the same point taken as origin, let us say $\boldsymbol{Q}'$, before to calculate their sum. The position of $\boldsymbol{Q}$ with respect to itself taken as origin being $x = 0$, its position with respect to the reference origin $\boldsymbol{Q}'$ is $x' = -dx$. According to the transformation law of torsor [2.10] for the infinitesimal spatial translation $dk = x - x' = dx$, the torsor of the external efforts $-F$ and $-M$ acting upon the slice through the upstream extremity cross-section with respect to the new position $\boldsymbol{Q}'$ is:

$$F' = -F, \qquad M' = -M + (-F) \times dx = -M + dx \times F$$

– the balance equation [2.12] of the slide reads:

$$F' + (F + dF) + f\,ds = dF + f\,ds = 0,$$

$$M' + (M + dM) + m\,ds = dx \times F + dM + m\,ds = 0.$$

Dividing by ds leads to the *local equilibrium equations of arches*:

$$\boxed{\frac{dF}{ds} + f = 0} \qquad [4.2]$$

$$\boxed{\frac{dM}{ds} + U \times F + m = 0} \qquad [4.3]$$

It is worth noting that the transport of the moment of the external forces from the barycenter to x' is not necessary because it generates in the moment equilibrium equation an additional infinitesimal term of the second order which can be neglected.

In the spirit of Chapter 3, we now hope to find again these equations by introducing the *covariant differential* of the torsor of internal efforts defined as:

$$\mathbf{d}\breve{\tau} = d(\breve{P}\,\breve{\tau}'\breve{P}^T)|_{x'=x} = (\breve{P}\,d\breve{\tau}'\breve{P}^T + d\breve{P}\,\breve{\tau}'\breve{P}^T + \breve{P}\,\breve{\tau}'d\breve{P}^T)\,|_{x'=x}\,.$$

When x' approaches x, $\breve{\tau}'$ approaches $\breve{\tau}$ and $\breve{P}$ approaches the identity:

$$\mathbf{d}\breve{\tau} = d\breve{\tau} + d\breve{P}\,\breve{\tau} + \breve{\tau}\,d\breve{P}^T. \qquad [4.4]$$

Considering as previously an infinitesimal translation $dk = dx$,

$$d\breve{P} = d\begin{pmatrix} 1 & 0 \\ k & R \end{pmatrix} = \begin{pmatrix} 0 & 0 \\ dx & 0 \end{pmatrix},$$

and owing to [7.11], the covariant differential of the torsor of the internal efforts reads:

$$\mathbf{d}\breve{\tau}^{int} = \begin{pmatrix} 0 & \mathbf{d}F^T \\ -\mathbf{d}F & -j(\mathbf{d}M) \end{pmatrix}, \qquad [4.5]$$

with:

$$\mathbf{d}F = dF, \qquad \mathbf{d}M = dM + dx \times F.$$

It is easy to verify that the local equilibrium equations [4.2] and [4.3] can be recast in the following compact form:

$$\frac{d\underset{\smile}{\mathcal{T}}^{int}}{ds} + \frac{d\underset{\smile}{\mathcal{T}}^{ext}}{ds} = 0. \qquad [4.6]$$

It is now possible to determine the distribution of internal efforts with respect to the external ones and the reactions at the ends of any part $\boldsymbol{AB}$ of the arch. Integrating [4.2] and owing to $F(s_A) = -R_A$ gives the internal force distribution:

$$F(s) = -\int_{s_A}^{s} f(s')\, ds' - R_A. \qquad [4.7]$$

According to the previous result and owing to $M(s_A) = -M_A$, the integration of [4.3] provides the distribution of the internal moment with respect to $\boldsymbol{A}$ taken as origin:

$$M(s) = -\int_{s_A}^{s} \left(m(s') + U(s') \times \int_{s_A}^{s'} f(s'')\, ds'' \right) ds' \\ - \left((x(s) - x(s_A)) \times R_A + M_A \right). \qquad [4.8]$$

Taking into account the conditions $F(s_B) = R_B$ and $M(s_B) = M_B$, the global equilibrium of the part $\boldsymbol{AB}$ reads:

$$\int_{s_A}^{s_B} f(s')\, ds' + R_A + R_B = 0.$$

$$\int_{s_A}^{s_B} \left(m(s') + U(s') \times \int_{s_A}^{s_B} f(s'')\, ds'' \right) ds' + (x(s_B) \\ - x(s_A)) \times R_A + M_A + M_B = 0$$

4.1.3. *Corotational equilibrium equations of arches*

In the previous section, we considered a moving coordinate system by translation of the origin along the mean line without rotation. We may also change the point of view by considering both translation and rotation. In other words, we assign to each point $\boldsymbol{Q}$ of the mean line a coordinate system x of which the origin is $\boldsymbol{Q}$, the corresponding coordinate system x' assigned to the downstream neighbor point $\boldsymbol{Q}'$ on the mean line being obtained from x by an infinitesimal Euclidean transformation $d\breve{P}$. According to [3.24], we have:

$$d\breve{P} = d\begin{pmatrix} 1 & 0 \\ k & R \end{pmatrix} = \begin{pmatrix} 0 & 0 \\ dx & j(d\psi) \end{pmatrix}.$$

Calculating the covariant differential of the torsor of the internal efforts by [4.4] gives:

$$\mathbf{d}\breve{\tau}^{int} = \begin{pmatrix} 0 & (dF + j(d\psi)F)^T \\ -(dF + j(d\psi)F) & -j(dM) + dr\, F^T - F\, dr^T - j(d\psi)j(M) + j(M)j(d\psi) \end{pmatrix}.$$

Introducing the column:

$$\Omega = \frac{d\psi}{ds}, \qquad [4.9]$$

taking into account [7.11], [7.14], the linearity of j, [4.5] and [4.6], the *corotational equilibrium equations of arches* reads:

$$\boxed{\frac{dF}{ds} + \Omega \times F + f = 0} \qquad [4.10]$$

$$\boxed{\frac{dM}{ds} + \Omega \times M + U \times F + m = 0} \qquad [4.11]$$

It is worth noting that these equations are true for any choice of Galilean coordinate system x moving along the curve and the associate moving orthonormal basis S. The assignment $s \mapsto S(s)$ is called a *moving frame.*

4.1.4. *Equilibrium equations of arches in Fresnet's moving frame*

Now, we specialize the corotational equilibrium equations to a particular moving frame due to Fresnet. We construct the moving orthonormal basis by adjoining to $\vec{\mathbf{U}}$ two vectors, $\vec{\mathbf{V}}$ and $\vec{\mathbf{W}}$, respectively, represented in the reference basis by the columns V and W . Differentiating $\| U \|^2 = U \cdot U = 1$ leads to:

$$U \cdot \frac{dU}{ds} = 0.$$

The normed vector V in the direction of dU/ds, called the *normal,* is orthogonal to U. The *curvature* κ of the mean line is defined by:

$$\frac{dU}{ds} = \kappa V. \qquad [4.12]$$

Next, we define the *binormal*:

$$W = U \times V, \qquad [4.13]$$

orthogonal to U and V. As V and U are orthogonal, owing to [7.18], W is a unit vector. The orthonormal basis $(\vec{\mathbf{U}}, \vec{\mathbf{V}}, \vec{\mathbf{W}})$ is called *Fresnet's basis*. Differentiating $\| W \|^2 = 1$ shows that dW/ds is orthogonal to W. On the other hand, differentiating [4.13] and taking into account [4.12] shows that $dW\,/\,ds$ is also orthogonal to U. The *torsion* θ of the mean line is defined by:

$$\frac{dW}{ds} = \theta V. \qquad [4.14]$$

As the basis (U, V, W) is orthonormal, its variation when running along the mean line of a length ds is an infinitesimal rotation. Owing to [4.12] and [4.14], it reads:

$$(dU, dV, dW) = (U, V, W) \begin{pmatrix} 0 & -\kappa & 0 \\ \kappa & 0 & \theta \\ 0 & -\theta & 0 \end{pmatrix} ds.$$

Then, [4.9] is:

$$\Omega = \begin{pmatrix} -\theta \\ 0 \\ \kappa \end{pmatrix}. \qquad [4.15]$$

In Fresnet's basis, the force acting onto the cross-section is represented by the columns:

$$F = \begin{pmatrix} N \\ T_n \\ T_b \end{pmatrix},$$

where N is the normal force and T_n, T_b are, respectively, the shear forces with respect to the normal and binormal, while the moment is represented by:

$$M = \begin{pmatrix} M_t \\ M_n \\ M_b \end{pmatrix},$$

where M_t is the torque and M_n, M_b are, respectively, the bending moments with respect to the normal and binormal. Taking into account [4.15], the corotational equilibrium equations [4.10] and [4.11] in Fresnet's moving frame read:

$$\frac{dN}{ds} + \kappa T_n + f_t = 0, \frac{dT_n}{ds} - \kappa N - \theta T_b + f_n = 0, \frac{dT_b}{ds} + \theta T_n + f_b = 0, \quad [4.16]$$

$$\frac{dM_t}{ds} + \kappa M_n + m_t = 0, \frac{dM_n}{ds} - \kappa M_t - \theta M_b - T_b + m_n = 0, \frac{dM_b}{ds} + \theta M_n + T_n + m_b = 0. \quad [4.17]$$

As an application, let us consider a *helical coil spring* (Figure 4.4). The mean line is a helix of radius a and pitch $2\pi h$, parametrized in a reference coordinate system $\boldsymbol{O}x'y'z'$ by:

$$x'(s) = \begin{pmatrix} a\cos\vartheta \\ a\sin\vartheta \\ h\vartheta \end{pmatrix}.$$

where $\vartheta = s/b$ with $b = \sqrt{a^2 + h^2}$. Its curvature and torsion are:

$$\kappa = \frac{a}{b^2}, \qquad \theta = -\frac{h}{b^2},$$

and the representation of Fresnet's basis in the reference basis is given by the rotation:

$$R = (U, V, W) = \begin{pmatrix} -\frac{a}{b}\sin\vartheta & -\cos\vartheta & \frac{h}{b}\sin\vartheta \\ \frac{a}{b}\cos\vartheta & -\sin\vartheta & -\frac{h}{b}\cos\vartheta \\ \frac{h}{b} & 0 & \frac{a}{b} \end{pmatrix}. \quad [4.18]$$

The spring is compressed through horizontal arms $\boldsymbol{OA}$ and $\boldsymbol{BC}$ by two opposite vertical forces R_O and R_C of same intensity P_0 acting, respectively, at the extremities $\boldsymbol{O}$ and $\boldsymbol{C}$. Cutting through the arch along the cross-section $\mathcal{S}_Q$, we draw the free body diagram of the upstream part $\boldsymbol{OAQ}$ of the body (Figure 4.5). The global force equilibrium equation [4.7] of this part gives in the reference basis the internal force F acting through the cross-section $\mathcal{S}_Q$:

$$F = -R_O = -\begin{pmatrix} 0 \\ 0 \\ P_0 \end{pmatrix}.$$

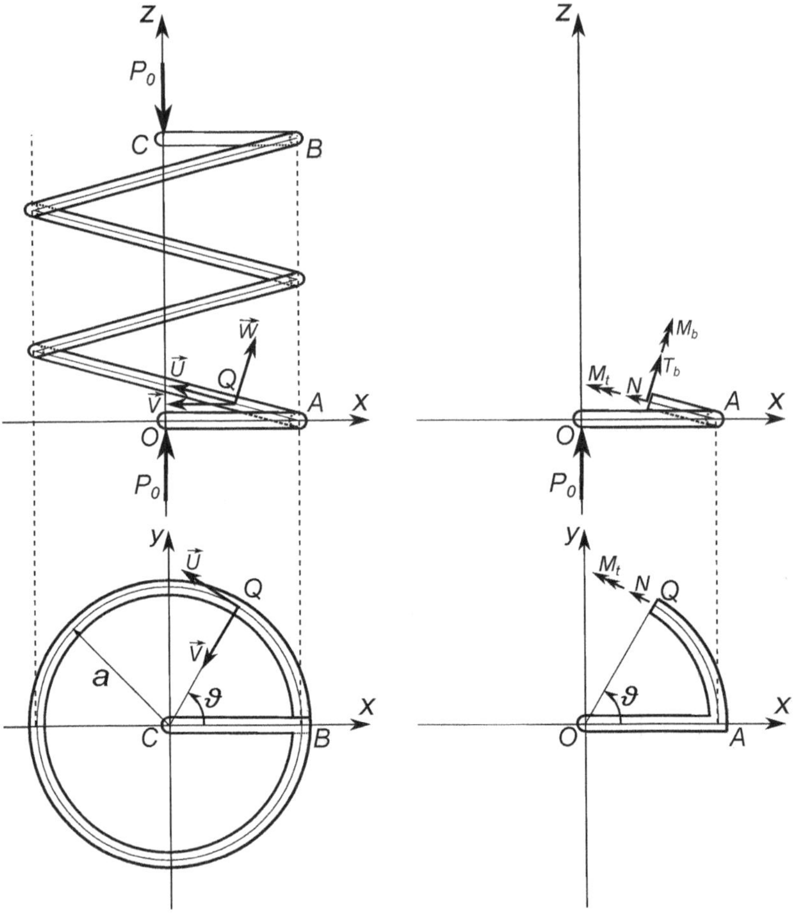

Figure 4.4. *Helical coil spring*

The shear force T_n is obviously null because the external force applied at O is vertical and the normal $\vec{\mathbf{V}}$ is horizontal. It is not represented in the figure. Using the rotation matrix [4.18], we obtain the internal force $\bar{F} = R^T F$ in Fresnet's basis:

$$\bar{F} = \begin{pmatrix} N \\ T_n \\ T_b \end{pmatrix} = -\frac{P_0}{b} \begin{pmatrix} h \\ 0 \\ a \end{pmatrix}. \qquad [4.19]$$

The global moment equilibrium equation [4.9] of $\boldsymbol{OAQ}$ gives in the reference basis the internal moment M acting through the cross-section $\mathcal{S}_Q$ with respect to point $\boldsymbol{Q}$:

$$M = x \times R_O = aP_0 \begin{pmatrix} \sin\vartheta \\ -\cos\vartheta \\ 0 \end{pmatrix},$$

from which we deduce the internal moment $\bar{M} = R^T M$ in Fresnet's basis:

$$\bar{M} = \begin{pmatrix} M_t \\ M_n \\ M_b \end{pmatrix} = \frac{aP_0}{b} \begin{pmatrix} -a \\ 0 \\ h \end{pmatrix}. \qquad [4.20]$$

It can easily be verified that the internal efforts [4.19] and [4.20] satisfy the corotational equilibrium equations [4.17] and [5.28].

4.2. Statics of cables

DEFINITION 4.1.– A *cable* is an arch so thin that it cannot resist moments:

$$M = m = 0.$$

Cable is a generic word for bodies such as *threads* and *strings* which exhibit such a behavior. Moment equilibrium equation [4.3] reads:

$$U \times F = 0,$$

from which it results the internal force F is proportional to the unit tangent vector U:

$$F = N\,U.$$

A cable is in equilibrium only in tension ($N > 0$). Its value and the shape of the cable are determined by solving the remaining equilibrium equation [4.2]:

$$\frac{d}{ds}(N\,U) + f = 0. \qquad [4.21]$$

It is worth noting that, if there is no distributed external force f on a part of a cable, $N\,U$ is constant. Hence, this part is straight and the tension N is constant, in agreement with the experience according to which a taut cable is straight.

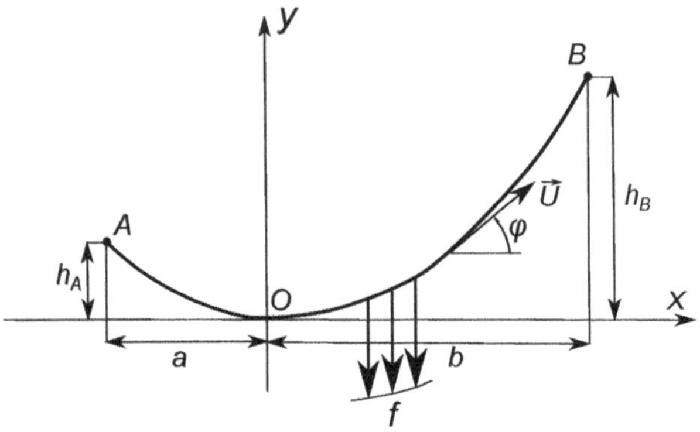

Figure 4.5. *Suspension bridge*

As an example, let us consider a cable $\boldsymbol{AB}$ in the $\boldsymbol{Oxy}$ plane (Figure 4.5). The horizontal distance between the extremities $\boldsymbol{A}$ and $\boldsymbol{B}$ is L and their heights with respect to the $\boldsymbol{Ox}$ axis are, respectively, h_A and h_B. The lack of moment at the extremities is symbolized by hinges. Under a vertical load f of intensity q by length unit and in opposite direction to $\boldsymbol{Oy}$, the cable is supposed to be tangent to the horizontal axis at the origin. φ being the angle of $\vec{U}$ with respect to the horizontal axis, the force equilibrium reads:

$$\frac{d}{ds}(N\,\cos\varphi) = 0, \qquad [4.22]$$

$$\frac{d}{ds}(N\,\sin\varphi) = q. \qquad [4.23]$$

As an application, let us consider a suspension cable of a *suspension bridge*. Introducing p such that:

$$p(x)\,dx = q(s)\,ds,$$

the last equation reads:

$$\frac{d}{dx}(N\,\sin\varphi) = p. \qquad [4.24]$$

integrating equations [4.22] and [4.24] gives:

$$N\,\cos\varphi = C_0, \qquad N\,\sin\varphi = \int_0^x p(x')\,dx' + C_1,$$

where C_0 and C_1 are constants. Eliminating N between these equations leads to:

$$\frac{dy}{dx} = \tan\varphi = \frac{1}{C_0}\left(\int_0^x p(x')\,dx' + C_1\right),$$

and by a new integration, to the shape of the cable:

$$y(x) = \frac{1}{C_0}\left(\int_0^x dx'' \int_0^{x''} p(x')\,dx' + C_1 x\right) + C_2,$$

where C_2 is constant. The proper weight of the cable being negligeable, it is subjected only to the weight of the deck hung on vertical suspenders. This load can be modeled by vertical forces of uniform intensity p by unit of horizontal length, which gives, taking into account the conditions $y(0) = 0$ and $\frac{dy}{dx}(0) = 0$:

$$y(x) = \frac{p}{2C_0}x^2.$$

The three unknowns, the distances a and b of $\boldsymbol{A}$ and $\boldsymbol{B}$ to the origin $\boldsymbol{O}$ and C_0, are determined by the additional conditions:

$$a + b = L, \qquad h_A = \frac{p}{2C_0}a^2, \qquad h_B = \frac{p}{2C_0}b^2.$$

4.3. Statics of trusses and beams

4.3.1. *Traction of trusses*

DEFINITION 4.2.– A *truss* is a straight arch axially loaded and ended by hinges.

They are not subjected to moments, and then the internal force F is proportional to the unit tangent vector U, as for the cables. However, unlike the latter that are in static equilibrium only in tension, trusses exhibit an internal force F in tension or compression. This is determined by equilibrium equation [4.21]. As an application, let us consider a *drilling riser* hung to an offshore platform and loaded by its own weight of intensity p (Figure 4.6). The truss is fixed to the platform with a hinge, while the other extremity is effort free.

Assuming the truss is deformed along its axis, the tangent unit vector is vertical. The force equilibrium equation [4.21] is reduced to:

$$\frac{dN}{ds} + p = 0.$$

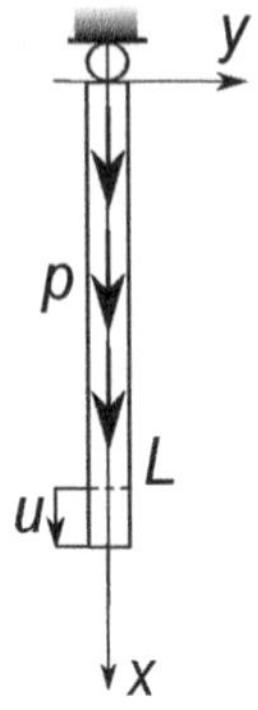

Figure 4.6. *Drilling riser hung to an offshore platform*

Integrating it with null value at the free extremity $x = L$ gives:

$$N(s) = p(L - s). \quad [4.25]$$

To determine the displacement of the points of the mean line, we need some additional assumptions concerning the behavior of the material constituting the body. If the deformation is small, many materials such as metals are *elastic*. They obey *Hooke's law* (1678):

LAW 4.1.– The effort is proportional to the corresponding deformation ("*ut tensio, sic vis*").

Under a normal force N, the material undergoes a deformation measured by the *extension:*

$$\varepsilon_x = \frac{ds - dx}{dx} = \frac{ds}{dx} - 1. \quad [4.26]$$

The *elasticity* law reads:

$$N = K_t \, \varepsilon_x, \quad [4.27]$$

where the *truss stiffness* K_t *a priori* depends on the material properties and the cross-section geometry. For small extensions as in elasticity, the displacements are so small that we can approximate the normal force [4.25] by:

$$N(s(x)) \cong p(L - x).$$

Combining with [4.26] and [4.27] leads to the equation:

$$\frac{ds}{dx} = 1 + \frac{p}{K_t}(L - x).$$

of which, owing to the condition $s(0) = 0$, the solution is for a uniform stiffness K_t:

$$s(x) = x + \frac{p}{K_t}\left(Lx - \frac{x^2}{2}\right).$$

The displacement at the free extremity of the truss is:

$$u = s(L) - L = \frac{p\,L^2}{2K_t}.$$

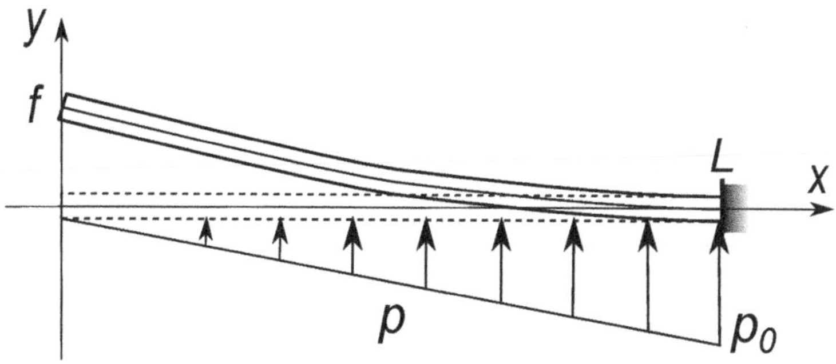

Figure 4.7. *Cantilever beam*

4.3.2. *Bending of beams*

DEFINITION 4.3.– A *beam* is a straight arch transversely loaded.

As an application, we consider a *Cantilever beam* subjected to a linearly distributed load, built-in at the left-hand end, effort free at the other (Figure 4.7). Intuitively, the transversal load induces a vertical *deflexion* $y(x)$ in the direction of the load. For easiness, we use simplified notations for derivatives:

$$y' = \frac{dy}{dx}, \qquad y'' = \frac{d^2y}{dx^2}, \qquad \ldots$$

The arclength element is:

$$ds = \sqrt{dx^2 + dy^2} = dx\sqrt{1 + (y')^2},$$

Fresnet's basis is represented in the reference frame by the rotation matrix:

$$R = (U, V, W) = \begin{pmatrix} \frac{1}{\sqrt{1+(y')^2}} & -\frac{y'}{\sqrt{1+(y')^2}} & 0 \\ \frac{y'}{\sqrt{1+(y')^2}} & \frac{1}{\sqrt{1+(y')^2}} & 0 \\ 0 & 0 & 1 \end{pmatrix}. \qquad [4.28]$$

The torsion is null and the curvature is:

$$\kappa = \frac{y''}{(1+(y')^2)^{3/2}}.$$

Using rotation matrix [4.28], we obtain the internal force $F' = RF$ and internal moment $M' = RM$ in the reference frame:

$$T_x = \frac{1}{\sqrt{1+(y')^2}}(N - y'T_n), \quad T_y = \frac{1}{\sqrt{1+(y')^2}}(y'N + T_n), \quad T_z = T_b,$$

$$M_x = \frac{1}{\sqrt{1+(y')^2}}(M_t - y'M_n),$$

$$M_y = \frac{1}{\sqrt{1+(y')^2}}(y'M_t + M_n), \quad M_z = M_b.$$

For elastic beams, the deflexions are so small that we can assume the sloop y' is negligeable with respect to the unity, leading to the following approximations:

$$ds \cong dx, \quad \kappa \cong y'', \quad T_x \cong N, \quad T_y \cong T_n, \quad M_x \cong M_t, \quad M_y \cong M_n.$$

Hence, force equilibrium equation [4.2] in the reference frame gives:

$$\frac{dN}{dx} = 0, \qquad \frac{dT_n}{dx} + p = 0, \qquad \frac{dT_b}{dx} = 0. \qquad [4.29]$$

Integrating with null values at the free extremity $x = L$, we conclude that N and T_b are identically null. The force equilibrium equation [4.3] in the reference frame gives:

$$\frac{dM_t}{dx} = 0, \qquad \frac{dM_n}{dx} = 0, \qquad \frac{dM_b}{dx} + T_n = 0. \qquad [4.30]$$

Integrating with null values at the free extremity $x = 0$, we conclude that M_t and M_n are identically null. Intuitively, under a bending moment M_b, the straight beam is curved. Applying Hooke's law 4.1, we assume that:

$$M_b = K_b \, \kappa, \qquad [4.31]$$

where the *flexural stiffness* K_b *a priori* depends on the material properties and the cross-section geometry. According to the above approximation on the curvature, combining this relation with the last condition in [4.30], the shear force is for a uniform stiffness K_b:

$$T_n = -K_b y''',$$

and the second condition in [4.29] gives:

$$y'''' = \frac{p}{K_b}.$$

Integrating it with the conditions $y'''(0) = -T_n(0)/K_b = 0$, $y''(0) = M_b(0)/K_b = 0$ at the free extremity and $y(L) = y'(L) = 0$ at the built-in support, we obtain the shape of the deformed beam:

$$y(x) = \frac{p_0}{120 K_b L}(x^5 - 5L^4 x + 4).$$

The deflexion at the free extremity $x = 0$ is:

$$f = \frac{p_0 L^4}{30 K_b}.$$

5

Dynamics of Rigid Bodies

5.1. Kinetic co-torsor

5.1.1. *Lagrangian coordinates*

A rigid body is such that all material lengths and angles remain unchanged by its motion. The Galilean coordinate systems are natural tools to model the rigid body motions, especially those X' in which every point of the body of position s' at any time t' is at rest:

$$v' = \frac{ds'}{dt'} = 0. \qquad [5.1]$$

We say the s'^j are the *Lagrangian coordinates* or the *material coordinates* of the point. Of course, this Lagrangian or material representation is not unique. For a given body, the change of Lagrangian coordinate systems is the time independent Euclidean transformation. Although it is more or less obvious, let us give a proof. If $\bar{s}'$ is the position of the points at time $\bar{t}'$ in another system $\bar{X}'$ of Lagrangian coordinates, it is related to X' by a change [3.27] composed of a rigid motion and a change clock:

$$s' = Q(t)\,\bar{s}' + s_0(t), \qquad \bar{t}' = t' + \tau.$$

Owing to the velocity addition formula [1.13], the corresponding velocity of transport must vanish:

$$u = v' - Q\,\bar{v}' = 0.$$

Taking into account [3.29], it holds for any s and t:

$$v = u = \dot{s}_0(t) + j(\varpi(t))(s' - s_0(t)) = 0,$$

where $\dot{Q} = j(\varpi)Q$. This affine map of s' is identically null if $\dot{s}_0(t) - j(\varpi(t))s_0(t)$ and $j(\varpi(t))$ vanish, then $\varpi(t)$ and $\dot{s}_0(t)$ so are. Taking into account [3.25], Q and s_0 are time independent, which achieves the proof:

$$s' = Q\,\bar{s}' + s_0. \qquad [5.2]$$

5.1.2. *Eulerian coordinates*

In an arbitrary Galilean coordinate system X, the particle of Lagrangian coordinates s' has at time t:

– the position of which the components x^i are called *Eulerian coordinates* or *spatial coordinates*:

$$x = R(t)s' + x_0(t), \qquad [5.3]$$

– and, owing to [1.13] and [5.1], velocity [3.29]:

$$v = u = \dot{x}_0(t) + \varpi(t) \times (x - x_0(t)). \qquad [5.4]$$

where ϖ is Poisson's vector defined by [3.25]:

$$\dot{R} = j(\varpi)\,R. \qquad [5.5]$$

DEFINITION 5.1.– A *vector field* is a distribution of vector in a given region. Vector fields can be graphically represented drawing for each point $\boldsymbol{Q}$ of coordinates x the corresponding bound vector as an arrow of origin $\boldsymbol{Q}$.

For instance, we could display the velocity field v of a rigid body at each time t but this Eulerian representation is blurred by the motion of the body itself. To avert this drawback, it is convenient to pull back the velocity onto the body at rest by drawing for each point $\boldsymbol{Q}'$ of Lagrangian coordinates s' the bound vector $R^T v$. The velocity field [5.4] is determined by the velocity $\dot{x}_0$ of the origin x_0 and Poisson's vector ϖ pulled back as:

$$\dot{x}'_0 = R^T \dot{x}_0, \qquad \varpi' = R^T \varpi. \qquad [5.6]$$

5.1.3. *Co-torsor*

Let us now consider any other Lagrangian coordinate system $\bar{s}'$ such that:

$$x = \bar{R}(t)\bar{s}' + \bar{x}_0(t),$$

Readers can easily deduce from [5.3] and the previous relation that s' and $\bar{s}'$ are related by [5.2] with:

$$Q = R^T \bar{R}, \qquad [5.7]$$

$$s_0 = R^T(\bar{x}_0 - x_0). \qquad [5.8]$$

First, let us observe that Poisson's vector is independent of the choice of the Lagrangian coordinate system because Q is time independent:

$$j(\bar{\varpi}) = \dot{\bar{R}}\bar{R}^T = (\dot{R}Q)(RQ)^T = \dot{R}R^T = j(\varpi),$$

hence, $\bar{\varpi} = \varpi$ because the map j is regular. By pulling back onto the new Lagrangian coordinate system, we have:

$$\dot{\bar{x}}_0' = \bar{R}^T \dot{\bar{x}}_0, \qquad \bar{\varpi}' = \bar{R}^T \varpi. \qquad [5.9]$$

From the previous relation, [5.6] and [5.7], we easily deduce:

$$\bar{\varpi}' = Q^T \varpi'. \qquad [5.10]$$

Next, taking into account that s_0 is time independent, [5.7] leads to:

$$\dot{\bar{x}}_0 = \dot{R}\, s_0 + \dot{x}_0.$$

Combining with [5.9] gives:

$$\dot{\bar{x}}_0' = \bar{R}^T(\dot{R}\, s_0 + \dot{x}_0),$$

and because of [5.7] and [5.6]:

$$\dot{\bar{x}}_0' = Q^T(\dot{x}_0' + R^T \dot{R}\, s_0),$$

but we have, owing to [5.5] and [7.23]:

$$R^T \dot{R} = R^T \dot{R} R^T R = R^T j(\varpi) R = j(R^T \varpi),$$

hence, taking into account [5.6], we obtain:

$$\dot{\bar{x}}_0' = Q^T(\dot{x}_0' + \varpi' \times s_0). \qquad [5.11]$$

As in statics, we work temporarily with the 4-column:

$$\breve{X} = \begin{pmatrix} 1 \\ s' \end{pmatrix},$$

and the 4×4 matrix:

$$\breve{P} = \begin{pmatrix} 1 & 0 \\ s_0 & Q \end{pmatrix}, \qquad [5.12]$$

so Euclidean transformation [5.2] looks like a simple regular linear transformation:

$$\breve{X} = \breve{P}\breve{\breve{X}}. \qquad [5.13]$$

DEFINITION 5.2.– A *co-torsor* $\breve{\gamma}$ is an object represented in a coordinate system by a skew-symmetric 4×4 matrix:

$$\boxed{\breve{\gamma} = \begin{pmatrix} 0 & v^T \\ -v & -j(\omega) \end{pmatrix},} \qquad [5.14]$$

where $v \in \mathbb{R}^3$ is its *velocity*, $\omega \in \mathbb{R}^3$ is its *spin* and of which the components, under Euclidean transformation [5.13], are modified according to the transformation law:

$$\breve{\gamma} = \breve{P}^{-T}\breve{\breve{\gamma}}\breve{P}^{-1}. \qquad [5.15]$$

Applying the rules of the matrix calculus, it is worth noting that if $\breve{\breve{\gamma}}$ is skew-symmetric, $\breve{\gamma}$ given by [5.15] also is. Under an Euclidean transformation (and more generally under an affine transformation), the skew-symmetry property is preserved, which ensures the consistency of the definition. By inversion of [5.15], we have:

$$\breve{\breve{\gamma}} = \breve{P}^{T}\breve{\gamma}\breve{P}. \qquad [5.16]$$

Taking into account [5.12] and [5.14], readers can easily verify the transformation law of the co-torsor components:

$$\bar{v} = Q^T(v + \omega \times s_0), \qquad \bar{\omega} = Q^T\omega. \qquad [5.17]$$

It is also easy to find two invariants under Euclidean transformations:

– the norm of the spin: $\| \omega \|$;

– the dot product of the velocity and the spin: $v \cdot \omega$.

The linear space $\mathbb{M}_{44}^{skew}$ of the 4×4 skew-symmetric matrices is of dimension 6. Let $\mathbf{T}_s^*$ be the set of co-torsors γ, in one-to-one correspondence with the skew-symmetric 4×4 matrices [5.14]. Due to this map, $\mathbf{T}_s^*$ is a linear space of 6 dimensions if we define by structure transport the addition of co-torsors and the multiplication of a co-torsor by a scalar.

It is worth observing the similarity and discrepancy between co-torsors and torsors. Formally comparing the transformation laws of their respective components, namely [2.10] and [5.17], we observe that the velocity v is analogous of the moment M and the spin ϖ of the force F but, comparing the positions of v and ϖ in $\breve{\gamma}$ to those of M and F in $\breve{\tau}$, the analogy is no more relevant. In fact, the co-torsors are objects distinct from the torsors [Comment 1].

On these grounds, the previous study shows that the velocity field of a given rigid body $\mathcal{B}$ can be characterized in the Lagrangian coordinates s' by a co-torsor of velocity $\dot{x}'_0$ and spin ϖ':

$$\breve{\gamma}'(\mathcal{B}) = \begin{pmatrix} 0 & \dot{x'}_0^T \\ -\dot{x}'_0 & -j(\varpi') \end{pmatrix}, \qquad [5.18]$$

because their transformation laws [5.11] and [5.10] are just those [5.17]) of a co-torsor. We call it the *kinetic co-torsor of a rigid body* because – x_0 being fixed – it contains the essential information to know, through [5.4], the velocity field of the body in its motion around it.

Let us now consider the rigid body at a fixed time then, for the sake of easiness, we increase it as variable. To determine the components of the co-torsor in the Lagrangian coordinates, we push them forward (the inverse operation of the pull back [5.6]) by applying [5.16] to [5.18] with $Q = R^T$ and $s_0 = 0$:

$$\breve{\gamma}(\mathcal{B}) = \breve{P}^T \breve{\gamma}'(\mathcal{B})\, \breve{P} = \begin{pmatrix} 0 & \dot{x}_0^T \\ -\dot{x}_0 & -j(\varpi) \end{pmatrix}.$$

If $v(x)$ is the velocity of the point x at the considered time, the elementary power provided by the elementary force $dF(x) = f(x)\, d\mathcal{V}(x)$ acting at the same time on the elementary volume $d\mathcal{V}(x)$ around x is:

$$d\mathcal{P}(x) = v(x) \cdot dF(x) = v(x) \cdot f(x)\, d\mathcal{V}(x).$$

As the power is an extensive quantity, the total power provided by the distribution of external forces f is:

$$\mathcal{P} = \iiint_{\mathcal{V}} d\mathcal{P}(x) = \iiint_{\mathcal{V}} v \cdot f\, d\mathcal{V}(x),$$

where $\mathcal{V}$ is the domain occupied by the body at the considered time. Taking into account [5.4] and that $\dot{x}_0, \varpi$ do not depend on x, it holds:

$$\mathcal{P} = \dot{x}_0 \cdot F + \varpi \cdot M,$$

where:

$$F = \iiint_{\mathcal{V}} f \, d\mathcal{V}(x), \qquad M = \iiint_{\mathcal{V}} (x - x_0) \times f \, d\mathcal{V}(x).$$

It is worth noting that the resultant torsor is:

$$\check{\tau}(\mathcal{B}) = \begin{pmatrix} 0 & F^T \\ -F & -j(M) \end{pmatrix}.$$

Then, owing to [7.3] and [7.13], the total power reads (see Comment 2, section 5.1):

$$\mathcal{P} = -\frac{1}{2} Tr \left(\check{\gamma}(\mathcal{B}) \, \check{\tau}(\mathcal{B}) \right).$$

5.2. Dynamical torsor

5.2.1. *Total mass and mass-center*

According to the experimental observations, we claim that:

LAW 5.1.– The mass is an extensive quantity.

On this ground, we define the dynamical torsor of a body occupying a continuous domain $\mathcal{B}$ in a given Lagrangian coordinate system. The *total mass* is:

$$m_{\mathcal{B}} = \iiint_{\mathcal{B}} dm(s').$$

As noted in section 3.1.2, the elementary particles as an electron *a priori* have a spin – even in classical mechanics – but, as far as we are concerned here, it will be neglected. According to [3.15], the elementary dynamical torsor of an elementary volume $d\mathcal{B}(s')$ around s' of infinitesimal mass $dm(s')$ has the reduced form in the Lagrangian coordinate system (because $d\mathcal{B}(s')$ is at rest in it):

$$d\tilde{\tau}'(s') = \begin{pmatrix} 0 & dm(s') & 0 \\ -dm(s') & 0 & 0 \\ 0 & 0 & 0 \end{pmatrix}, \qquad [5.19]$$

Let us now consider the point of coordinates:

$$s'_{\mathcal{B}} = \frac{1}{m_{\mathcal{B}}} \iiint_{\mathcal{B}} s' \, dm(s').$$

Owing to the linearity of the integral and [5.2], it is represented in another Lagrangian coordinate system by:

$$\bar{s}'_{\mathcal{B}} = Q^T (s'_{\mathcal{B}} - s_0),$$

hence, its coordinates change as the components of a point called the *mass-center* of $\mathcal{B}$. Its definition does not depend on the choice of the Lagrangian coordinate system. The Lagrangian coordinate systems X' of which the space origin is the mass-center are called *barycentric coordinate systems* and, of course, we have in such coordinate systems:

$$s'_{\mathcal{B}} = \frac{1}{m_{\mathcal{B}}} \iiint_{\mathcal{B}} s' \, dm(s') = 0. \qquad [5.20]$$

The changes of barycentric coordinate systems are time independent rotations.

5.2.2. *The rigid body as a particle*

According to Newton's third law 2.1, the resultant torsor is an extensive quantity. On these grounds, we claim that:

LAW 5.2.– The dynamical torsor is an extensive quantity.

This law must be applied with caution. Indeed, we wish to calculate the resultant dynamical torsor of the rigid body. In order to be summed, the elementary dynamical torsors of the elementary volumes must be represented in a common coordinate system X (in the same spirit as in section 2.3.1). For this aim, we apply the boost method of section 3.1.2 to the reduced form [5.19], given in a particular coordinate system X' for which the elementary volume $d\mathcal{B}(s')$, assimilated to a particle, is at rest at the position $x' = 0$. Let us consider the other coordinate system $X = PX' + C$ with the Galilean boost:

$$v = \frac{dx}{dt}, \qquad [5.21]$$

and a translation k of the origin $x' = 0$ at $x = k$ (hence, $\tau = 0$ and $R = 1_{\mathbb{R}^3}$). Performing the matrix product [3.2] applied to [5.19] gives the new components of the elementary dynamical torsor:

$$d\tilde{\tau}(s') = \tilde{\tau}(\mathcal{B}) = \iiint_{\mathcal{B}} d\tilde{\tau}(s') = \begin{pmatrix} 0 & m_{\mathcal{B}} & p_{\mathcal{B}}^T \\ -m_{\mathcal{B}} & 0 & -q_{\mathcal{B}}^T \\ -p_{\mathcal{B}} & q_{\mathcal{B}} & -j(l_{\mathcal{B}}) \end{pmatrix}, \qquad [5.22]$$

where:

$$dp(s') = v\, dm(s'),\ dq(s') = x\, dm(s'),\ dl(s') = x \times v\, dm(s'). \qquad [5.23]$$

We are now allowed to calculate the sum of these elementary contributions because they are given with respect to a common coordinate system X of arbitrary origin. This leads to the following definition:

DEFINITION 5.3.– The *dynamical torsor of a body* $\mathcal{B}$ is the integral of the elementary dynamical torsors of every infinitesimal parts:

$$\tilde{\tau}(\mathcal{B}) = \iiint_{\mathcal{B}} d\tilde{\tau}(s') = \begin{pmatrix} 0 & m_{\mathcal{B}} & p_{\mathcal{B}}^T \\ -m_{\mathcal{B}} & 0 & -q_{\mathcal{B}}^T \\ -p_{\mathcal{B}} & q_{\mathcal{B}} & -j(l_{\mathcal{B}}) \end{pmatrix},$$

where:

$$p_{\mathcal{B}} = \iiint_{\mathcal{B}} v\, dm(s'), \quad q_{\mathcal{B}} = \iiint_{\mathcal{B}} x\, dm(s'), \quad l_{\mathcal{B}} = \iiint_{\mathcal{B}} x \times v\, dm(s')$$

In the arbitrary Galilean coordinate system X, the particle of barycentric coordinates s' has at time t the position:

$$x = R(t)s' + x_{\mathcal{B}}(t), \qquad [5.24]$$

and velocity [5.21]. Using the mass-center of Eulerian coordinates $x^i_{\mathcal{B}}$, it is possible to decompose the dynamic of a rigid body into the overall motion of the body with the mass concentrated at its mass-center and the motion of the body around it. This is the purpose of *König's first theorem*:

THEOREM 5.1.– The motion of a rigid body is equivalent to that of a particle of mass $m_{\mathcal{B}}$, position $x_{\mathcal{B}}$, velocity $\dot{x}_{\mathcal{B}}$ and a spin angular momentum $l_{0\mathcal{B}}$ linearly depending on Poisson's vector ϖ, the components of its dynamical torsor being:

– the mass: $m_{\mathcal{B}}$;

– the linear momentum: $p_{\mathcal{B}} = m_{\mathcal{B}}\dot{x}_{\mathcal{B}}$;

– the passage: $q_{\mathcal{B}} = m_{\mathcal{B}} x_{\mathcal{B}}$;

– the angular momentum: $l_{\mathcal{B}} = x_{\mathcal{B}} \times m_{\mathcal{B}} \dot{x}_{\mathcal{B}} + \mathcal{J}_{\mathcal{B}} \varpi$.

when working in barycentric coordinates.

PROOF.–

Owing to the velocity addition formula [5.13] and formula [5.29] for the velocity of transport, the velocity in the considered Eulerian coordinate system X is:

$$v = u + R v' = u = \dot{x}_{\mathcal{B}}(t) + \varpi(t) \times (x - x_{\mathcal{B}}(t)).$$

Combining with [5.24] gives:

$$v = \dot{x}_{\mathcal{B}}(t) + \varpi(t) \times (R(t) s'). \quad [5.25]$$

Introducing this expression into that of the linear momentum $p_{\mathcal{B}}$, one has:

$$p_{\mathcal{B}} = \dot{x}_{\mathcal{B}}(t) \iiint_{\mathcal{B}} dm(s') + \varpi(t) \times \left(R(t) \iiint_{\mathcal{B}} s' \, dm(s') \right).$$

Because we are working in barycentric coordinates, [5.20] leads to:

$$p_{\mathcal{B}} = m_{\mathcal{B}} \dot{x}_{\mathcal{B}}. \quad [5.26]$$

Introducing expression [5.24] into that of the passage $q_{\mathcal{B}}$, one has:

$$q_{\mathcal{B}} = x_{\mathcal{B}}(t) \iiint_{\mathcal{B}} dm(s') + R(t) \iiint_{\mathcal{B}} s' \, dm(s').$$

Owing to [5.20] leads to:

$$q_{\mathcal{B}} = m_{\mathcal{B}} x_{\mathcal{B}}. \quad [5.27]$$

Once again introducing expressions [5.24] and [5.26] into that of the angular momentum $l_{\mathcal{B}}$, one has:

$$\begin{aligned} l_{\mathcal{B}} = {} & x_{\mathcal{B}} \times \dot{x}_{\mathcal{B}}(t) \iiint_{\mathcal{B}} dm(s') + x_{\mathcal{B}} \times \left(\varpi(t) \times \left(R(t) \iiint_{\mathcal{B}} s' \, dm(s') \right) \right) \\ & + \left(R(t) \iiint_{\mathcal{B}} s' \, dm(s') \right) \times \dot{x}_{\mathcal{B}}(t) \\ & + \iiint_{\mathcal{B}} (R(t) s') \times (\varpi(t) \times (R(t) s')) \, dm(s'). \end{aligned} \quad [5.28]$$

Owing to [5.20], the second and third terms of the left hand side vanish and:

$$l_{\mathcal{B}} = l_{0\mathcal{B}} + x_{\mathcal{B}} \times m_{\mathcal{B}} \dot{x}_{\mathcal{B}},$$

where the second term of the right hand side is the orbital angular momentum and the spin angular momentum:

$$l_{0\mathcal{B}} = \iiint_{\mathcal{B}} (R(t)s') \times (\varpi(t) \times (R(t)s'))\, dm(s'). \quad [5.29]$$

is a linear function of Poisson's vector $\varpi(t)$:

$$l_{0\mathcal{B}} = \mathcal{J}_{\mathcal{B}}(t)\, \varpi(t), \quad [5.30]$$

that achieves the proof. ■

In short, the dynamical torsor of a body depends on:

– the whole motion of the body through the mass-center position $x_{\mathcal{B}}$,

– the motion of the body arround it through $\dot{x}_{\mathcal{B}}$ and ϖ or, equivalently, through the kinetic co-torsor which describes this motion,

– the geometrical and material characteristics of the body, the total mass and the moment of inertia matrix.

5.2.3. *The moment of inertia matrix*

Now we would like to calculate explicitly the 3×3 matrix $\mathcal{J}_{\mathcal{B}}(t)$.

THEOREM 5.2.– The linear map from Poisson's vector ϖ onto the spin angular momentum $l_{0\mathcal{B}}$ is:

$$\mathcal{J}_{\mathcal{B}}(t) = R(t)\mathcal{J}'_{\mathcal{B}}(R(t))^T \quad [5.31]$$

with the time-independent *moment of inertia matrix*:

$$\mathcal{J}'_{\mathcal{B}} = \iiint_{\mathcal{B}} (\| s' \|^2\, 1_{\mathbb{R}^3} - s' s'^T)\, dm(s') \quad [5.32]$$

PROOF.–

Using [5.6] and [7.22], we have:

$$(R\, s') \times (\varpi \times (R\, s')) = (R\, s') \times ((R\, \varpi') \times (R\, s')) = R\, (s' \times (\varpi' \times s')).$$

Taking into account the vector triple product [7.16] and [5.6], one has:

$$(R\,s') \times (\varpi \times (R\,s')) = R\,(\| \, s' \, \|^2 \, \varpi' - (s' \cdot \varpi')s')$$
$$= R(t)\,(\| \, s' \, \|^2 \, 1_{\mathbb{R}^3} - s's'^T)\,(R(t))^T \varpi(t).$$

Introducing this expression into [5.29] and identifying with [5.30], we obtain [5.31] with [5.32], that achieves the proof. ■

Let us prove that $\mathcal{J}_{\mathcal{B}}$ is positive definite, according to [7.9]:

$$\forall \varpi \neq 0, \qquad \varpi \cdot (\mathcal{J}_{\mathcal{B}} \varpi) > 0.$$

Taking into account [5.6], it is equivalent to show that $\mathcal{J}'_{\mathcal{B}}$ is positive definite, which it is true because of [5.31] and [7.18]:

$$\iiint_{\mathcal{B}} (\| \, s' \, \|^2 \, \| \, \varpi' \, \|^2 - (s' \cdot \varpi')^2) \, dm(s') = \iiint_{\mathcal{B}} \| \, s' \times \varpi' \, \|^2 \, dm(s') > 0.$$

Denoting x', y', z' the Lagrangian coordinates, the moment of inertia matrix is:

$$\mathcal{J}'_{\mathcal{B}} = \begin{pmatrix} A' & -H' & -G' \\ -H' & B' & -F' \\ -G' & -F' & C' \end{pmatrix},$$

where the moments of inertia with respect to the coordinate axes are:

$$A' = \iiint_{\mathcal{B}} (y'^2 + z'^2) \, dm, \quad B' = \iiint_{\mathcal{B}} (z'^2 + x'^2) \, dm,$$
$$C' = \iiint_{\mathcal{B}} (x'^2 + y'^2) \, dm,$$

and the product of inertia are:

$$F' = \iiint_{\mathcal{B}} y'z' \, dm, \quad G' = \iiint_{\mathcal{B}} z'x' \, dm, \quad H' = \iiint_{\mathcal{B}} x'y' \, dm.$$

The moments of inertia matrix $\mathcal{J}'_{\mathcal{B}}$ is symmetric then diagonalizable with real eigenvalues and the corresponding matrix $P = (V_1, V_2, V_3)$ is orthogonal. The eigenvectors V_1, V_2, V_3 are mutually orthogonal and of unit norm. The corresponding axes are called *principal axis of inertia*. The eigenvalues are the *principal moments of inertia* A, B, C.

Using [5.24], it is also worth noting that the moment of inertia matrix [5.31] can be recast as:

$$\mathcal{J}_{\mathcal{B}} = \iiint_{\mathcal{V}} (\| x - x_{\mathcal{B}} \|^2 \, 1_{\mathbb{R}^3} - (x - x_{\mathcal{B}})(x - x_{\mathcal{B}})^T) \, dm(x), \qquad [5.33]$$

where $\mathcal{V}$ is the image of $\mathcal{B}$ through the map $s' \longmapsto x$. Then we define the matrix of inertia at the origin $x = 0$ as:

$$\mathcal{J}_{\mathcal{B}O} = \iiint_{\mathcal{V}} (\| x \|^2 \, 1_{\mathbb{R}^3} - x \, x^T) \, dm(x), \qquad [5.34]$$

and we prove *Huygens' theorem*:

THEOREM 5.3.– The matrix of inertia $\mathcal{J}_{\mathcal{B}}$ is transported at the origin according to:

$$\mathcal{J}_{\mathcal{B}O} = \mathcal{J}_{\mathcal{B}} + m_{\mathcal{B}}(\| x_{\mathcal{B}} \|^2 \, 1_{\mathbb{R}^3} - x_{\mathcal{B}} x_{\mathcal{B}}^T). \qquad [5.35]$$

PROOF.–

Expanding [5.33] and taking into account [5.34], it holds:

$$\begin{aligned} \mathcal{J}_{\mathcal{B}} = {} & \mathcal{J}_{\mathcal{B}O} + m_{\mathcal{B}}(\| x_{\mathcal{B}} \|^2 \, 1_{\mathbb{R}^3} - x_{\mathcal{B}} x_{\mathcal{B}}^T) \\ & + \left(\iiint_{\mathcal{V}} x \, dm(x) \right) x_{\mathcal{B}}^T + x_{\mathcal{B}} \left(\iiint_{\mathcal{V}} x \, dm(x) \right)^T \\ & - 2 \left(\left(\iiint_{\mathcal{V}} x \, dm(x) \right) \cdot x_{\mathcal{B}} \right) 1_{\mathbb{R}^3} \end{aligned} \qquad [5.36]$$

But, owing to [5.24] and [5.20], it holds:

$$\iiint_{\mathcal{V}} x \, dm(x) = R(t) \iiint_{\mathcal{B}} s' \, dm(s') + m_{\mathcal{B}} x_{\mathcal{B}} = m_{\mathcal{B}} x_{\mathcal{B}}.$$

Using this last relation to simplify [5.36] leads to:

$$\mathcal{J}_{\mathcal{B}} = \mathcal{J}_{\mathcal{B}O} - m_{\mathcal{B}}(\| x_{\mathcal{B}} \|^2 \, 1_{\mathbb{R}^3} - x_{\mathcal{B}} x_{\mathcal{B}}^T),$$

that achieves the proof. ■

Considering a change of coordinate system $x = Rx'$, the reader can easily verify that the moment of inertia matrix is transformed according to a formula similar to [5.31]:

$$\mathcal{J}_{\mathcal{B}O} = R \, \mathcal{J}'_{\mathcal{B}O} \, R^T. \qquad [5.37]$$

5.2.4. *Kinetic energy of a body*

The kinetic energy [3.73] of the elementary volume $d\mathcal{B}(s')$ around s' is:

$$de(s') = \frac{1}{2} \parallel v \parallel^2 \, dm(s').$$

Assuming – as the mass – the kinetic energy is an extensive quantity, the *kinetic energy of a body* is:

$$e_{\mathcal{B}} = \iiint_{\mathcal{B}} \frac{1}{2} \parallel v \parallel^2 \, dm(s'). \qquad [5.38]$$

König's second theorem is a straighforward extension of the first theorem:

THEOREM 5.4.– The kinetic energy of the body is decomposed into the energy of the body with the mass $m_{\mathcal{B}}$ concentrated at its mass-center and that relative to the motion of the body around it:

$$e_{\mathcal{B}} = \frac{1}{2} m_{\mathcal{B}} \parallel \dot{x}_{\mathcal{B}} \parallel^2 + \frac{1}{2} \varpi \cdot (\mathcal{J}_{\mathcal{B}} \, \varpi), \qquad [5.39]$$

when working in barycentric coordinates.

PROOF.–

Taking into account expression [5.25] of the velocity in the Eulerian coordinate system, one has:

$$e_{\mathcal{B}} = \iiint_{\mathcal{B}} \frac{1}{2} \parallel \dot{x}_{\mathcal{B}}(t) + \varpi(t) \times (R(t)s') \parallel^2 \, dm(s').$$

Owing to [5.20] leads to:

$$e_{\mathcal{B}} = \frac{1}{2} m_{\mathcal{B}} \parallel \dot{x}_{\mathcal{B}} \parallel^2 + \iiint_{\mathcal{B}} \frac{1}{2} \parallel \varpi(t) \times (R(t)s') \parallel^2 \, dm(s'). \qquad [5.40]$$

Taking into account [7.18], one has:

$$\begin{aligned}
\parallel \varpi \times (R\,s') \parallel^2 &= \parallel \varpi \parallel^2 \parallel R\,s' \parallel^2 - (\varpi \cdot (R\,s'))^2 \\
&= \varpi \cdot \left[\parallel R\,s' \parallel^2 \varpi - R\,s'((R\,s') \cdot \varpi) \right] \\
&= \varpi \cdot \left[\parallel s' \parallel^2 1_{\mathbb{R}^3} - R\,s'(R\,s')^T \right] \varpi \\
&= \varpi \cdot R \left[\parallel s' \parallel^2 1_{\mathbb{R}^3} - s's'^T \right] R^T \varpi.
\end{aligned}$$

Introducing it into the last term of [5.40] and owing to [5.31] and [5.32] achieves the proof. ■

5.3. Generalized equations of motion

5.3.1. *Resultant torsor of the other forces*

As in section 3.3.2, we suppose given some linear map:

$$dX \mapsto d\tilde{P} = \tilde{\Gamma}(dX).$$

By analogy with what was done in Chapter 4 with the torsor of internal efforts, we introduce the *covariant differential* of the dynamical torsor defined as:

$$\mathbf{d}\tilde{\tau} = d(\tilde{P}\,\tilde{\tau}'\tilde{P}^T)|_{X'=X} = (\tilde{P}\,d\tilde{\tau}'\tilde{P}^T + d\tilde{P}\,\tilde{\tau}'\tilde{P}^T + \tilde{P}\,\tilde{\tau}'d\tilde{P}^T)\,|_{X'=X}\,.$$

$$\mathbf{d}\tilde{\tau} = (\tilde{P}\,d\tilde{\tau}'\tilde{P}^T + \tilde{\Gamma}(dX)\,\tilde{\tau}'\tilde{P}^T + \tilde{P}\,\tilde{\tau}'(\tilde{\Gamma}(dX))^T)\,|_{X'=X}\,.$$

When X' approaches X, $\tilde{\tau}'$ approaches $\tilde{\tau}$ and $\tilde{P}$ approaches the identity:

$$\mathbf{d}\tilde{\tau} = d\tilde{\tau} + \tilde{\Gamma}(dX)\,\tilde{\tau} + \tilde{\tau}\,(\tilde{\Gamma}(dX))^T. \qquad [5.41]$$

Taking into account the structure of the dynamical torsor [3.1] and:

$$d\tilde{P} = d\begin{pmatrix} 1 & 0 \\ C & P \end{pmatrix} = \begin{pmatrix} 0 & 0 \\ \Gamma_A(dX) & \Gamma(dX) \end{pmatrix}, \qquad [5.42]$$

where Γ is a Galilean gravitation and Γ_A is a new object that will be studied further, the covariant differential of the dynamical torsor reads:

$$\mathbf{d}\tilde{\tau} = \begin{pmatrix} 0 & \mathbf{d}T^T \\ -\mathbf{d}T & \mathbf{d}J \end{pmatrix},$$

with:

$$\mathbf{d}T = dT + \Gamma(dX)\,T, \qquad [5.43]$$

$$\mathbf{d}J = dJ + \Gamma(dX)\,J + J\,(\Gamma(dX))^T + \Gamma_A(dX)\,T^T - T\,(\Gamma_A(dX))^T,$$

where the first relation is nothing other than [3.37]. Dividing by dt and using the same notation as in [3.43], the covariant derivative of the dynamical torsor reads:

$$\mathring{\tilde{\tau}} = \begin{pmatrix} 0 & \mathring{T}^T \\ -\mathring{T} & \mathring{J} \end{pmatrix}, \qquad [5.44]$$

with:

$$\overset{\circ}{T} = \dot{T} + \Gamma(U)\,T, \overset{\circ}{J} = \dot{J} + \Gamma(U)\,J + J\,(\Gamma(U))^T + \Gamma_A(U)\,T^T - T\,(\Gamma_A(U))^T \quad [5.45]$$

On the other hand, let us introduce the following definition:

DEFINITION 5.4.– The *resultant torsor of the other forces* (i.e. different from the gravitation) is represented by:

$$\tilde{\tau}^\star = \begin{pmatrix} 0 & H^T \\ -H & G \end{pmatrix}, \quad [5.46]$$

where the component $H \in \mathbb{R}^4$ represents the resultant of the other forces and is given by [3.76] (or [3.84] for the special case of thrust). $G \in \mathbb{M}_{44}^{skew}$ represent the resultant moment of the other forces.

Thus we can generalized law 3.4 to the rigid bodies:

LAW 5.3.– For any rigid body subjected to a Galilean gravitation and other forces, the motion is governed by the equation:

$$\overset{\circ}{\tilde{\tau}} = \tilde{\tau}^\star.$$

5.3.2. *Transformation laws*

Once again, guided by Galileo's principle of relativity 1.1, we claim this law is the same in all the Galilean coordinate systems. Thus in another Galilean coordinate system X', we must have:

$$\mathbf{d}\tilde{\tau}' = d\tilde{\tau}' + \tilde{\Gamma}'(dX')\,\tilde{\tau}' + \tilde{\tau}'\,(\tilde{\Gamma}'(dX'))^T.$$

Introducing [3.30] into [5.41], differentiating the products and taking into account [3.28] gives:

$$\mathbf{d}\tilde{\tau} = \tilde{P}\mathbf{d}\tilde{\tau}'\tilde{P}^T, \quad [5.47]$$

provided:

$$\tilde{\Gamma}'(dX') = \tilde{P}^{-1}(\tilde{\Gamma}(\tilde{P}\,dX')\,\tilde{P} + d\tilde{P}). \quad [5.48]$$

Dividing both members of [5.47] by $dt = dt'$ leads to:

$$\overset{\circ}{\tilde{\tau}} = \tilde{P}\,\overset{\circ}{\tilde{\tau}}{}'\,\tilde{P}^T. \quad [5.49]$$

To be consistent with Galileo's principle of relativity 1.1, taking into account [5.49], $\tilde{\tau}^\star$ must be transformed under a change of Galilean coordinate systems $X' \mapsto X$ according to transformation law [3.2] of a torsor, that justifies the definition 5.4. Using equivalent formulae [3.30] and decomposition [5.46], we obtain:

$$H = P\,H', \qquad G = P\,G'P^T + C(P\,H')^T - (P\,H')\,C^T, \tag{5.50}$$

where the first relation is nothing else [3.77]. For H of form [3.76] and G of the form:

$$G = \begin{pmatrix} 0 & 0 \\ 0 & -j(M) \end{pmatrix}, \tag{5.51}$$

applying [5.50] with a translation k and a rotation R leaves the null components of H and G while we recover transformation law [2.10] of the statical torsor. In fact, the resultant torsor $\tilde{\tau}^\star$ of the other forces is nothing else the expansion of their resultant torsor $\check{\tau}$ when recovering the extra dimension of time. The reader can also verify the relevancy of definition 5.4 to model rocket thrust. With respect to a Galilean coordinate system X' in which the rocket of mass m is at rest, H is given by [3.84] and G is null. In a coordinate system X obtained from X' by a boost v and a translation x, the reader can verify that the dynamical torsor of the thrust is given by [3.85] and:

$$G' = \begin{pmatrix} 0 & -\dot{m}x^T \\ \dot{m}x & -j(x \times (\dot{m}(v + w))) \end{pmatrix}, \tag{5.52}$$

where w is the velocity of the exhaust gases with respect to a Galilean coordinate system X'.

Let us examine now in more detail transformation law [5.48] of $\tilde{\Gamma}$. Taking into account decomposition [1.16] of $\tilde{P}$ and [5.42] of $\tilde{\Gamma}$ allows us to recover transformation law [3.49] of the gravitation and reveal that of the new object Γ_A:

$$\Gamma'_A(dX') = P^{-1}(\Gamma_A(P\,dX') + dC + \Gamma(P\,dX')C). \tag{5.53}$$

Unlike the gravitation Γ, the new object Γ_A has no deep physical meaning but we can link it to physical features of the motion according to the following reasoning. As a rigid body can be considered as a particle of mass $m_{\mathcal{B}}$ and spin $l_{0\mathcal{B}}$, let us consider two proper coordinate systems X' and $\bar{X}'$ of this particle (defined in section 3.1.2). As the change of coordinate systems $X' \mapsto \bar{X}' = P\,X'$ is linear, the translation C vanishes and:

$$\Gamma'_A(dX') = P^{-1}\,\bar{\Gamma}'_A(P\,dX'). \tag{5.54}$$

As the map Γ'_A is linear, the most simple choice for the proper coordinate systems is the identity:

$$\Gamma'_A(dX') = dX', \qquad [5.55]$$

for which [5.54] is the transformation law for the components of the vector $\overrightarrow{\mathbf{dX}'}$. Next we can deduce the expression of Γ_A in any other coordinate system $X = P\,X' + C$ thanks to its transformation law [5.53] which reads by inversion:

$$\Gamma_A(dX) = P\,\Gamma'_A(dX') - (dC + \Gamma(dX)\,C) = P\,dX' - (dC + \Gamma(dX)\,C),$$

or in short:

$$\boxed{\Gamma_A(dX) = dX - \mathbf{d}C.} \qquad [5.56]$$

If the observer is located at the moving origin, the new object Γ_A represents her or his infinitesimal motion.

5.3.3. *Equations of motion of a rigid body*

Let us consider a proper coordinate system X' of the rigid body considered as a particle. In another coordinate system $X = PX' + C$ obtained by applying a Galilean boost $\dot{x}_\mathcal{B}$ and a translation of the origin at $k = x_\mathcal{B}$ (hence $\tau = 0$ and $R = 1_{\mathbb{R}^3}$), the dynamical torsor is given by König's first theorem 5.1. In the coordinate system:

$$X = \begin{pmatrix} t \\ x_\mathcal{B} \end{pmatrix}, \qquad C = \begin{pmatrix} 0 \\ x_\mathcal{B} \end{pmatrix}.$$

Owing to expression [3.38] of the Galilean gravitation, Γ_A is calculated by the transformation law [5.56]:

$$\begin{aligned} \Gamma_A &= d\begin{pmatrix} t \\ x_\mathcal{B} \end{pmatrix} - d\begin{pmatrix} 0 \\ x_\mathcal{B} \end{pmatrix} - \begin{pmatrix} 0 & 0 \\ j(\Omega)\,dx_\mathcal{B} - g\,dt & j(\Omega)\,dt \end{pmatrix}\begin{pmatrix} 0 \\ x_\mathcal{B} \end{pmatrix} \qquad [5.57] \\ &= \begin{pmatrix} dt \\ -\Omega \times x_\mathcal{B}\,dt \end{pmatrix}. \end{aligned}$$

Taking into account [5.44] and [5.46], law 5.3 of the rigid body motion reads:

$$\mathring{T} = H, \qquad \mathring{J} = G. \qquad [5.58]$$

The first equation is nothing other than law 3.4 of the linear momentum that has been studied in section 3.5.1. With the notations concerning the rigid body, equation of motion [3.78] reads:

$$\dot{m}_{\mathcal{B}} = 0, \qquad \dot{p}_{\mathcal{B}} = m_{\mathcal{B}}\,(g - 2\,\Omega \times \dot{x}_{\mathcal{B}}) + F. \qquad [5.59]$$

Let us now detail the second one. With some abusive notations again, we write:

$$\mathring{J} = \begin{pmatrix} 0 & -\mathring{q}_{\mathcal{B}}^{T} \\ \mathring{q}_{\mathcal{B}} & -j(\mathring{l}_{\mathcal{B}}) \end{pmatrix}.$$

Also note that:

$$U = \begin{pmatrix} 1 \\ \dot{x}_{\mathcal{B}} \end{pmatrix}, \qquad \Gamma_A(U) = \begin{pmatrix} 1 \\ -\Omega \times x_{\mathcal{B}} \end{pmatrix}.$$

Owing to [3.44], the second relation of [5.45] leads to:

$$\mathring{q}_{\mathcal{B}} = \dot{q}_{\mathcal{B}} + \Omega \times (q_{\mathcal{B}} - m_{\mathcal{B}}\, x_{\mathcal{B}}) - p_{\mathcal{B}}, \qquad [5.60]$$

$$\mathring{l}_{\mathcal{B}} = \dot{l}_{\mathcal{B}} + \Omega \times l_{\mathcal{B}} + q_{\mathcal{B}} \times (\Omega \times \dot{x}_{\mathcal{B}} - g) - (\Omega \times x_{\mathcal{B}}) \times p_{\mathcal{B}}. \qquad [5.61]$$

Applying Jacobi's identity [7.17] to the last term of the expression of $\mathring{l}_{\mathcal{B}}$, owing to definition 3.13 of the spin and the expression of the passage $q_{\mathcal{B}} = m_{\mathcal{B}} x_{\mathcal{B}}$ stemming from König's first theorem 5.1, we obtain:

$$\mathring{q}_{\mathcal{B}} = \dot{q}_{\mathcal{B}} - p_{\mathcal{B}}, \qquad \mathring{l}_{\mathcal{B}} = \dot{l}_{\mathcal{B}} + \Omega \times l_{0\mathcal{B}} - q_{\mathcal{B}} \times (g - 2\,\Omega \times \dot{x}_{\mathcal{B}}). \qquad [5.62]$$

Once again using the expression of the passage, the second equation in [5.58] reveals – in addition to [5.59] – new *equations of motion* (see Comment 3, section 5.6):

$$\boxed{\dot{q}_{\mathcal{B}} = p_{\mathcal{B}},} \qquad [5.63]$$

$$\boxed{\dot{l}_{\mathcal{B}} + \Omega \times l_{0\mathcal{B}} = x_{\mathcal{B}} \times m_{\mathcal{B}}(g - 2\,\Omega \times \dot{x}_{\mathcal{B}}) + M.} \qquad [5.64]$$

The first relation is obvious taking into account the expressions of the passage and the linear momentum given by König's first theorem 5.1, and the time independance of mass [5.59]. The second one is relevant for the study of the motion of the body around it. For the particular case of no spinning ($\Omega = 0$), we deduce the *theorem of the angular momentum*:

THEOREM 5.5.– For a rigid body subjected to gravity g without spinning and other forces of resultant moment M, the time derivative of the angular momentum is equal to the resultant moment of the gravity and the other forces:

$$\dot{l}_{\mathcal{B}} = x_{\mathcal{B}} \times m_{\mathcal{B}} g + M.$$

5.4. Motion of a free rigid body around it

Let us consider a rigid body free of gravitation and other forces. A typical application is a satellite at so large a distance from the Earth that the gravitation effects are negligeable in a suitable space-time window as discussed in section 3.2.2. Within this window, the mass-center is in uniform straight motion. For convenience, we use an Eulerian coordinate system in which it is at rest:

$$\dot{x}_{\mathcal{B}} = 0, \qquad [5.65]$$

hence the angular momentum given by König's first theorem 5.1 is reduced to the spin:

$$l_{\mathcal{B}} = l_{0\mathcal{B}}.$$

We hope to study the motion of the body around it, starting from the new equation of motion [5.64]. Because $g = \Omega = M = 0$ in the Eulerian representation, it reads:

$$\dot{l}_{\mathcal{B}} = \dot{l}_{0\mathcal{B}} = 0,$$

Owing to [5.30] and [5.31], we obtain three integral of the motion:

$$R(t)\mathcal{J}'_{\mathcal{B}}(R(t))^T\,\varpi(t) = l_{0\mathcal{B}} = C^{te}.$$

It is worth pulling back the spin onto the body at rest by working in the Lagrangian representation but we must take care that transformation law [3.51] leads to a non vanishing spinning:

$$\Omega' = R^T\varpi = \varpi'.$$

where we use [5.6]. Thus, in the Lagrangian representation, the equation of motion [5.64] leads to *Euler's equation of motion* of a rigid body:

$$\dot{l}'_{0\mathcal{B}} + \varpi' \times l'_{0\mathcal{B}} = 0. \qquad [5.66]$$

Because of transformation law [3.14] of the spin, [5.30], [5.31] and [5.6]:

$$l'_{0\mathcal{B}} = R^T l_{0\mathcal{B}} = R^T\mathcal{J}_{\mathcal{B}}\,\varpi = R^T R\mathcal{J}'_{\mathcal{B}}R^T\varpi = \mathcal{J}'_{\mathcal{B}}\varpi'$$

Because the moment of inertia matrix $\mathcal{J}'_{\mathcal{B}}$ is time independent, the equation of motion becomes:

$$\mathcal{J}'_{\mathcal{B}}\dot{\varpi}' + \varpi' \times (\mathcal{J}'_{\mathcal{B}}\varpi') = 0.$$

Performing the dot product by ϖ' gives:

$$\varpi' \cdot (\mathcal{J}'_{\mathcal{B}}\dot{\varpi}') = 0, \qquad [5.67]$$

and we obtain a new integral of the motion:

$$\varpi' \cdot (\mathcal{J}'_{\mathcal{B}}\varpi') = C^{te} = 2\, e_{\mathcal{B}}, \qquad [5.68]$$

where, according to König's second theorem 5.4 and [5.65], $e_{\mathcal{B}}$ is the kinetic energy of the body. The geometrical interpretation of this relation – due to Poinsot – is that Poisson's vector ϖ' lies on an ellipsoid of equation [5.68] (Figure 5.1). Because the moment of inertia matrix is symmetric, [5.67] reads:

$$\dot{\varpi}' \cdot (\mathcal{J}'_{\mathcal{B}}\varpi') = 0,$$

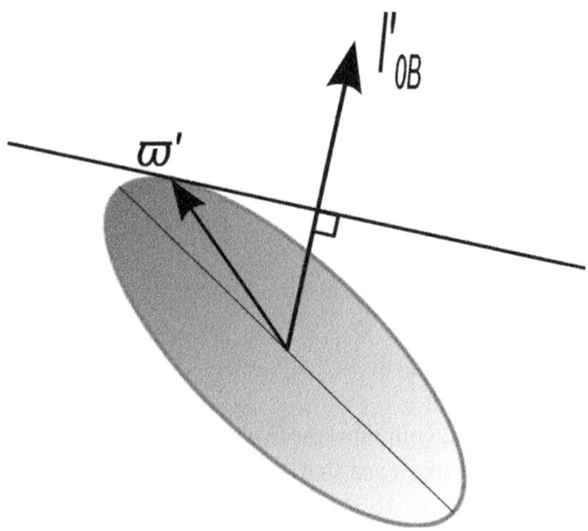

Figure 5.1. *Poinsot's construction*

thus the tangent plane at ϖ' to *Poinsot's ellipsoid* is perpendicular to the spin angular momentum $l'_{0\mathcal{B}}$. Its distance to the origin:

$$\varpi' \cdot \frac{l'_{0\mathcal{B}}}{\| \, l'_{0\mathcal{B}} \, \|} = \frac{2\, e_{\mathcal{B}}}{\| \, l_{0\mathcal{B}} \, \|},$$

is time independent. Pulling back this construction into the Eulerian coordinates, the motion is described by saying that the ellipsoid rolls on the invariable plane drawn perpendicular to the time independent vector $l_{0\mathcal{B}}$ at the invariable distance $2\,e_{\mathcal{B}}/ \parallel l_{0\mathcal{B}} \parallel$ of the origin. The vector drawn from the origin to the contact point is Poisson's vector ϖ'. The curve traced by this point of contact on the ellipsoid and the plane are respectively called the polhode and the herpolhode.

Let us now examine the particular case of a body with rotational symmetry around an axis. In the principal axis of inertia with the third one being the rotational symmetry axis, $A = B$ and the spin angular momentum reads:

$$l'_{0\mathcal{B}} = \mathcal{J}'_{\mathcal{B}}\varpi' = \begin{pmatrix} A & 0 & 0 \\ 0 & A & 0 \\ 0 & 0 & C \end{pmatrix} \begin{pmatrix} \varpi'_1 \\ \varpi'_2 \\ \varpi'_3 \end{pmatrix} = \begin{pmatrix} A\varpi'_1 \\ A\varpi'_2 \\ C\varpi'_3 \end{pmatrix}.$$

Hence Euler's equation of motion [5.66] reads:

$$A\dot{\varpi}'_1 + (C - A)\,\varpi'_2\varpi'_3 = 0,$$
$$A\dot{\varpi}'_2 + (C - A)\,\varpi'_3\varpi'_1 = 0,$$
$$C\dot{\varpi}'_3 = 0,$$

where A and C are time independent. From the last equation, we find a first integral of the motion, the *spin* ϖ'_3 of the body around its rotational symmetry axis. Next, adding the first equation multiplied by ϖ'_1 and the second one multiplied by ϖ'_2 gives the second integral of the motion:

$$\varpi'^2_1 + \varpi'^2_2 = C^{te},$$

and, consequently:

$$\parallel \varpi' \parallel^2 = \varpi'^2_1 + \varpi'^2_2 + \varpi'^2_3 = C^{te}.$$

The herpolhode is a circle of time independent radius:

$$\varrho = \sqrt{\parallel \varpi' \parallel^2 - \left(\frac{2\,e_{\mathcal{B}}}{\parallel l_{0\mathcal{B}} \parallel}\right)^2}.$$

5.5. Motion of a rigid body with a contact point (Lagrange's top)

The toy spinning top is a solid of revolution subjected to the gravity and placed in contact with a horizontal plane (Figure 5.2). In absence of spinning, the top naturally tumbles because of the gravity but if it is set spinning about its axis of revolution at a sufficient rotation velocity, it stands upright. In this section, we would like to explain why the spinning motion prevents the top taking a tumble. For this aim, the top is modeled as a rigid body subjected in a suitable Eulerian coordinate system x to a uniform gravity g without spinning ($\Omega = 0$), in punctual contact with a rough surface at a point $\boldsymbol{O}$ then no sliding is allowed. For convenience, let us pick z's axis directed vertically upward, as determined by a plumb line, and the origin at $\boldsymbol{O}$. Drawing the free body diagram of the top, we remove the support at $\boldsymbol{O}$ and we draw a reaction force F (Figure 5.2) acting at $\boldsymbol{O}$. As the gravity is uniform, we can model its action upon the top by its resultant $m_{\mathcal{B}}g$ acting at the mass-center $\boldsymbol{G}$, and obtained thanks to the equation of motion [5.59] in absence of spinning:

$$F = \dot{p}_{\mathcal{B}} - m_{\mathcal{B}}\, g,$$

after determining the trajectory. For this aim, we use equation of the motion [5.64] of the body around it. As the contact point $\boldsymbol{O}$ is at rest in the considered Eulerian coordinate system, taking into account [5.4], the no sliding condition reads:

$$\dot{x}_{\mathcal{B}} = \varpi \times x_{\mathcal{B}}.$$

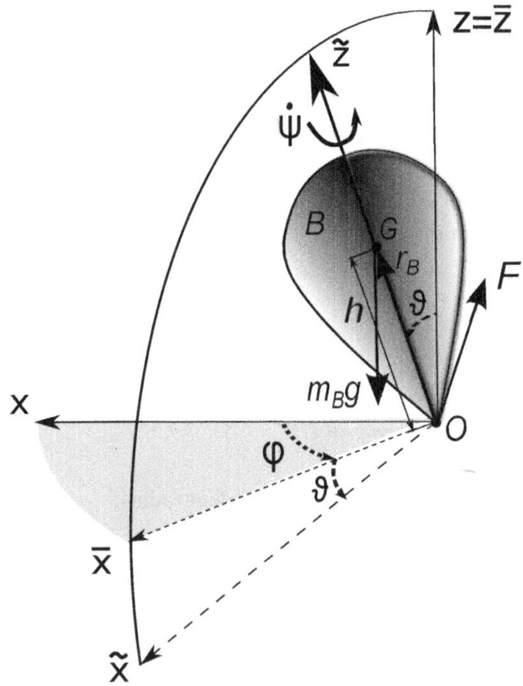

Figure 5.2. *Lagrange's top*

Hence, starting from the expression of the angular momentum in König's first theorem 5.1 and owing to [7.12], it holds:

$$l_{\mathcal{B}} = m_{\mathcal{B}} x_{\mathcal{B}} \times (\varpi \times x_{\mathcal{B}}) + \mathcal{J}_{\mathcal{B}}\varpi = (-m_{\mathcal{B}}\, j(x_{\mathcal{B}}) j(x_{\mathcal{B}}) + \mathcal{J}_{\mathcal{B}})\varpi \qquad [5.69]$$

$$= (m_{\mathcal{B}}(\| x_{\mathcal{B}} \|^2\, 1_{\mathbb{R}^3} - x_{\mathcal{B}} x_{\mathcal{B}}^T) + \mathcal{J}_{\mathcal{B}})\varpi.$$

By Huygens' theorem 5.3, the angular momentum of the top reads:

$$l_{\mathcal{B}} = \mathcal{J}_{\mathcal{B}O}\varpi. \qquad [5.70]$$

In a similar way, starting from the expression of the kinetic energy in König's second theorem 5.4 and owing to [7.18], it holds:

$$e_{\mathcal{B}} = \frac{1}{2}\, m_{\mathcal{B}}\,(\| x_{\mathcal{B}} \|^2 \| \varpi \|^2 - (x_{\mathcal{B}} \cdot \varpi)^2) + \frac{1}{2}\,\varpi \cdot (\mathcal{J}_{\mathcal{B}}\,\varpi).$$

Because of Huygens' theorem 5.3, the kinetic energy of the top reads:

$$e_{\mathcal{B}} = \frac{1}{2}\,\varpi \cdot (\mathcal{J}_{\mathcal{B}O}\varpi). \qquad [5.71]$$

On this ground, we consider a Lagrangian coordinate system x' with origin at the contact point $\boldsymbol{O}$, the rotational symmetry axis being the one of z', and we model the body motion by the rotation matrix R of the map from the reference Eulerian coordinate system x onto $x' = R^T x$. To express ϖ in terms of Euler's angles, we differentiate [3.20] with respect to the time:

$$\dot{R} = \dot{R}_{\varphi} R_{\vartheta} R_{\psi} + R_{\varphi} \dot{R}_{\vartheta} R_{\psi} + R_{\varphi} R_{\vartheta} \dot{R}_{\psi}.$$

Hence, we have:

$$\dot{R}R^T = \dot{R}_{\varphi} R_{\varphi}^T + R_{\varphi}(\dot{R}_{\vartheta} R_{\vartheta}^T) R_{\varphi}^T + (R_{\varphi} R_{\vartheta})(\dot{R}_{\psi} R_{\psi}^T)(R_{\varphi} R_{\vartheta})^T.$$

ϖ_{φ}, ϖ_{ϑ} and ϖ_{ψ} being respectively Poisson's vectors of R_{φ}, R_{ϑ} and R_{ψ}, it holds, owing to [7.23]:

$$j(\varpi) = j(\varpi_{\varphi}) + j(R_{\varphi}\varpi_{\vartheta}) + j(R_{\varphi} R_{\vartheta}\varpi_{\psi}).$$

Because the map j is linear and regular, we obtain:

$$\varpi = \varpi_{\varphi} + R_{\varphi}\varpi_{\vartheta} + R_{\varphi} R_{\vartheta}\varpi_{\psi},$$

Introducing $R_{\varphi\vartheta} = R_\varphi R_\vartheta$, its pull back onto the coordinate system $\tilde{x} = R_{\varphi\vartheta}^T x$ is:

$$\tilde{\varpi} = R_\vartheta^T R_\varphi^T \varpi = R_{\varphi\vartheta}^T \varpi_\varphi + R_\vartheta^T \varpi_\vartheta + \varpi_\psi. \qquad [5.72]$$

Taking into account [3.21], it holds:

$$\tilde{\varpi} = \begin{pmatrix} \cos\varphi\cos\vartheta & \sin\varphi\cos\vartheta & -\sin\vartheta \\ -\sin\varphi & \cos\varphi & 0 \\ \cos\varphi\sin\vartheta & \sin\varphi\sin\vartheta & \cos\vartheta \end{pmatrix} \begin{pmatrix} 0 \\ 0 \\ \dot{\varphi} \end{pmatrix} + \begin{pmatrix} \cos\vartheta & 0 & -\sin\vartheta \\ 0 & 1 & 0 \\ \sin\vartheta & 0 & \cos\vartheta \end{pmatrix} \begin{pmatrix} 0 \\ \dot{\vartheta} \\ 0 \end{pmatrix} + \begin{pmatrix} 0 \\ 0 \\ \dot{\psi} \end{pmatrix},$$

and we obtain:

$$\tilde{\varpi} = \begin{pmatrix} -\dot{\varphi}\sin\vartheta \\ \dot{\vartheta} \\ \dot{\psi} + \dot{\varphi}\cos\vartheta \end{pmatrix}. \qquad [5.73]$$

According to [5.37], the moment of inertia matrix reads in the Lagrangian coordinate system $\tilde{x}$:

$$\tilde{\mathcal{J}}_{BO} = R_\psi \mathcal{J}'_{BO} R_\psi^T,$$

or, owing to [3.21]:

$$\tilde{\mathcal{J}}_{BO} = \begin{pmatrix} \cos\psi & -\sin\psi & 0 \\ \sin\psi & \cos\psi & 0 \\ 0 & 0 & 1 \end{pmatrix} \begin{pmatrix} A & 0 & 0 \\ 0 & A & 0 \\ 0 & 0 & C \end{pmatrix} \begin{pmatrix} \cos\psi & \sin\psi & 0 \\ -\sin\psi & \cos\psi & 0 \\ 0 & 0 & 1 \end{pmatrix},$$

that leads to:

$$\tilde{\mathcal{J}}_{BO} = \mathcal{J}'_{BO} \qquad [5.74]$$

which is nothing other than expression of the rotational symmetry of the body around the $\tilde{z}$'s axis. By putting $\varpi_\psi = 0$ in [5.72], Poisson's vector of the rotation matrix $R_{\varphi\vartheta}$ is found to be in the Eulerian coordinate system $\tilde{x}$:

$$\tilde{\varpi}_{\varphi\vartheta} = R_{\varphi\vartheta}^T \varpi_\varphi + R_\vartheta^T \varpi_\vartheta = \begin{pmatrix} -\dot{\varphi}\sin\vartheta \\ \dot{\vartheta} \\ \dot{\varphi}\cos\vartheta \end{pmatrix}.$$

Now, we are able to find three integrals of the motion:

– Let us calculate the time derivative of the total kinetic energy [5.38] and take into account the equation of motion of the elementary mass $dm(s')$ in the reference

Eulerian coordinate system x were the spinning Ω vanishes and the gravity g is uniform:

$$\dot{e}_{\mathcal{B}} = \iiint_{\mathcal{B}} v \cdot \dot{v}\, dm(s') = \iiint_{\mathcal{B}} v \cdot g\, dm(s') = \left(\iiint_{\mathcal{B}} v\, dm(s') \right) \cdot g.$$

Because of definition [5.5] and König's first theorem 5.1, one has:

$$\dot{e}_{\mathcal{B}} = p_{\mathcal{B}} \cdot g = m_{\mathcal{B}}\, g \cdot \dot{x}_{\mathcal{B}}.$$

Introducing the gravitational potential:

$$\phi = -g \cdot x_{\mathcal{B}},$$

we obtain a first integral of the motion, the total energy:

$$e_T = e_{\mathcal{B}} + m_{\mathcal{B}} \phi.$$

The total kinetic energy [5.71] becomes:

$$e_{\mathcal{B}} = \frac{1}{2} (R_{\varphi\vartheta} \tilde{\varpi})^T \cdot (\mathcal{J}_{\mathcal{B}O}(R_{\varphi\vartheta} \tilde{\varpi})) = \frac{1}{2} \tilde{\varpi} \cdot (\tilde{\mathcal{J}}_{\mathcal{B}O} \tilde{\varpi}),$$

where $\tilde{\mathcal{J}}_{\mathcal{B}O} = R^T_{\varphi\vartheta} \mathcal{J}_{\mathcal{B}O} R_{\varphi\vartheta}$, according to transformation law [5.37]. Owing to [5.73] and [5.74], one has:

$$e_{\mathcal{B}} = \frac{1}{2} \left(A\,(\dot{\vartheta}^2 + \dot{\varphi}^2 \sin^2 \vartheta) + C\,(\dot{\psi} + \dot{\varphi} \cos \vartheta)^2 \right).$$

In addition, h being the distance between the mass-center $\boldsymbol{G}$ and the contact point $\boldsymbol{O}$, one has:

$$x_{\mathcal{B}} = R_{\varphi\vartheta}\, \tilde{x}_{\mathcal{B}} = \begin{pmatrix} \cos\varphi \cos\vartheta & -\sin\varphi & \cos\varphi \sin\vartheta \\ \sin\varphi \cos\vartheta & \cos\varphi & \sin\varphi \sin\vartheta \\ -\sin\vartheta & 0 & \cos\vartheta \end{pmatrix} \begin{pmatrix} 0 \\ 0 \\ h \end{pmatrix} = h \begin{pmatrix} \cos\varphi \sin\vartheta \\ \sin\varphi \sin\vartheta \\ \cos\vartheta \end{pmatrix},$$

and the gravitational potential reads:

$$\phi = -\left(0\ 0\ -\parallel g \parallel \right) h \begin{pmatrix} \cos\varphi \sin\vartheta \\ \sin\varphi \sin\vartheta \\ \cos\vartheta \end{pmatrix} = \parallel g \parallel\ h \cos\vartheta.$$

The first integral of the motion, the kinetic energy reads:

$$e_T = \frac{1}{2} \left(A\,(\dot{\vartheta}^2 + \dot{\varphi}^2 \sin^2 \vartheta) + C(\dot{\psi} + \dot{\varphi} \cos\vartheta)^2 \right) + m_{\mathcal{B}} \parallel g \parallel h \cos\vartheta = C^{te}. \quad [5.75]$$

– The second integral of the motion is obtained considering equation [5.64] of motion in the Eulerian coordinate system $\tilde{x}$:

$$\dot{\tilde{l}}_{\mathcal{B}} + \tilde{\Omega} \times \tilde{l}_{0\mathcal{B}} = \tilde{x}_{\mathcal{B}} \times m_{\mathcal{B}}(\tilde{g} - 2\tilde{\Omega} \times \dot{\tilde{x}}_{\mathcal{B}}),$$

where the moment of the reaction force acting at the origin O vanishes. Because the position vector $\tilde{x}_{\mathcal{B}}$ of the mass-center is collinear to the basis vector $e_{\tilde{z}}$, we obtain by projection:

$$e_{\tilde{z}} \cdot \left(\dot{\tilde{l}}_{\mathcal{B}} + \tilde{\Omega} \times \tilde{l}_{0\mathcal{B}} \right) = 0. \qquad [5.76]$$

Owing to [5.70], [5.73] and [5.74], the angular momentum reads:

$$\tilde{l}_{\mathcal{B}} = \tilde{\mathcal{J}}_{\mathcal{B}O}\tilde{\varpi} = \begin{pmatrix} -A\dot{\varphi}\sin\vartheta \\ A\dot{\vartheta} \\ C\left(\dot{\psi} + \dot{\varphi}\cos\vartheta\right) \end{pmatrix}. \qquad [5.77]$$

On the other hand, the spin angular momentum reads:

$$\tilde{l}_{0\mathcal{B}} = \tilde{\mathcal{J}}_{\mathcal{B}}\tilde{\varpi}$$

Owing to Huygens' theorem 5.3, namely [5.35], one has:

$$\tilde{\mathcal{J}}_{\mathcal{B}} = \tilde{\mathcal{J}}_{\mathcal{B}O} - m_{\mathcal{B}}(\| \tilde{x}_{\mathcal{B}} \|^2 1_{\mathbb{R}^3} - \tilde{x}_{\mathcal{B}}\tilde{x}_{\mathcal{B}}^T) = \begin{pmatrix} A^* & 0 & 0 \\ 0 & A^* & 0 \\ 0 & 0 & C^* \end{pmatrix},$$

where $A^* = A - m_{\mathcal{B}}h^2$, $C^* = C$ and, taking into account [5.73]:

$$\tilde{l}_{0\mathcal{B}} = \begin{pmatrix} -A^*\dot{\varphi}\sin\vartheta \\ A^*\dot{\vartheta} \\ C^*\left(\dot{\psi} + \dot{\varphi}\cos\vartheta\right) \end{pmatrix}.$$

Moreover, using transformation law [3.51] of the spinning and because $\Omega = 0$ in the reference Eulerian coordinate system x, one has:

$$\tilde{\Omega} = \tilde{\varpi}_{\varphi\vartheta}.$$

After calculation, it holds:

$$\dot{\tilde{l}}_{\mathcal{B}} + \Omega \times \tilde{l}_{0\mathcal{B}} = \begin{pmatrix} -\frac{d}{dt}(A^*\dot{\varphi}\sin\vartheta) + (C^* - A^*)\dot{\vartheta}\dot{\varphi}\cos\vartheta + C^*\dot{\vartheta}\dot{\psi} \\ \frac{d}{dt}(A^*\dot{\vartheta}) + (C^* - A^*)\dot{\varphi}^2\sin\vartheta\cos\vartheta + C^*\dot{\varphi}\dot{\psi}\sin\vartheta \\ \frac{d}{dt}(C^*(\dot{\psi} + \dot{\varphi}\cos\vartheta)) \end{pmatrix}.$$

Because C is time independent, relation [5.76] leads to a second integral of the motion, the projection of the angular momentum onto the top symmetry axis:

$$\dot{\psi} + \dot{\varphi}\cos\vartheta = C^{te} = n, \qquad [5.78]$$

often called *spin* and denoted by n. It is worth noting that, as for the motion of a free rigid body around it, the spin integral of motion results from the rotational symmetry of the body.

– The third integral of motion is obtained considering once again equation [5.64] of motion but in the reference Eulerian coordinate system x. Hence, the assumptions of theorem 5.5 of the angular momentum are fulfilled and the equation of motion is reduced to:

$$\dot{l}_{\mathcal{B}} = x_{\mathcal{B}} \times m_{\mathcal{B}} g. \qquad [5.79]$$

The gravity g being collinear to the time-independent basis vector e_z along the vertical z's axis, it holds:

$$\dot{l}_{\mathcal{B}} \cdot e_z = \frac{d}{dt}(l_{\mathcal{B}} \cdot e_z) = \dot{l}_z = 0,$$

and the z-component of the angular momentum is an integral of the motion:

$$l_z = C^{te}. \qquad [5.80]$$

For the change of coordinate system $x = R_{\varphi\vartheta}\tilde{x}$, the velocity of transport [3.29] vanishes at $x = 0$, then transformation law [3.8] of the angular momentum gives :

$$l_{\mathcal{B}} = R_{\varphi\vartheta}\tilde{l}_{\mathcal{B}}.$$

Introducing:

$$\tilde{e}_z = R^T_{\varphi\vartheta} e_z = \begin{pmatrix} \cos\varphi\cos\vartheta & \sin\varphi\cos\vartheta & -\sin\vartheta \\ -\sin\varphi & \cos\varphi & 0 \\ \cos\varphi\sin\vartheta & \sin\varphi\sin\vartheta & \cos\vartheta \end{pmatrix} \begin{pmatrix} 0 \\ 0 \\ 1 \end{pmatrix} = \begin{pmatrix} -\sin\vartheta \\ 0 \\ \cos\vartheta \end{pmatrix},$$

and taking into account [5.77] and [5.78], the last integral of motion reads:

$$l_z = e^T_z l_{\mathcal{B}} = \tilde{e}^T_z \tilde{l}_{\mathcal{B}} = A\dot{\varphi}\sin^2\vartheta + C\,n\cos\vartheta. \qquad [5.81]$$

Setting aside the events $C = 0$ and $n = 0$ definitely, we define the adimensional constants:

$$\alpha = \frac{2e_T - C\,n^2}{2\,m_{\mathcal{B}} \parallel g \parallel h}, \qquad \beta = \frac{l_z}{C\,n}, \qquad b = \frac{C}{A}, \qquad n_* = n\sqrt{\frac{A}{2\,m_{\mathcal{B}} \parallel g \parallel h}},$$

and the adimensional variables:

$$x = \cos\vartheta, \qquad \varphi, \qquad \psi, \qquad \tau = \sqrt{\frac{2\,m_{\mathcal{B}}\parallel g \parallel h}{A}}\,t.$$

Denoting $'$ the derivative with respect to τ, the integral of motion [5.75], [5.78] and [5.81] read:

$$x'^2 + \varphi'^2(1-x^2)^2 = (\alpha - x)\,(1-x^2), \qquad [5.82]$$

$$\psi' + \varphi' x = n_*, \qquad [5.83]$$

$$\varphi'(1-x^2) = b\,n_*(\beta - x). \qquad [5.84]$$

Eliminating φ', we get the differential equation:

$$x'^2 = f(x) = (\alpha - x)\,(1-x^2) - b^2 n_*^2 (\beta - x)^2.$$

The motion is possible only if $f(x)$ is positive and $\mid x \mid \leq 1$ (since $x = \cos\vartheta$). For large values of $\mid x \mid$, $f(x)$ is dominated by the cubic term in x, then

$$\lim_{x\to\pm\infty} f(x) = \pm\infty.$$

As $f(-1) < 0$ and $f(1) < 0$, the cubic polynomial f has three real zeros x_1, x_2, x_3 such that:

$$-1 < x_1 < x_2 < 1 < x_3,$$

special cases of equality being disregarded here. The variable x oscillates within the interval $[x_1, x_2]$. Then the nutation angle ϑ varies between limit values corresponding to x_1 and x_2, which explains why the spinning top does not take a tumble. The azimutal angle φ is given by:

$$\varphi' = \frac{b\,n_*(\beta - x)}{1-x^2}.$$

It is clear that φ' has one sign throughout the motion if and only if β lies outside the interval $[x_1, x_2]$. The motion is most clearly followed by tracing the path of $e_{\bar{z}}$ on the unit sphere with coordinates ϑ, φ. This path is bounded by the two circles $x = x_1$ (above) and $x = x_2$ (below), and the path crosses itself if and only if φ' changes sign during the motion.

5.6. Comments for experts

COMMENT 1.– This partial analogy is a cause of misleading and confusion in literature where co-torsors are erroneously identified as torsors. In fact, co-torsors are skew-symmetric covariant affine tensors of rank 2.

COMMENT 2.– The total power is the opposite of half the contracted product of a torsor, a 2-contravariant affine tensor, and a co-torsor, a 2-covariant affine tensor, that reveals the duality between them.

COMMENT 3.– We will recover in Chapter 17 these equations of motion both as a generalization of Euler–Poincaré equations and as canonical equations deriving from Kirillov–Kostant–Souriau structure on a manifold.

6

Calculus of Variations

6.1. Introduction

For a long time, it has been noted that the laws of many natural phenomena can be obtained by realizing an extremum (minimum or maximum) of a certain physical quantity assigned to a function modeling the considered phenomenon. In other words, we use a real-valued map defined on a set of functions and called a functional (a function of functions). Such a formulation, called a variational principle, is used to obtain the physical laws of the considered phenomenon by means of the calculus of variations. The variational principles have over all a mnemonic value which allows deducing the physical laws in a consistent and systematic way.

In the dynamics of particles and rigid bodies, the considered functional is called the action and the corresponding variational principle is the principle of least action that allows us to deduce the equations of motion in a more abstract way as in Chapter 3.

The starting point is the *Lagrangian*, i.e. a differentiable real function:

$$\mathcal{L} : \mathbb{R} \times \mathbb{R}^n \times \mathbb{R}^n \longrightarrow \mathbb{R} : (t, y, z) \longmapsto \lambda = \mathcal{L}(t, y, z).$$

To a differentiable map $t \longmapsto y$, we associate the number:

$$\alpha[y] = \int_{t_0}^{t_1} \mathcal{L}(t, y, \dot{y})\, dt,$$

called the *action*, which defines the *functional* $y \longmapsto \alpha[y]$. Let us suppose that y depends on a parameter ϵ so that the map:

$$\begin{pmatrix} t \\ \epsilon \end{pmatrix} \longmapsto y, \qquad [6.1]$$

is twice continuously differentiable and:

$$\forall \epsilon, \qquad y(t_0, \epsilon) = y_0, \qquad y(t_1, \epsilon) = y_1. \tag{6.2}$$

The action depends now on ϵ:

$$\alpha(\epsilon) = \alpha[y].$$

If the action has a minimum for the map $t \longmapsto y(t, 0)$, the function $\epsilon \longmapsto \alpha$ has a minimum at $\epsilon = 0$ and thus:

$$\delta \alpha = \alpha'(0) = \int_{t_0}^{t_1} \left. \frac{\partial \mathcal{L}}{\partial \epsilon} \right|_{\epsilon=0} dt = 0. \tag{6.3}$$

The time derivative $\dot{y}$ is now a simplified notation for the partial derivative of y with respect to t and the partial derivative with respect to the parameter ϵ:

$$\delta y = \left. \frac{\partial y}{\partial \epsilon} \right|_{\epsilon=0},$$

is called the *variation* of y, hence the name of *calculus of variations*. As the Lagrangian depends on ϵ through y and $\dot{y}$, the chain rule provides:

$$\frac{\partial \mathcal{L}}{\partial \epsilon} = \frac{\partial \mathcal{L}}{\partial y} \frac{\partial y}{\partial \epsilon} + \frac{\partial \mathcal{L}}{\partial \dot{y}} \frac{\partial^2 y}{\partial \epsilon \partial t} = \frac{\partial \mathcal{L}}{\partial y} \frac{\partial y}{\partial \epsilon} + \frac{\partial \mathcal{L}}{\partial \dot{y}} \frac{\partial^2 y}{\partial t \partial \epsilon},$$

because y is twice continuously differentiable. With simplified notations, we have:

$$\left. \frac{\partial \mathcal{L}}{\partial \epsilon} \right|_{\epsilon=0} = \frac{\partial \mathcal{L}}{\partial y} \delta y + \frac{\partial \mathcal{L}}{\partial \dot{y}} \delta \left(\frac{dy}{dt} \right) = \frac{\partial \mathcal{L}}{\partial y} \delta y + \frac{\partial \mathcal{L}}{\partial \dot{y}} \frac{d}{dt} (\delta y), \tag{6.4}$$

where the derivative symbols d/dt and δ are permuted. Substituting this expression into the variation [6.3] leads to:

$$\delta \alpha = \int_{t_0}^{t_1} \left[\frac{\partial \mathcal{L}}{\partial y} \delta y + \frac{\partial \mathcal{L}}{\partial \dot{y}} \frac{d}{dt} (\delta y) \right] dt = 0.$$

Integrating by parts, it holds:

$$\delta \alpha = \left[\frac{\partial \mathcal{L}}{\partial \dot{y}} \delta y \right]_{t_0}^{t_1} + \int_{t_0}^{t_1} \left[\frac{\partial \mathcal{L}}{\partial y} - \frac{d}{dt} \left(\frac{\partial \mathcal{L}}{\partial \dot{y}} \right) \right] \delta y \, dt = 0.$$

Taking into account [6.2], the variation of y vanishes at $t = t_0$ and $t = t_1$ and we have for any map [6.1] hence for any variation δy:

$$\delta\alpha = \int_{t_0}^{t_1} \left[\frac{\partial\mathcal{L}}{\partial y} - \frac{d}{dt}\left(\frac{\partial\mathcal{L}}{\partial\dot{y}}\right)\right] \delta y\, dt = 0. \qquad [6.5]$$

We are in a situation where f being a given map defined on $[t_0, t_1]$ we have:

$$\int_{t_0}^{t_1} f(t)\, \delta y(t)\, dt = 0,$$

for every continuous function $t \longmapsto \delta y$, for instance $\delta y = f\, g$ where $g(t) = -(t - t_0)\,(t - t_1)$:

$$\int_{t_0}^{t_1} f(t)\, \delta y(t)\, dt = \int_{t_0}^{t_1} (f(t))^2\, g(t)\, dt = 0.$$

The integrand is non-negative so it must be zero. Applying this result to [6.5] leads to:

$$\frac{d}{dt}\left(\frac{\partial\mathcal{L}}{\partial\dot{y}}\right) - \frac{\partial\mathcal{L}}{\partial y} = 0, \qquad [6.6]$$

and, by transposition, to *Euler–Lagrange equations*:

$$\boxed{\frac{d}{dt}\,(grad_{\dot{y}}\,\mathcal{L}) - grad_{y}\,\mathcal{L} = 0.} \qquad [6.7]$$

The curve of $\mathbb{R}^n$ represented by the map $t \mapsto y$ realizing the minimum of the action is called *natural path*. If $\mathcal{L}$ does not depend explicitly on t, taking into account [6.6], we have along the natural path:

$$\frac{d\mathcal{L}}{dt} = \frac{\partial\mathcal{L}}{\partial y}\frac{dy}{dt} + \frac{\partial\mathcal{L}}{\partial\dot{y}}\frac{d\dot{y}}{dt} = \frac{d}{dt}\left(\frac{\partial\mathcal{L}}{\partial\dot{y}}\right)\dot{y} + \frac{\partial\mathcal{L}}{\partial\dot{y}}\frac{d\dot{y}}{dt},$$

that reads:

$$\frac{d}{dt}\left(\frac{\partial\mathcal{L}}{\partial\dot{y}}\,\dot{y} - \mathcal{L}\right) = 0,$$

hence, we obtain a quantity preserved along the natural path, Legendre's transform of the Lagrangian and called *Hamiltonian*:

$$\mathcal{H} = \frac{\partial \mathcal{L}}{\partial \dot{y}}\, \dot{y} - \mathcal{L} = C^{te}. \qquad [6.8]$$

Incidentally, it is worth introducing:

$$\pi = grad_{\dot{y}}\, \mathcal{L}.$$

As Legendre's transform of $\mathcal{L}$, the Hamiltonian:

$$\mathcal{H} = \pi \cdot \dot{y} - \mathcal{L},$$

is a function of t, y, π and:

$$grad_y\, \mathcal{H} = -grad_y\, \mathcal{L}.$$

Then, Euler–Lagrange equations [6.7] are broken into the system of *canonical equations*:

$$\dot{y} = grad_\pi\, \mathcal{H}, \qquad \dot{\pi} = -grad_y\, \mathcal{H}, \qquad [6.9]$$

where intermediate variables π are seen as independent of y. Hence, we double the number of unknowns and equations but they are now differential equations of rank 1 instead of Euler–Lagrange ones which are of rank 2. This alternative framework is known as the *Hamiltonian formalism*.

Moreover, let us examine the current situations in which the variable y is subjected to a constraint:

$$\forall t \in [t_0, t_1], \qquad f(t, y(t)) = 0.$$

We would like that:

$$\{\delta f = grad_y\, f \cdot \delta y = 0\} \qquad \Rightarrow$$
$$\left\{\delta \mathcal{L} = \left[\frac{d}{dt}\,(grad_{\dot{y}}\, \mathcal{L}) - grad_y\, \mathcal{L}\right] \cdot \delta y = 0\right\},$$

hence, there exists a Lagrange's multiplier $\lambda \in \mathbb{R}$ such that:

$$\boxed{\frac{d}{dt}\,(grad_{\dot{y}}\, \mathcal{L}) - grad_y\, \mathcal{L} = \lambda\, grad_y\, f.} \qquad [6.10]$$

6.2. Particle subjected to the Galilean gravitation

6.2.1. *Guessing the Lagrangian expression*

We would like to find the equations of motion of a particle by this method. Although at first glance it may seem attractive, to be honest we have to say that it is not so easy to know the expression of the corresponding Lagrangian. To get it, we use a heuristic way. To lay the ground, we first try to guess it in the simple case of a free particle, hence in uniform straight motion in some Galilean coordinate system X'. The studied phenomenon is the trajectory of a particle of mass m, modeled by the function $t \longmapsto x'$. In the absence of gravitation, the equation of motion [3.45] is reduced to:

$$\frac{dp'}{dt} = 0.$$

From the comparison with Euler–Lagrange equations [6.7] (where y is x' and $\dot{x}'$ is v'), we gather:

$$grad_{v'}\,\mathcal{L} = p' = m\,v', \qquad grad_{x'}\,\mathcal{L} = 0.$$

Modulo a constant, an obvious solution is:

$$\mathcal{L}(t, r', v') = \frac{1}{2} m \parallel v' \parallel^2,$$

which is nothing other then the kinetic energy [3.73].

Unfortunately, the calculus of variations is littered with traps (but we will learn to avert some of them). The problem of this Lagrangian is that it was found in a very peculiar situation where the particle is in uniform straight motion in X'. In fact, the previous expression is not general and our goal now is to find its generic form in any Galilean coordinate system X. For this aim, we use the boost method in the spirit of section 3.1.2. The coordinate change $X' \longmapsto X$ being characterized by a boost u and a rotation R, we use the velocity addition formula [1.13] to express the Lagrangian in terms of the velocity in the new Galilean coordinate system X:

$$\mathcal{L} = \frac{1}{2} m \parallel v' \parallel^2 = \frac{1}{2} m \parallel R\,v' \parallel^2 = \frac{1}{2} m \parallel v - u \parallel^2,$$

thus, expanding:

$$\mathcal{L} = \frac{1}{2} m \parallel v \parallel^2 + \frac{1}{2} m \parallel u \parallel^2 - m\,u \cdot v. \qquad [6.11]$$

For this new expression of the Lagrangian, Euler–Lagrange equations:

$$\frac{d}{dt}\left(grad_{\,v}\,\mathcal{L}\right) - grad_{\,x}\,\mathcal{L} = 0, \qquad [6.12]$$

give:

$$\frac{d}{dt}\left(m\,(v-u)\right) + m\,(grad_x u)\,(v-u) = 0.$$

Taking into account the expression [3.29] of the velocity of transport, we have:

$$m\,\dot{v} = m\left[\dot{u} + j(\varpi)\,(v-u)\right];$$

that, owing to the expression [3.53] of the acceleration of transport, leads to:

$$m\,\dot{v} = m\,a_t.$$

In the old coordinate system X', the particle is gravitation free hence $g' = \Omega' = 0$ and, taking into account the transformation law [3.52] of the gravitation, we recover Souriau's equation of motion [3.47]:

$$m\ddot{x} = m\,(g - 2\,\Omega \times v),$$

in any Galilean coordinate system X.

6.2.2. *The potentials of the Galilean gravitation*

Next, let us consider a more general case where there is not necessarily a particular Galilean coordinate system in which the particle is in uniform straight motion. Having a look at [6.11], we claim that the Lagrangian for a particle subjected to a Galilean gravitation has the following general form:

$$\boxed{\mathcal{L}(t,x,v) = \tfrac{1}{2}\,m \parallel v \parallel^2 + m\,A\cdot v - m\,\phi,} \qquad [6.13]$$

where $(t,x) \longmapsto \phi(t,x) \in \mathbb{R}$ and $(t,x) \longmapsto A(t,x) \in \mathbb{R}^3$ are given scalar and vector fields assumed to model the gravitation (for instance, the particular Lagrangian [6.11] is obtained with $\phi = - \| u \|^2 /2$ and $A = -u$). Corresponding Euler–Lagrange equation reads:

$$\frac{d}{dt}\,(grad_v\,\mathcal{L}) - grad_x\,\mathcal{L}$$

$$= \frac{d}{dt}\,[m\,(v + A)] + m\,[grad\,\phi - (grad\,A)\,v] = 0.$$

As the field A depends on x and t, we have:

$$\dot{p} + m\left[\frac{\partial A}{\partial t} + \frac{\partial A}{\partial x}\,v + grad\,\phi - (grad\,A)\,v\right] = 0.$$

$$\dot{p} + m\left[grad\,\phi + \frac{\partial A}{\partial t} + j(curl\,A)\,v\right] = 0.$$

We recover the equation of motion [3.45]:

$$\dot{p} = m\,(g - 2\Omega \times v),$$

as Euler–Lagrange variation equation of the least action principle, provided that we put:

$$\boxed{g = -grad\,\phi - \frac{\partial A}{\partial t}, \qquad \Omega = \tfrac{1}{2}\,curl\,A.} \qquad [6.14]$$

The fields ϕ and A are called the *potentials of the Galilean gravitation*. It is said that the components g, Ω of the Galilean gravitation *admit* or *are generated by the potentials* ϕ, A. In fact, we already know the scalar potential ϕ. We met it in the particular case of the Newtonian gravitation and it was given by [3.72] in Kepler's problem. By a straightforward calculation, the readers can verify that, for a given Galilean gravitation, a necessary condition for the existence of the gravitation potentials is:

$$curl\,g + 2\,\frac{\partial \Omega}{\partial t} = 0, \qquad div\,\Omega = 0. \qquad [6.15]$$

It is worth noting that there is not always a variational formulation because it is conditioned by the satisfaction of these two conditions (this is another trap of the calculus of variation).

Moreover, let us note that for a given gravitation field, the choice of the potentials ϕ and A is not unique. Indeed, let ϕ^* and A^* be potentials allowing us to recover the same gravity and spinning fields by [6.14]. Hence, $\Delta\phi = \phi^* - \phi$ and $\Delta A = A^* - A$ satisfy:

$$grad\,(\Delta\phi) + \frac{\partial}{\partial t}(\Delta A) = 0, \qquad curl\,(\Delta A) = 0.$$

Because of the last condition, there exists (at least within a simply connected subdomain of $\mathbb{R}^4$) a scalar field $(t, x) \longmapsto f(t, x)$ such that:

$$\Delta A = grad\, f,$$

and the first condition reads:

$$grad\left(\Delta\phi + \frac{\partial f}{\partial t}\right) = 0.$$

We can conclude that:

$$\boxed{\phi^* = \phi - \frac{\partial f}{\partial t}, \qquad A^* = A + grad\, f,} \qquad [6.16]$$

leads to the same gravitation field as ϕ and A. The arbitrary field f is called a *gauge function* and the previous condition is the *gauge transformation*.

In the Hamiltonian formalism, introducing the *generalized linear momentum*:

$$\pi = grad_v\, \mathcal{L} = m\,(v + A) = p + m\,A, \qquad [6.17]$$

and taking into account the expression [6.13], we obtain the *Hamiltonian of a particle*:

$$\mathcal{H} = \pi \cdot v - \mathcal{L} = \frac{1}{2}m \parallel v \parallel^2 + m\,\phi, \qquad [6.18]$$

which is nothing other then the total energy [3.74], an important integral of the motion already encountered about Kepler's problem in section 3.4 but considered now in a more general context. Also, eliminating v, it reads:

$$\mathcal{H} = \frac{1}{2m} \parallel \pi - m\,A \parallel^2 + m\,\phi.$$

Explicitly, the *canonical equations* [6.9] read:

$$\dot{x} = v = grad_{\pi} \mathcal{H} = \frac{\pi}{m} - A, \quad [6.19]$$

$$\dot{\pi} = -grad_x \mathcal{H} = m\,(grad\, A)\left(\frac{\pi}{m} - A\right) - m\, grad\, \phi \quad [6.20]$$

$$= m\,[(grad\, A)\, v - grad\, \phi]. \quad [6.21]$$

Before going further, let us have a look once again at the example of section 3.3.1. In the Galilean coordinate system X' such that $g' = \Omega' = 0$, the particle is at rest. For the observer rotating at $x_0 = 0$ at constant rotation velocity ϖ and working with the Galilean coordinate system X, the potentials of the gravitation are, under the conditions [3.63], given by:

$$\phi = -\frac{1}{2} \parallel u \parallel^2 = -\frac{1}{2} \parallel \varpi \times x \parallel^2, \qquad A = -u = -\varpi \times x.$$

As an exercise, the readers can verify that [6.14] allows us to recover the expression [3.64] of the spinning and gravity fields.

6.2.3. *Transformation law of the potentials of the gravitation*

Introducing into the Lagrangian [6.13] the expression of v given by the velocity addition formula [1.13], we obtain after straightforward calculations:

$$\mathcal{L}(t, x', v') = \frac{1}{2} m \parallel v' \parallel^2 - m\, \phi' + m\, A' \cdot v',$$

where:

$$\boxed{\phi' = \phi - A \cdot u - \frac{1}{2} \parallel u \parallel^2, \qquad A' = R^T(A + u).} \quad [6.22]$$

Let us prove it is the *transformation law of the Galilean gravitation potentials* in the following sense:

THEOREM 6.1.– If the components g, Ω of the Galilean gravitation in the Galilean coordinate system X are generated by the potentials ϕ, A, according to [6.14], then the corresponding components g', Ω' in another Galilean coordinate system X' admit the potentials ϕ', A' given by [6.22]:

$$g' = -grad_{x'}\, \phi' - \frac{\partial A'}{\partial t'}, \qquad \Omega' = \frac{1}{2} curl_{x'}\, A'. \quad [6.23]$$

The quantity:

$$I_0 = \phi + \parallel A \parallel^2 / 2 \qquad [6.24]$$

is a Galilean invariant.

PROOF.–

The calculus is decomposed into four steps:

– Step 1: *recasting the transformation law of the gravity.* Let us note by time differentiating [3.29] at constant x that [3.60] is nothing else:

$$a_t^* = \frac{\partial u}{\partial t}.$$

Hence, [3.61] reads:

$$\boxed{g = \frac{\partial u}{\partial t} + \varpi \times u + 2\,\Omega \times u + R\,g',} \qquad [6.25]$$

which is a more compact way to write the transformation law of the gravity [3.62].

– Step 2: *establishing the transformation law for the derivatives of a column field.* Owing to [3.28], it holds for any column field $X \longmapsto v'(X) \in \mathbb{R}^n$:

$$dv' = \frac{\partial v'}{\partial X}\,dX = \frac{\partial v'}{\partial X}\,P\,dX'.$$

On the other hand, v' being seen as a function of X' through the coordinate change $X \longmapsto X'$, we have:

$$dv' = \frac{\partial v'}{\partial X'}\,dX'.$$

dX' being arbitrary, we obtain by comparing the previous relations:

$$\frac{\partial v'}{\partial X'} = \frac{\partial v'}{\partial X}\,P.$$

Taking into account [1.9], we have:

$$\frac{\partial v'}{\partial t'} = \frac{\partial v'}{\partial t} + \frac{\partial v'}{\partial x}\,u, \qquad \frac{\partial v'}{\partial x'} = \frac{\partial v'}{\partial x}\,R. \qquad [6.26]$$

– Step 3: *demonstrating the second condition* [6.23]. Applying this formula to the potential field A and owing to its transformation law [6.22], we have:

$$\frac{\partial A'}{\partial x'} = \frac{\partial}{\partial x}(R^T(A+u))\,R = R^T\frac{\partial}{\partial x}(A+u)\,R,$$

because R is independent of x. Owing to the definition [7.43] of $curl$ and [7.23], it holds:

$$j(curl_{x'}\,A') = \frac{\partial A'}{\partial x'} - \left(\frac{\partial A'}{\partial x'}\right)^T = R^T\left[\frac{\partial}{\partial x}(A+u) - \left(\frac{\partial}{\partial x}(A+u)\right)^T\right]R,$$

$$j(curl_{x'}\,A') = R^T j(curl_x\,(A+u))\,R = j(R^T curl_x\,(A+u)).$$

Because the map j is regular, we obtain:

$$curl_{x'}\,A' = R^T(curl_x\,A + curl_x\,u).$$

Differentiating [3.29] with respect to x gives:

$$\frac{\partial u}{\partial x} = j(\varpi), \qquad grad_x\,u = -j(\varpi), \tag{6.27}$$

and using once again [7.43]:

$$curl_x\,u = 2\varpi, \tag{6.28}$$

that proves, owing to [6.14], the second condition [6.23]:

$$\frac{1}{2}\,curl_{x'}\,A' = R^T(\frac{1}{2}\,curl_x\,A + \varpi) = \Omega'.$$

– Step 4: *demonstrating the first condition* [6.23]. Applying the first condition [6.26] to A and taking into account [6.22], we have:

$$\frac{\partial A'}{\partial t'} = \frac{\partial A'}{\partial t} + \frac{\partial A'}{\partial x}\,u = \dot{R}^T(A+u) + R^T\left(\frac{\partial}{\partial t}(A+u) + \frac{\partial}{\partial x}(A+u)\,u\right),$$

or, taking into account [3.25]:

$$\frac{\partial A'}{\partial t'} = R^T\left(\frac{\partial}{\partial t}(A+u) + \frac{\partial}{\partial x}(A+u)\,u - \varpi\times(A+u)\right).$$

On the other hand, applying the second condition [6.26] to ϕ, transposing and owing to [6.22] and [7.39], we have:

$$grad_{x'}\,\phi' = R^T\left(grad_x\,\phi - (grad_x\,(A+u))\,u - (grad_x\,u)\,A\right).$$

Combining the last two results and owing to [6.14] gives:

$$-grad_{x'}\,\phi' - \frac{\partial A'}{\partial t'} = R^T [g - \frac{\partial u}{\partial t} + \left(grad_x\,(A+u) - \frac{\partial}{\partial x}(A+u) \right)\, u$$

$$+ (grad_x\, u)\, A + \varpi \times (A+u)].$$

Owing to the definition [7.43] of $curl$, [6.14], [6.28] and [6.27], we have:

$$-grad_{x'}\,\phi' - \frac{\partial A'}{\partial t'} = R^T \left[g - \frac{\partial u}{\partial t} - \varpi \times u - 2\,\Omega \times u \right],$$

that, taking into account [6.25], proves the first relation [6.23].

Moreover, by a straightforward calculation resulting from [6.22], it is easy to verify that the quantity [6.24] is invariant.■

6.2.4. *How to manage holonomic constraints?*

Due to Lagrange's multipliers, some modifications can be done to model other forces as the gravitation. For instance, let us consider Foucault's pendulum (section 3.5.2). The bob moving on the sphere of center $\boldsymbol{P}$, radius l, its position is subjected to the holonomic constraint:

$$f(x) = 1 - \frac{1}{l}\sqrt{x^2 + y^2 + (z-l)^2} = 0,$$

hence:

$$grad\, f = \frac{1}{l}\begin{pmatrix} -x \\ -y \\ l-z \end{pmatrix}.$$

The gravity g and the spinning Ω given by [3.79] being uniform, considering the gravitation potentials:

$$\phi = m\ \parallel g \parallel\ z, \qquad A = \Omega \times x,$$

and Lagrange's multiplier S, the modified Euler–Lagrange equations [6.10] with $y = x$ allow us to recover the generalized equations of motion [3.78] with the tension force along the thread:

$$F = S\, grad\, f,$$

where Lagrange's multiplier S is physically interpreted as the intensity of the tension force. Finally, we recover the equations [3.80], [3.81] and [3.82] of Foucault's pendulum motion by a variational formulation.

7

Elementary Mathematical Tools

7.1. Maps

We call a *map* an assignment $f : x \mapsto y = f(x)$ of elements of a set into elements of another set. It is sometimes denoted by $x \mapsto y$. The existence of a map f entails the one of its *definition set*:

$$x \in def(f) \qquad \Leftrightarrow \qquad f(x) \quad \text{exists},$$

and of its *value set*:

$$y \in val(f) \qquad \Leftrightarrow \qquad \exists x \quad \text{such that} \quad y = f(x).$$

It is useful to consider the *identity* of a set E, denoted by 1_E and the *impotent map* of which the definition set is empty, which we denote by $1_\emptyset$. The *composition* of functions is defined by:

$$(f \circ g)(x) = f(g(x)),$$

every time it exists. If there is no confusion, the composition $f \circ g$ is simply denoted by fg. The composition of two maps is always a map, even if it is impotent. The composition is associative but no commutative. A map f is *regular* or *one-to-one* if:

$$f(x) = f(y) \qquad \Leftrightarrow \qquad x = y.$$

f being regular, the *inverse* map f^{-1} is defined by:

$$f^{-1}(y) = x \qquad \Leftrightarrow \qquad y = f(x).$$

We verify immediately that f^{-1} is regular and:

$$(f^{-1})^{-1} = f, \quad def(f^{-1}) = val(f), \quad val(f^{-1}) = def(f),$$
$$f^{-1}f = 1_{def(f)}, \quad ff^{-1} = 1_{val(f)}.$$

If f and g are regular, fg is regular and:

$$(fg)^{-1} = g^{-1}f^{-1}.$$

Two maps f and g being defined on the same set, there exists a map h such that $f = hg$ if and only if:

$$\{g(x) = g(y)\} \qquad \Rightarrow \qquad \{f(x) = f(y)\}\,.$$

Moreover, the map h is unique if $def(h) = val(g)$. Hence, it is denoted by f/g and is called the *quotient* of f by g.

If a map has two variables, for instance $(x, y) \mapsto f(x, y)$, the partial map $y \mapsto f(x, y)$ is sometimes denoted by $f(x, \bullet)$.

7.2. Matrix calculus

If you need to brush up on matrix calculus and linear algebra, this would be a good time to consult, for instance [SOU 92].

7.2.1. *Columns*

An *n-column* or simply a *column* is an element of $\mathbb{R}^n$ and is denoted by:

$$V = \begin{pmatrix} V^1 \\ V^2 \\ \vdots \\ V^n \end{pmatrix}.$$

V^i are its *components*. $\mathbb{R}^n$ has two operations, the addition of vectors and the multiplication by a scalar:

$$\begin{pmatrix} U^1 \\ U^2 \\ \vdots \\ U^n \end{pmatrix} + \begin{pmatrix} V^1 \\ V^2 \\ \vdots \\ V^n \end{pmatrix} = \begin{pmatrix} U^1 + V^1 \\ U^2 + V^2 \\ \vdots \\ U^n + V^n \end{pmatrix} \quad \text{and} \quad \lambda \begin{pmatrix} V^1 \\ V^2 \\ \vdots \\ V^n \end{pmatrix} = \begin{pmatrix} \lambda V^1 \\ \lambda V^2 \\ \vdots \\ \lambda V^n \end{pmatrix}.$$

The addition is associative and commutative. The multiplication by a scalar is distributive over the column addition. We call *key-columns* the following elements of $\mathbb{R}^n$:

$$e_1 = \begin{pmatrix} 1 \\ 0 \\ \vdots \\ 0 \end{pmatrix}, \quad e_2 = \begin{pmatrix} 0 \\ 1 \\ \vdots \\ 0 \end{pmatrix}, \quad \ldots, \quad e_n = \begin{pmatrix} 0 \\ 0 \\ \vdots \\ 1 \end{pmatrix}.$$

Thus, any n-column is a linear combination of the key-columns:

$$V = V^1 e_1 + V^1 e_2 + \cdots + V^n e_n.$$

The n-columns $V_1, V_2, \cdots, V_p$ are *linearly independent* if any linear combination $\lambda_1 V_1 + \lambda_2 V_2 + \cdots + \lambda_p V_p$ is zero if and only if $\lambda_1 = \lambda_2 = \cdots = \lambda_p = 0$.

7.2.2. *Rows*

An *n-row* or simply a *row* is a numerical linear function on $\mathbb{R}^n$:

$$\Phi(\lambda U + \mu V) = \lambda \Phi(U) + \mu \Phi(V).$$

It is denoted by:

$$\Phi = \begin{pmatrix} \Phi_1 & \Phi_2 & \cdots & \Phi_n \end{pmatrix}.$$

Φ_i are its *components* and the value of Φ for the n-column V is:

$$\Phi(V) = \Phi_1 V^1 + \Phi_1 V^1 + \cdots + \Phi_n V^n.$$

The set $(\mathbb{R}^n)^*$ of the rows has two operations, the addition:

$$\begin{pmatrix} \Phi_1 & \Phi_2 & \cdots & \Phi_n \end{pmatrix} + \begin{pmatrix} \Theta_1 & \Theta_2 & \cdots & \Theta_n \end{pmatrix} = \begin{pmatrix} \Phi_1 + \Theta_1 & \Phi_2 + \Theta_2 & \cdots & \Phi_n + \Theta n \end{pmatrix},$$

and the multiplication by a scalar:

$$\lambda \begin{pmatrix} \Phi_1 & \Phi_2 & \cdots & \Phi_n \end{pmatrix} = \begin{pmatrix} \lambda\Phi_1 & \lambda\Phi_2 & \cdots & \lambda\Phi_n \end{pmatrix}.$$

The addition is associative and commutative. The multiplication by a scalar is distributive over the row addition. We call *key-rows* the following rows:

$$e^1 = \begin{pmatrix} 1 & 0 & \cdots & 0 \end{pmatrix}, \quad e^2 = \begin{pmatrix} 0 & 1 & \cdots & 0 \end{pmatrix}, \quad \ldots, \quad e^n = \begin{pmatrix} 0 & \cdots & 0 & 1 \end{pmatrix}.$$

Thus, any n-row is a linear combination of the key-rows:

$$\Phi = \Phi_1 e^1 + \Phi_2 e^2 + \cdots + \Phi_n e^n.$$

7.2.3. *Matrices*

An $n \times p$ *matrix* is a linear map from $\mathbb{R}^p$ into $\mathbb{R}^n$:

$$M(\lambda U + \mu V) = \lambda M(U) + M\Phi(V).$$

It is denoted by:

$$M = \begin{pmatrix} M_1^1 & M_2^1 & \cdots & M_p^1 \\ M_1^2 & M_2^2 & \cdots & M_p^2 \\ \vdots & \vdots & \ddots & \vdots \\ M_1^n & M_2^n & \cdots & M_p^n \end{pmatrix}.$$

M_j^i are its *elements*. The set $\mathbb{M}_{np}$ of the $n \times p$ matrices has two basic operations, the addition:

$$\begin{pmatrix} M_1^1 & \cdots & M_p^1 \\ \vdots & \ddots & \vdots \\ M_1^n & \cdots & M_p^n \end{pmatrix} + \begin{pmatrix} N_1^1 & \cdots & N_p^1 \\ \vdots & \ddots & \vdots \\ N_1^n & \cdots & N_p^n \end{pmatrix} = \begin{pmatrix} M_1^1 + N_1^1 & \cdots & M_p^1 + N_p^1 \\ \vdots & \ddots & \vdots \\ M_1^n + N_1^n & \cdots & M_p^n + N_p^n \end{pmatrix},$$

and the multiplication by a scalar:

$$\lambda \begin{pmatrix} M_1^1 & \cdots & M_p^1 \\ \vdots & \ddots & \vdots \\ M_1^n & \cdots & M_p^n \end{pmatrix} = \begin{pmatrix} \lambda M_1^1 & \cdots & \lambda M_p^1 \\ \vdots & \ddots & \vdots \\ \lambda M_1^n & \cdots & \lambda M_p^n \end{pmatrix}.$$

The matrix addition is associative and commutative. The multiplication by a scalar is distributive over the matrix addition. The j-th column of M is the column M_j of which the components are M_j^i, which allows us to write in a more compact way:

$$M = \begin{pmatrix} M_1 & \cdots & M_n \end{pmatrix}.$$

The i-th row of M is the row M^i of which the components are M_j^i, which allows us to write:

$$M = \begin{pmatrix} M^1 \\ \vdots \\ M^n \end{pmatrix}.$$

The composition or product of the $n \times p$ matrix M by the $p \times q$ matrix N is the $n \times q$ matrix obtained performing products "rows by columns":

$$MN = \begin{pmatrix} M^1N_1 & M^1N_2 & \cdots & M^1N_q \\ M^2N_1 & M^2N_2 & \cdots & M^2N_q \\ \vdots & \vdots & \ddots & \vdots \\ M^nN_1 & M^nN_2 & \cdots & M^nN_q \end{pmatrix}.$$

The matrix product is associative but not commutative. The *identity matrix*:

$$1_{\mathbb{R}^n} = \begin{pmatrix} 1 & 0 & \cdots & 0 \\ 0 & 1 & \cdots & 0 \\ \vdots & \vdots & \ddots & \vdots \\ 0 & 0 & \cdots & 1 \end{pmatrix},$$

is such that for any $n \times n$ matrix M:

$$M\,1_{\mathbb{R}^n} = 1_{\mathbb{R}^n} M = M$$

The components of the identity matrix are denoted by δ^i_j and are called *Kronecker's symbols*. An $n \times n$ matrix M is *diagonal* if it has the form:

$$M = \begin{pmatrix} M^1_1 & 0 & \cdots & 0 \\ 0 & M^2_2 & \cdots & 0 \\ \vdots & \vdots & \ddots & \vdots \\ 0 & 0 & \cdots & M^n_n \end{pmatrix}.$$

We denote it by:

$$M = diag(M^1_1, M^2_2, \cdots, M^n_n).$$

The *transpose matrix* of the $n \times p$ matrix M is the $p \times n$ matrix:

$$M^T = \begin{pmatrix} M^1_1 & M^2_1 & \cdots & M^n_1 \\ M^1_2 & M^2_2 & \cdots & M^n_2 \\ \vdots & \vdots & \ddots & \vdots \\ M^1_p & M^2_p & \cdots & M^n_p \end{pmatrix}.$$

We verify that:

$$(M + M')^T = M^T + M'^T, \qquad (M\,N)^T = N^T M^T, \qquad (M^T)^T = M.$$

An $n \times n$ matrix M is *symmetric* if $M^T = M$ and *skew-symmetric* if $M^T = -M$.

Of course, as particular cases:

– n-columns are $n \times 1$ matrices (then linear maps from $\mathbb{R}$ into $\mathbb{R}^n$) and $\mathbb{R}^n = \mathbb{M}_{n1}$;

– n-rows are $1 \times n$ matrices and $(\mathbb{R}^n)^* = \mathbb{M}_{1n}$. Moreover: $\Phi(V) = \Phi V$,;

– scalars are 1×1 matrices and commute with any other matrices, in particular with columns and rows.

The *dot product* of two n-columns is the scalar:

$$U \cdot V = U^T V.$$

The dot product is commutative. The *norm* of an n-column is:

$$\| U \| = \sqrt{U \cdot U}.$$

The norm is non-negative. It vanishes if and only if the column vanishes. For any n-columns U, V and scalar λ :

$$\| \lambda U \| = | \lambda | \, \| U \|, \qquad [7.1]$$

$$\| U + V \| \leq \| U \| + \| V \| . \qquad [7.2]$$

The *trace* of an $n \times n$ matrix M is the sum of its diagonal elements:

$$Tr(M) = M_1^1 + M_2^2 + \cdots + M_n^n.$$

Of course, we have:

$$Tr(M^T) = Tr(M),$$

and if M is skew-symmetric, its trace vanishes. We verify that for an n-row Φ and n-columns U, V :

$$Tr(V\Phi) = \Phi V \qquad \text{and} \qquad Tr(VU^T) = U \cdot V, \qquad [7.3]$$

and that for any $n \times p$ matrix M and any $p \times n$ matrix N:

$$Tr(MN) = Tr(NM). \qquad [7.4]$$

The *determinant* of an $n \times n$ matrix M is the unique numerical function det of M, linear with respect to each of its columns $M_1, M_2, \cdots M_n$, completely skew-symmetric with respect to them and such that:

$$\det(1_{\mathbb{R}^n}) = 1.$$

The determinant of a 2×2 matrix is:

$$\det(M) = M_1^1 M_2^2 - M_1^2 M_2^1,$$

and the one of a 3×3 matrix is:

$$\begin{aligned}\det(M) &= M_1^1 M_2^2 M_3^3 + M_1^2 M_2^3 M_3^1 + M_1^3 M_2^1 M_3^2 \\ &- M_1^3 M_2^2 M_3^1 - M_1^2 M_2^1 M_3^3 - M_1^1 M_2^3 M_3^2.\end{aligned}$$

If M, N are $n \times n$ matrix, we verify that:

$$\det(-M) = (-1)^n \det(M) \qquad \text{and} \qquad \det(MN) = \det(M)\det(N).$$

An $n \times n$ matrix M is regular if and only if its determinant is not null. Then:

$$\det(M^{-1}) = (\det(M))^{-1}.$$

The *dot product* of two $n \times n$ matrices M and N is the scalar:

$$M : N = Tr(M^T N). \qquad [7.5]$$

The dot product is commutative. The *norm* of an $n \times n$ matrix is:

$$\| M \| = \sqrt{M : M}. \qquad [7.6]$$

The matrix norm has similar properties to the ones of the vector norm, in particular [7.1] and [7.2].

The *inverse* of an $n \times n$ matrix M is –if there exists– the unique matrix M^{-1} such that:

$$M M^{-1} = M^{-1} M = 1_{\mathbb{R}^n}. \qquad [7.7]$$

Then, M is said to be *regular*. A matrix is regular if and only if its determinant is non-zero.

The (i, j) *minor* of the $n \times n$ matrix M, denoted by $minor(i, j, M)$, is the determinant of the $(n-1) \times (n-1)$ matrix that results from deleting the i-th row and j-th column of M. The *adjugate* of M is the $n \times n$ matrix $adj(M)$ of which the element $(adj(M))^i_j$ is the (j, i) *cofactor* of M:

$$(adj(M))^i_j = (-1)^{i+j} minor(j, i, M).$$

If M is regular, its inverse can be obtained due to *Cramer's rule:*

$$M^{-1} = \frac{adj(M)}{\det(M)}. \qquad [7.8]$$

An $n \times n$ matrix M has a real or complex *eigenvalue* λ if there is a non-zero *eigenvector* $V \in \mathbb{C}^n$ such that

$$M\,V = \lambda\,V.$$

The eigenvalues of M are roots of the *characteristic equation*:

$$\det(M - \lambda 1_{\mathbb{R}^n}) = 0.$$

The matrix M is *diagonalizable* if there is a family of n linearly independent eigenvectors $V_1, V_2, \cdots, V_n$ of respective eigenvalues $\lambda_1, \lambda_2, \cdots, \lambda_n$. Then, considering the $n \times n$ matrix $P = (V_1, V_2, \cdots, V_n)$, we have:

$$P^{-1} M\,P = diag(\lambda_1, \lambda_2, \cdots, \lambda_n).$$

We say that P *diagonalizes* M. A symmetric matrix M is diagonalizable, its eigenvalues are real numbers and its eigenvectors are mutually orthogonal. As they are defined to within a scalar factor, they may be chosen as orthonormal.

An $n \times n$ matrix M is *positive definite* if for all non-vanishing $x \in \mathbb{R}^n$:

$$x \cdot (M\,x) > 0. \qquad [7.9]$$

7.2.4. *Block matrix*

A *block matrix* or a partitioned matrix is a matrix which is interpreted as having been broken into sections called *blocks* or *submatrices*. For instance, the matrix:

$$M = \begin{pmatrix} 1\,1\,2\,2\,2 \\ 1\,1\,2\,2\,2 \\ 3\,3\,4\,4\,4 \end{pmatrix},$$

can be partitioned into four blocks:

$$M_1^1 = \begin{pmatrix} 1\,1 \\ 1\,1 \end{pmatrix}, \quad M_1^2 = \begin{pmatrix} 2\,2\,2 \\ 2\,2\,2 \end{pmatrix},$$
$$M_2^1 = \begin{pmatrix} 3\,3 \end{pmatrix}, \qquad M_2^2 = \begin{pmatrix} 4\,4\,4 \end{pmatrix}.$$

The partitioned matrix can then be written as:

$$M = \begin{pmatrix} M_1^1\; M_2^1 \\ M_1^2\; M_2^2 \end{pmatrix}.$$

Sums and products can be extended to block matrices provided that the partitions of terms and factors are compatible.

7.3. Vector calculus in $\mathbb{R}^3$

With any 3-column u a unique 3×3 skew-symmetric matrix is associated:

$$j(u) = \begin{pmatrix} 0 & -u^3 & u^2 \\ u^3 & 0 & -u^1 \\ -u^2 & u^1 & 0 \end{pmatrix}. \qquad [7.10]$$

u is sometimes called the *axial vector* of $j(u)$. The map j is regular, linear and satisfies the following identities:

$$j(j(u)v) = vu^T - uv^T, \qquad [7.11]$$

$$j(u)j(v) = vu^T - (u \cdot v)1_{\mathbb{R}^3}, \qquad [7.12]$$

from which we deduce:

$$Tr(j(u)j(v)) = -2u \cdot v, \qquad [7.13]$$

and:

$$j(u)j(v) - j(v)j(u) = j(j(u)v). \qquad [7.14]$$

The *cross-product* of two 3-columns u and v is the 3-column $u \times v$ defined by:

$$u \times v = j(u)v.$$

Considering the components, we have:

$$\begin{pmatrix} u^1 \\ u^2 \\ u^3 \end{pmatrix} \times \begin{pmatrix} v^1 \\ v^2 \\ v^3 \end{pmatrix} = \begin{pmatrix} u^2v^3 - u^3v^2 \\ u^3v^1 - u^1v^3 \\ u^1v^2 - u^2v^1 \end{pmatrix}.$$

The cross-product is anticommutative: $u \times v = -v \times u$ and is not associative. Of course, $u \times u = 0$ and the cross-product of two colinear columns is null. The cross-product is distributive over the addition. From [7.11] and [7.12], we deduce the expressions of the *vector triple products*:

$$(u \times v) \times w = (u \cdot w)v - (v \cdot w)u, \qquad [7.15]$$

$$u \times (v \times w) = (u \cdot w)v - (v \cdot u)w. \qquad [7.16]$$

Also, condition [7.14] leads to *Jacobi's identity*:

$$u \times (v \times w) + v \times (w \times u) + w \times (u \times v) = 0. \qquad [7.17]$$

The oriented volumes are measured by the *scalar triple product*, symmetric by circular permutation of its arguments:

$$(u \times v) \cdot w = (w \times u) \cdot v = (v \times w) \cdot u,$$

and skew-symmetric with respect to any couple of arguments:

$$(u \times v) \cdot w = -(v \times u) \cdot w = -(w \times v) \cdot u = (u \times w) \cdot v.$$

Hence, the scalar triple product vanishes if two of its arguments are identical. From [7.12], we deduce the relation:

$$\| u \times v \|^2 + (u \cdot v)^2 = \| u \|^2 \| v \|^2 . \qquad [7.18]$$

For any 3×3 matrix M and any 3-columns u, v, w, we have:

$$Tr(M)\, u \cdot (v \times w) = (M\, u) \cdot (v \times w) + u \cdot ((M\, v) \times w) + u \cdot (v \times (M\, w)),$$

from which we deduce:

$$(M\, v) \times w + v \times (M\, w) = (Tr(M)\, 1_{\mathbb{R}^3} - M)^T\, (v \times w). \qquad [7.19]$$

An *orthogonal matrix* R is a 3×3 matrix preserving the dot products of two 3-columns:

$$(R\,u) \cdot (R\,v) = u \cdot v. \qquad [7.20]$$

We verify that the inverse of R is its transpose matrix:

$$R^T R = R\,R^T = 1_{\mathbb{R}^3}, \qquad [7.21]$$

and that:

$$(R\,u) \times (R\,v) = R(u \times v). \qquad [7.22]$$

Using the map j, this relation reads:

$$j(R^T u) = R^T j(u) R \qquad [7.23]$$

A *rotation* is an orthogonal transformation preserving the oriented volumes, then its determinant is equal to 1. An *Euclidean transformation* is an affine transformation of $\mathbb{R}^3$ preserving the dot product of two vectors and the oriented volumes, then composed of a rotation R and a translation $k \in \mathbb{R}^3$:

$$x = R\,x' + k.$$

A 3×3 symmetric matrix M is diagonalizable, its eigenvalues are real numbers and the corresponding matrix $P = (V_1, V_2, V_3)$ is orthogonal. The vectors V_1, V_2, V_3 are mutually orthogonal and of unit norm.

7.4. Linear algebra

7.4.1. *Linear space*

A *linear space* (or *vector space*) $\mathcal{T}$ is a non-empty set with two operations, the addition and the multiplication by a scalar. Its elements are called *vectors* and denoted with an arrow: $\vec{\mathbf{U}}, \vec{\mathbf{V}}, \cdots$. The addition is associative, commutative and has a zero $\vec{\mathbf{0}}$:

$$\vec{\mathbf{U}} + \vec{\mathbf{0}} = \vec{\mathbf{U}}.$$

Each vector has an opposite one:

$$\vec{\mathbf{U}} + (-\vec{\mathbf{U}}) = \vec{\mathbf{0}}.$$

The multiplication by a scalar is distributive over the vector addition and:

$$\lambda(\mu\vec{\mathbf{U}}) = (\lambda\mu)\vec{\mathbf{U}},$$
$$1\vec{\mathbf{U}} = \vec{\mathbf{U}}.$$

If the scalars are real numbers, $\mathcal{T}$ is a real linear space. Unless otherwise specified, linear spaces considered in this book are real. The set $\mathbb{R}^n$ of n-columns, the set $(\mathbb{R}^n)^*$ of n-rows and, more generally, the set $\mathbb{M}_{np}$ of the $n \times p$ matrices are examples of linear spaces.

A *linear subspace* of $\mathcal{T}$ is a subset that is closed under taking linear combinations. The set $\mathbb{M}_{nn}^{symm}$ of the $n \times n$ symmetric matrices and the set $\mathbb{M}_{nn}^{skew}$ of the $n \times n$ skew-symmetric matrices are linear subspaces of $\mathbb{M}_{nn}$.

Let $\mathcal{R}$ be another linear space. A map $\mathbf{A} : \mathcal{T} \to \mathcal{R}$ is a *linear map* if it preserves linear combinations:

$$\mathbf{A}\left(\lambda_1\vec{\mathbf{U}}_1 + \lambda_2\vec{\mathbf{U}}_2 + \cdots + \lambda_p\vec{\mathbf{U}}_p\right) = \lambda_1\mathbf{A}\left(\vec{\mathbf{U}}_1\right) + \lambda_2\mathbf{A}\left(\vec{\mathbf{U}}_2\right) + \cdots + \lambda_p\mathbf{A}\left(\vec{\mathbf{U}}_p\right).$$

A linear space $\mathcal{T}$ has a finite *dimension* n if there exists a linear regular map S of which the definition set is $\mathbb{R}^n$ and the value set is $\mathcal{T}$. S is called a *basis* or *linear frame* of $\mathcal{T}$. The $\vec{\mathbf{e}}_i = S(e_i)$ are called the **basis** *vectors.* Thus, any vector can be decomposed into the unique linear combination of the basis vectors:

$$\vec{\mathbf{V}} = V^1\vec{\mathbf{e}}_1 + V^2\vec{\mathbf{e}}_2 + \cdots + V^n\vec{\mathbf{e}}_n,$$

where the V^i are called the *components* of $\vec{\mathbf{V}}$ with respect to the basis. We denote it indifferently by S or $(\vec{\mathbf{e}}_1, \vec{\mathbf{e}}_2, \cdots, \vec{\mathbf{e}}_n)$ or more simply $(\vec{\mathbf{e}}_i)$, allowing us to write the previous relation:

$$\vec{\mathbf{V}} = S(V) = (\vec{\mathbf{e}}_1, \vec{\mathbf{e}}_2, \cdots, \vec{\mathbf{e}}_n)V,$$

the last expression being understood as a product of a row by a column. If we change the basis $(\vec{\mathbf{e}}_i)$ for a new one $(\vec{\mathbf{e}}'_i)$, the corresponding *transformation matrix* is the regular $n \times n$ matrix $P = S^{-1}S'$ such that:

$$(\vec{\mathbf{e}}'_1, \vec{\mathbf{e}}'_2, \cdots, \vec{\mathbf{e}}'_n) = (\vec{\mathbf{e}}_1, \vec{\mathbf{e}}_2, \cdots, \vec{\mathbf{e}}_n)\,P,$$

that also reads:

$$\vec{\mathbf{e}}'_i = P_i^1\vec{\mathbf{e}}_1 + P_i^2\vec{\mathbf{e}}_2 + \cdots + P_n^n\vec{\mathbf{e}}_n.$$

In the new basis, the vector $\vec{\mathbf{V}}$ is represented by the column:

$$V' = P^{-1}V. \quad [7.24]$$

Conversely, if there exists a regular map S of which the definition set is $\mathbb{R}^n$, its value set $\mathcal{T}$ is a linear space of dimension n, defining by *structure transport* the vector addition:

$$\vec{\mathbf{U}} + \vec{\mathbf{V}} = S(S^{-1}(\vec{\mathbf{U}}) + S^{-1}(\vec{\mathbf{V}})),$$

and the multiplication by a scalar:

$$\lambda\vec{\mathbf{U}} = S(\lambda S^{-1}(\vec{\mathbf{U}})).$$

Let $\mathcal{F}$ be the set of the maps f of which the definition set is a given set A and the value set is a given linear space $\mathcal{T}$. Defining for $f, g \in \mathcal{F}$ the vector addition by:

$$\forall x \in A, \qquad (f+g)(x) = f(x) + g(x), \quad [7.25]$$

and the multiplication by a scalar:

$$\forall \lambda \in \mathbb{R}, \qquad \forall x \in A, \qquad (\lambda f)(x) = \lambda f(x), \quad [7.26]$$

the set $\mathcal{F}$ is a linear space.

7.4.2. *Linear form*

The set $\mathcal{T}^*$ of the linear maps from $\mathcal{T}$ into $\mathbb{R}$ is a linear space and is called the *dual space* of $\mathcal{T}$. Its elements $\mathbf{\Phi}$ are called *linear forms* or *covectors*. If $\mathcal{T}$ has a finite dimension n, then its dual space has the same dimension. S or $(\vec{\mathbf{e}}_i)$ being a basis of $\mathcal{T}$, we have:

$$\mathbf{\Phi}(\vec{\mathbf{V}}) = \mathbf{\Phi}(S\,V) = (\mathbf{\Phi}\,S)\,V.$$

Hence, $\Phi = \mathbf{\Phi}\,S$ is the unique n-row such that $\mathbf{\Phi} = \Phi\,S^{-1}$. The map S^{-1} is called a *cobasis* or a *linear coframe* and the components Φ_i of Φ are the *components* of $\mathbf{\Phi}$ with respect to this cobasis. The $\mathbf{e}^i = e^i S^{-1}$ are called the *basis covectors*. The value of the form $\mathbf{e}^i$ for the vector $\vec{\mathbf{U}}$ is its i-th component in the basis S:

$$\mathbf{e}^i(\vec{\mathbf{U}}) = U^i.$$

We denote the cobasis indifferently by S^{-1} or $(\mathbf{e}^i)$, allowing us to write the previous relation:

$$\mathbf{\Phi} = \Phi\, S^{-1} = \Phi \begin{pmatrix} \mathbf{e}^1 \\ \mathbf{e}^2 \\ \vdots \\ \mathbf{e}^n \end{pmatrix},$$

the last expression being understood as a product of a row by a column. Because:

$$\mathbf{e}^i(\vec{\mathbf{e}}_j) = \delta^i_j,$$

the cobasis S^{-1} is called the *dual basis* or *dual linear frame* of S. We deduce:

$$\mathbf{\Phi}(\vec{\mathbf{e}}_i) = \Phi_i. \qquad [7.27]$$

In a new cobasis S'^{-1}, the linear form $\mathbf{\Phi}$ is represented by the row:

$$\Phi' = \Phi\, P, \qquad [7.28]$$

where occurs the transformation matrix $P = S^{-1}S'$. The map:

$$\mathcal{T}^* \times \mathcal{T} \to \mathbb{R} : (\mathbf{\Phi}, \vec{\mathbf{V}}) \mapsto \left\langle \mathbf{\Phi}, \vec{\mathbf{V}} \right\rangle = \mathbf{\Phi}(\vec{\mathbf{V}}) = \Phi\, V,$$

is linear with respect to each of its arguments and is called the *dual pairing*.

7.4.3. *Linear map*

Let $\mathcal{R}$ be another linear space of finite dimension p and $(\vec{\boldsymbol{\eta}}_i)$ be one of its basis. Let $\mathbf{A} : \mathcal{T} \to \mathcal{R} : \vec{\mathbf{U}} \mapsto \vec{\mathbf{V}} = \mathbf{A}\,(\vec{\mathbf{U}})$ be a linear map. If in the basis $(\vec{\boldsymbol{\eta}}_i)$, the vector $\vec{\mathbf{V}}$ is represented by the column V:

$$\vec{\mathbf{V}} = (\vec{\boldsymbol{\eta}}_1, \dots, \vec{\boldsymbol{\eta}}_p)V = \hat{S}(V),$$

and each vector $\mathbf{A}(\vec{\mathbf{e}}_j)$ by the column A_j of components A^i_j, the linear map $\mathbf{A}$ is represented by the matrix:

$$A = (A_1, \cdots, A_p)\,, \qquad V = AU.$$

The linear map $\mathbf{A}$ and the matrix A representing it are linked by:

$$\mathbf{A} = \hat{S}AS^{-1}, \quad A = \hat{S}^{-1}\mathbf{A}\, S.$$

Let Q be the transformation matrix of the change between $(\vec{\eta}_i)$ and a new basis $(\vec{\eta}'_i)$. Then, in the basis $(\vec{e}'_j)$ and $(\vec{\eta}'_i)$, the linear map is represented by the *equivalent matrix*:

$$A' = Q^{-1}AP. \qquad [7.29]$$

When $\mathcal{T} = \mathcal{R}$, $\mathbf{A}$ is represented by the *similar matrix*:

$$A' = P^{-1}AP,$$

and:

$$Tr(A') = Tr(P^{-1}AP) = Tr(APP^{-1}) = Tr(A),$$

does not depend on the choice of the basis but only on $\mathbf{A}$. We call it the *trace* of $\mathbf{A}$ and denote it by $Tr(\mathbf{A})$. The element A^i_j of the matrix A representing $\mathbf{A}$ in the basis S is given by:

$$A^i_j = \mathbf{e}^i(\mathbf{A}(\vec{\mathbf{e}}_j)),$$

Hence:

$$Tr(\mathbf{A}) = \sum_{i=1}^{n} \mathbf{e}^i(\mathbf{A}(\vec{\mathbf{e}}_i)). \qquad [7.30]$$

The set $Hom(\mathcal{T}, \mathcal{R})$ of the linear maps from $\mathcal{T}$ into $\mathcal{R}$ is a linear space of dimension np. In particular, $Hom(\mathcal{T}, \mathbb{R}) = \mathcal{T}^*$ and $Hom(\mathbb{R}^p, \mathbb{R}^n) = \mathbb{M}_{np}$.

The *transpose* of the linear map $\mathbf{A} : \mathcal{T} \to \mathcal{R}$ is the linear map ${}^t\mathbf{A} : \mathcal{R}^* \to \mathcal{T}^*$ such that:

$$\forall \mathbf{\Phi} \in \mathcal{R}^*, \quad \forall \vec{\mathbf{V}} \in \mathcal{T}, \qquad \left\langle \mathbf{\Phi}, \mathbf{A}(\vec{\mathbf{V}}) \right\rangle = \left\langle {}^t\mathbf{A}(\mathbf{\Phi}), \vec{\mathbf{V}} \right\rangle.$$

The transpose map ${}^t\mathbf{A}$ is represented by the transpose matrix A^T of A representing $\mathbf{A}$.

If two maps $\mathbf{A}$ and $\mathbf{B}$ are linear and if:

$$\left\{\mathbf{B}(\vec{\mathbf{U}}) = \vec{\mathbf{0}}\right\} \quad \Rightarrow \quad \left\{\mathbf{A}(\vec{\mathbf{U}}) = \vec{\mathbf{0}}\right\},$$

the quotient of $\mathbf{A}$ by $\mathbf{B}$ exists and is linear (see section 7.1). The map $\boldsymbol{\lambda} = \mathbf{A}/\mathbf{B}$ is called a *Lagrange's multiplier*, hence:

$$\mathbf{A} = \boldsymbol{\lambda}\mathbf{B}.$$

7.5. Affine geometry

An *affine space* is a non-empty set $A\mathcal{T}$ of *points*, associated with a linear space $\mathcal{T}$ through a map:

$$A\mathcal{T} \times \mathcal{T} \rightarrow A\mathcal{T} : (\mathbf{a}, \vec{\mathbf{U}}) \mapsto \mathbf{a} + \vec{\mathbf{U}},$$

satisfying the following three conditions:

– for any $\mathbf{a} \in A\mathcal{T}$, $\mathbf{a} + \vec{\mathbf{0}} = \mathbf{a}$;

– for any $\mathbf{a} \in A\mathcal{T}$ and $\vec{\mathbf{U}}, \vec{\mathbf{V}} \in \mathcal{T}$, $(\mathbf{a} + \vec{\mathbf{U}}) + \vec{\mathbf{V}} = \mathbf{a} + (\vec{\mathbf{U}} + \vec{\mathbf{V}})$;

– for any points $\mathbf{a}, \mathbf{b} \in A\mathcal{T}$, there is a unique $\vec{\mathbf{U}} \in \mathcal{T}$ such that $\mathbf{a} + \vec{\mathbf{U}} = \mathbf{b}$.

The unique vector $\vec{\mathbf{U}}$ such that $\mathbf{a} + \vec{\mathbf{U}} = \mathbf{b}$ is denoted by $\overrightarrow{\mathbf{ab}}$. By taking $\mathbf{a}$ as the *origin* in $A\mathcal{T}$, we identify $A\mathcal{T}$ with $\mathcal{T}$ through the regular map $\mathbf{b} \mapsto \overrightarrow{\mathbf{ab}}$.

For any family of m points $(\mathbf{a}_i)$ in $A\mathcal{T}$, for any family of m scalars (λ_i) such that $\lambda_1 + \lambda_2 + \cdots + \lambda_m = 1$, and for any $\mathbf{a} \in A\mathcal{T}$, the point:

$$\mathbf{a} + \lambda_1 \, \overrightarrow{\mathbf{a}\mathbf{a}_1} + \lambda_2 \, \overrightarrow{\mathbf{a}\mathbf{a}_2} + \cdots + \lambda_m \, \overrightarrow{\mathbf{a}\mathbf{a}_m},$$

signed with the weights λ_i, does not depend on the choice of $\mathbf{a}$ and is called *barycenter* and is denoted by:

$$\lambda_1 \mathbf{a}_1 + \lambda_2 \mathbf{a}_2 + \cdots + \lambda_m \mathbf{a}_m.$$

An *affine subspace* of $A\mathcal{T}$ is a subset that is closed under taking barycenters. For any point $\mathbf{a} \in A\mathcal{T}$ and any subset $W \subset \mathcal{T}$, let $\mathbf{a} + W$ denote the following subset of $A\mathcal{T}$:

$$\mathbf{a} + W = \left\{ \mathbf{a} + \vec{\mathbf{V}} \mid \vec{\mathbf{V}} \in W \right\}.$$

A non-empty subset $A \subset \mathcal{T}$ is an affine subspace if and only if for every point $\mathbf{a} \in A$, the set:

$$W_{\mathbf{a}} = \left\{ \overrightarrow{\mathbf{ab}} \mid \mathbf{b} \in A \right\},$$

is a subspace of $\mathcal{T}$. Consequently, $A = \mathbf{a} + W_{\mathbf{a}}$.

Let $A\mathcal{R}$ be another affine space. A map $\boldsymbol{\alpha} : A\mathcal{T} \rightarrow A\mathcal{R}$ is an *affine map* if it preserves barycenters:

$$\boldsymbol{\alpha}(\lambda_1 \mathbf{a}_1 + \lambda_2 \mathbf{a}_2 + \cdots + \lambda_p \mathbf{a}_p) = \lambda_1 \boldsymbol{\alpha}(\mathbf{a}_1) + \lambda_2 \boldsymbol{\alpha}(\mathbf{a}_2) + \cdots + \lambda_p (\mathbf{a}_p).$$

An affine map preserves affine subspaces and parallelotopes. There exists a unique linear map $\mathbf{A} : \mathcal{T} \to \mathcal{R}$ such that:

$$\alpha\left(\mathrm{a} + \vec{\mathbf{U}}\right) = \alpha\left(\mathrm{a}\right) + \mathbf{A}\left(\vec{\mathbf{U}}\right), \qquad \mathbf{A}\left(\overrightarrow{\mathrm{ab}}\right) = \alpha\left(\mathrm{b}\right) - \alpha\left(\mathrm{a}\right).$$

It is called the *linear part* of α and is denoted by $\mathbf{A} = lin\left(\alpha\right)$. For instance, the map $a : A\mathbb{R}^n \to A\mathbb{R}^n : V \mapsto V' = C + PV$, which may be identified to the couple (C, P), is affine and P is its linear part.

If $\mathcal{T}$ has a finite *dimension* n, we say that $A\mathcal{T}$ has the dimension n. Let $(\mathrm{a}_0, \mathrm{a}_1, \ldots, \mathrm{a}_n)$ be a set of $(n+1)$ points such that the set of the vectors $\vec{\mathbf{e}}_i = \mathrm{a}_i - \mathrm{a}_0$ is a linear frame. We say that: $(\mathrm{a}_0, (\vec{\mathbf{e}}_i))$ is an affine frame of $A\mathcal{T}$ of origin a_0. For any $\mathrm{a} \in A\mathcal{T}$, the decomposition:

$$\mathrm{a} = \mathrm{a}_0 + V^1\vec{\mathbf{e}}_1 + V^2\vec{\mathbf{e}}_2 + \cdots + V^n\vec{\mathbf{e}}_n, \qquad [7.31]$$

is unique. We call V^i the *(affine) components* of a. In other words, the correspondence between $\mathrm{a} \in A\mathcal{T}$ and the column V collecting the components V^i is one-to-one. This defines a one-to-one affine map: $A\mathbb{R}^n \to A\mathcal{T} : V \mapsto \mathrm{a} = f\left(V\right)$. We say it is an *affine frame map*. Conversely, let f be a given affine frame map. It defines an affine frame by:

$$\mathrm{a}_0 = f\left(0\right), \quad \mathrm{a}_i = f\left(e_i\right), \quad \vec{\mathbf{e}}_i = f\left(e_i\right) - \mathrm{a}_0,$$

and [7.31] reads:

$$\mathrm{a} = f\left(V\right) = \mathrm{a}_0 + S(V), \qquad [7.32]$$

where the basis $S = lin\left(f\right)$ is the linear part of f. If we change the affine frame $(\mathrm{a}_0, (\vec{\mathbf{e}}_i))$ for a new one $(\mathrm{a}'_0, (\vec{\mathbf{e}}'_i))$, the corresponding transformation matrix being P and $C' = S'^{-1}(\overrightarrow{\mathrm{a}'_0\mathrm{a}_0})$ being the n-column gathering the components of $\overrightarrow{\mathrm{a}'_0\mathrm{a}_0}$ in the new basis, the decomposition [7.32] leads to:

$$\mathrm{a} = \mathrm{a}'_0 + \overrightarrow{\mathrm{a}'_0\mathrm{a}_0} + S(V) = \mathrm{a}'_0 + S'(V')$$

with the transformation law for the components of a point:

$$V' = C' + P^{-1}V. \qquad [7.33]$$

Compare it to the corresponding relation [7.24] for vectors: points are not vectors.

Conversely, introducing $C = -P\,C'$, we have:

$$V = C + P\,V'. \qquad [7.34]$$

There is a useful trick to convert this relation into what looks like a linear relation. We add 1 as the $(n+1)$-th component to the vectors V and V', and form the $(n+1) \times (n+1)$ matrix $\tilde{P}$:

$$\tilde{V} = \begin{pmatrix} 1 \\ V \end{pmatrix}, \qquad \tilde{P} = \begin{pmatrix} 1 & 0 \\ C & P \end{pmatrix}, \qquad \tilde{V}' = \begin{pmatrix} 1 \\ V' \end{pmatrix}, \qquad [7.35]$$

so that [7.34] is equivalent to:

$$\tilde{V} = \tilde{P}\tilde{V}'.$$

The affine maps $\boldsymbol{\Psi}$ from $A\mathcal{T}$ into $\mathbb{R}$ are called *affine forms* and their set is denoted by $A^*\mathcal{T}$. If $A\mathcal{T}$ has a finite dimension n and f is an affine frame map, $\Psi = \boldsymbol{\Psi} \circ f$ is a affine function from $\mathbb{R}^n$ into $\mathbb{R}$. Hence, it holds:

$$\boldsymbol{\Psi}(\boldsymbol{a}) = \Psi(V) = \chi + \Phi\,V, \qquad [7.36]$$

where $\chi = \Psi(0) = \boldsymbol{\Psi}(\boldsymbol{a}_0)$ and $\Phi = lin(\Psi)$ is an n-row. We call $\Phi_1, \Phi_2, \cdots, \Phi_n, \chi$ the *(affine) components* of $\boldsymbol{\Psi}$. It is convenient to gather the components into an $(n+1)$-row $\tilde{\Psi}$ according to:

$$\Psi(V) = \tilde{\Psi}\,\tilde{V} = (\chi \quad \Phi) \begin{pmatrix} 1 \\ V \end{pmatrix}.$$

The set $A^*\mathcal{T}$ is a linear space of dimension $(n+1)$ called the *vector dual* of $A\mathcal{T}$. If we change the affine frame $(\boldsymbol{a}_0, (\vec{\mathbf{e}}_i))$ for a new one $(\boldsymbol{a}'_0, (\vec{\mathbf{e}}'_i))$, the components of an affine form are modified according to $\tilde{\Psi}' = \tilde{\Psi}\,\tilde{P}$, which leads to:

$$\chi' = \chi + \Phi\,C, \qquad \Phi' = \Phi\,P.$$

It is easy to verify that the inverse transformation law:

$$\tilde{\Psi} = \tilde{\Psi}'\,\tilde{P}^{-1}, \qquad [7.37]$$

reads:

$$\chi = \chi' - \Phi'\,P^{-1}C, \qquad \Phi = \Phi'\,P^{-1}.$$

7.6. Limit and continuity

Let $t \mapsto U(t)$ be a map from an open interval I of $\mathbb{R}$ into $\mathbb{R}^n$. If $t_0 \in I$, we say that $U(t)$ approaches the **limit** $U_0 \in \mathbb{R}^n$ as t approaches t_0 if for any $\varepsilon > 0$ we can find $\eta > 0$ such that:

$$| t - t_0 |< \eta \qquad \Rightarrow \qquad \| U(t) - U_0 \|< \varepsilon,$$

that reads:

$$\lim_{t \to t_0} U(t) = U_0.$$

Moreover, let V be a function valued in $\mathbb{R}^n$ and f be a scalar function, both defined on I, and:

$$\lim_{t \to t_0} V(t) = V_0, \qquad \lim_{t \to t_0} f(t) = f_0,$$

then:

$$\lim_{t \to t_0} (U(t) + V(t)) = U_0 + V_0, \qquad \lim_{t \to t_0} (f(t)U(t)) = f_0 U_0,$$

$$\lim_{t \to t_0} (U(t) \cdot V(t)) = U_0 \cdot V_0, \qquad \lim_{t \to t_0} (U(t) \times V(t)) = U_0 \times V_0.$$

The function U is said to be *continuous* at t_0 if:

$$\lim_{t \to t_0} U(t) = U(t_0).$$

It is continuous on the interval $J \subset I$ if it is continuous at every $t \in J$.

The extension of the previous considerations to functions M valued in $\mathbb{M}_{nn}$ is straightforward due to the matrix norm [7.6]. If U is a function valued in $\mathbb{R}^n$ and:

$$\lim_{t \to t_0} M(t) = M_0,$$

we have:

$$\lim_{t \to t_0} (M(t)\, U(t)) = M_0\, U_0,$$

every time the right-hand side exists.

7.7. Derivative

We said that U is *differentiable* at t_0 if:

$$\dot{U}(t) = \frac{dU}{dt}(t) = \lim_{t \to t_0} \frac{U(t) - U(t_0)}{t - t_0}.$$

exists. This limit is called the *derivative* of U at t_0. We said that U is differentiable on I if it is differentiable at every $t_0 \in I$. Thus, we have:

$$\frac{d}{dt}(fU) = \frac{df}{dt}U + f\frac{dU}{dt}, \qquad \frac{d}{dt}(U \cdot V) = \frac{dU}{dt} \cdot V + u \cdot \frac{dU}{dt},$$
$$\frac{d}{dt}(U \times V) = \frac{dU}{dt} \times V + U \times \frac{dV}{dt}.$$

The extension of the derivative to functions M valued in $\mathbb{M}_{nn}$ is straightforward. Then, we have:

$$\frac{d}{dt}(MU) = \frac{dM}{dt}U + M\frac{dU}{dt}.$$

Multiplying by dt, we can adopt the language of differentials:

$$d(fU) = df\,U + f\,dU, \qquad d(U \cdot V) = dU \cdot V + U \cdot dV,$$
$$d(U \times V) = dU \times V + U \times dV, \qquad d(MU) = dM\,U + M\,dU.$$

If a scalar function f is differentiable at t_0 and U is differentiable at $f(t_0)$, then the composition $V = Uf$ is differentiable at t_0 with:

$$\frac{d}{dt}(U(f(t))) = \frac{dU}{df}\frac{df}{dt}.$$

This is the *chain rule*.

7.8. Partial derivative

Let f be a map from a subset of $\mathbb{R}^n$ into $\mathbb{R}^m$ and v be a n-column. The *directional derivative of f at $x \in def(f)$ in the direction of v* is defined by:

$$Df(x)(v) = \lim_{t \to 0} \frac{f(x + t\,v) - f(x)}{t},$$

if the limit exists. That is $Df(x)(v)$ is the ordinary derivative of the function $t \mapsto f(x+tv)$ at $t=0$. If the directional derivative of f at x exists in any direction and $Df(x)$ is linear, we say that f is differentiable at x. The map $Df(x)$ is called the *derivative* of f and, if there is no confusion, we often denote its value by $Df(x)\,v$. If f is differentiable at every $x \in def(f)$, we said that f is differentiable. Then, it is continuous on $def(f)$. We say that f is *continuously differentiable* (or *of class* C^1) if the map $x \mapsto Df(x)$ is continuous.

Let us now consider *scalar fields* f, i.e. such that $val(f) \subset \mathbb{R}$. The *partial derivative of f with respect to x^i at x* is:

$$\frac{\partial f}{\partial x^i}(x) = \partial_i f(x) = Df(x)\,e_i.$$

If the function $x \mapsto \partial_i f(x)$ has a partial derivative with respect to x^j, we denote it by:

$$\frac{\partial}{\partial x^j}\left(\frac{\partial f}{\partial x^i}\right) = \frac{\partial^2 f}{\partial x^j \partial x^i}.$$

If a function f has continuous partial derivatives up to the order p, we said that it is *of class* C^p. If a function is at least of class C^2, the partial derivatives commute:

$$\frac{\partial^2 f}{\partial x^i \partial x^j} = \frac{\partial^2 f}{\partial x^j \partial x^i}.$$

7.9. Vector analysis

7.9.1. *Gradient*

The *derivative* of a scalar field at x, denoted by:

$$Df(x) = \frac{\partial f}{\partial x},$$

is a linear map from $\mathbb{R}^n$ into $\mathbb{R}$, which is an n-row and:

$$\frac{\partial f}{\partial x} = \left(\frac{\partial f}{\partial x^1}\ \frac{\partial f}{\partial x^2} \cdots \frac{\partial f}{\partial x^n}\right).$$

The *gradient* of the scalar field f is the n-column:

$$grad\, f = \left(\frac{\partial f}{\partial x}\right)^T = \begin{pmatrix} \frac{\partial f}{\partial x^1} \\ \frac{\partial f}{\partial x^2} \\ \vdots \\ \frac{\partial f}{\partial x^n} \end{pmatrix}.$$

For any scalar fields λ and μ, it holds:

$$grad\,(\lambda\mu) = \lambda\, grad\, \mu + \mu\, grad\, \lambda.$$

Let v be a *vector field*, i.e. such that $val(v) \subset \mathbb{R}^p$. Its *derivative* at x, denoted by:

$$Dv(x) = \frac{\partial v}{\partial x},$$

is a linear map from $\mathbb{R}^n$ into $\mathbb{R}^p$, which is a $p \times n$ matrix. Its *gradient* is the $n \times p$ matrix:

$$grad\, v = \left(\frac{\partial v}{\partial x}\right)^T.$$

For any scalar field λ and any vector fields u, v, we have:

$$grad\,(\lambda v) = grad\, \lambda\, v^T + \lambda\, grad\, v. \qquad [7.38]$$

$$grad\,(u \cdot v) = (grad\, u)\, v + (grad\, v)\, u. \qquad [7.39]$$

If $n = p$, the *symmetric gradient* of v is the symmetric $n \times n$ matrix:

$$grad_s\, v = \frac{1}{2}\left[\frac{\partial v}{\partial x} + \left(\frac{\partial v}{\partial x}\right)^T\right].$$

The *skew-symmetric gradient* of v is the skew-symmetric $n \times n$ matrix:

$$grad_a\, v = \frac{1}{2}\left[\frac{\partial v}{\partial x} - \left(\frac{\partial v}{\partial x}\right)^T\right],$$

If the map:

$$\mathbb{R}^n \to \mathbb{R}^n : x \mapsto y = (grad\, f)\,(x),$$

is regular, we define *Legendre's transform* of f (with respect to x) as the scalar function:

$$f^*(y) = x \cdot ((grad\, f)\,(x)) - f(x), \qquad [7.40]$$

where $x = (grad\, f)^{-1}(y)$. Then, the inverse map is:

$$\mathbb{R}^n \to \mathbb{R}^n : y \mapsto x = (grad\, f^*)\,(y).$$

With some abusive notations, Legendre's transform reads:

$$f^* = x \cdot grad\, f - f = \frac{\partial f}{\partial x}\, x - f.$$

7.9.2. *Divergence*

Let v be a *vector field*, such that $val(v), def(v) \subset \mathbb{R}^n$. Its *divergence* is the scalar field:

$$div\, v = Tr\left(\frac{\partial v}{\partial x}\right),$$

and for any scalar field λ:

$$div\,(\lambda v) = \lambda\, div\, v + \frac{\partial \lambda}{\partial x}\, v = \lambda\, div\, v + grad\, \lambda \cdot v. \qquad [7.41]$$

7.9.3. *Vector analysis in $\mathbb{R}^3$ and curl*

For every vector fields $u, v \in \mathbb{R}^3$, we have:

$$\frac{\partial}{\partial x}(u \times v) = j(u)\, \frac{\partial v}{\partial x} - j(v)\, \frac{\partial u}{\partial x}. \qquad [7.42]$$

For any column field $x \longmapsto v(x) \in \mathbb{R}^3$ of class C^1, we call *curl* of v the unique 3-column field $curl\, v$ associated with the skew-symmetric gradient of v by the map j:

$$j(curl\, v) = \frac{\partial v}{\partial x} - \left(\frac{\partial v}{\partial x}\right)^T. \qquad [7.43]$$

For any scalar field λ and any vector field v, we have:

$$curl\,(grad\, \lambda) = 0, \qquad div\,(curl\, v) = 0.$$

PART 2

Continuous Media

8

Statics of 3D Continua

8.1. Stresses

8.1.1. *Stress tensor*

The aim of this chapter is to study the static equilibrium of a bulky body occupying an open domain $\mathcal{V}$ of the three-dimensional (3D) affine space (Figure 8.1). First, we have to model the internal and external forces, defined in a global way in section 2.3.3. We would like to give a more accurate description of them. To identify the internal forces, we isolate a part $\mathcal{V}_-$ of the body by cutting it along a smooth surface $\mathcal{S}$ of equation $f(x) = 0$ where f is a level-set function of class C^1 and conventionally negative in $\mathcal{V}_-$.

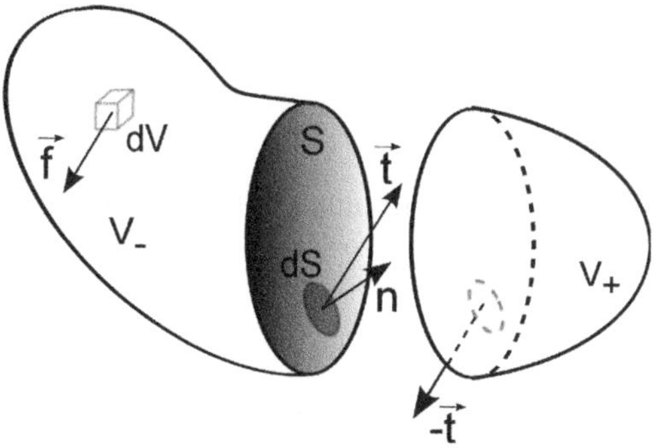

Figure 8.1. *Stress vector*

Drawing the free body diagram of the part $\mathcal{V}_-$, the distribution of internal forces acting upon it through the surface will be assumed at least continuous. At a given point $\boldsymbol{P}$ of the interior of the surface, let us consider an infinitesimal *surface element* completely defined by its area dS and orientation. Working in a Galilean coordinate system x, the surface element is perpendicular to the gradient of f, the components of which are the partial derivative of f. In a Galilean coordinate system, the components of the unit normal to the surface, pointing away from the considered part, are:

$$n_i = \frac{1}{\| \frac{\partial f}{\partial x} \|} \frac{\partial f}{\partial x^i},$$

which are the components of a covector $\mathbf{n}$, which is a 1-covariant tensor, called *unit normal* to $\mathcal{S}$ at $\boldsymbol{P}$. According to the transformation law [2.10], we saw that forces are vectors, i.e. 1-contravariant tensors. Let dF^j be the components of the elementary internal force vector $\overrightarrow{\boldsymbol{dF}}_s$ acting at $\boldsymbol{P}$ upon $\mathcal{V}_-$ through the surface element dS.

DEFINITION 8.1.– The *stress vector* acting at the point $\boldsymbol{P}$ through the surface $\mathcal{S}$ (from $\mathcal{V}_+$ to $\mathcal{V}_-$) is:

$$\vec{t} = \frac{\overrightarrow{\boldsymbol{dF}_s}}{dS},$$

value of a map:

$$(\boldsymbol{P}, \boldsymbol{n}) \mapsto \vec{t} = \boldsymbol{T}(\boldsymbol{P}, \boldsymbol{n}).$$

In other words, if $\overrightarrow{\boldsymbol{\Delta F}}$ is the resultant of forces acting upon $\mathcal{V}_-$ through the small portion $\Delta\mathcal{S}$ of $\mathcal{S}$ around $\boldsymbol{P}$, it is the limit:

$$\vec{t} = \lim_{\Delta\mathcal{S} \to 0} \frac{\overrightarrow{\boldsymbol{\Delta F}}}{\Delta\mathcal{S}}.$$

By Newton's third law [2.1], the resultant of forces $\overrightarrow{\boldsymbol{\Delta F'}}$ acting upon $\mathcal{V}_+$ through $\Delta\mathcal{S}$ is such that:

$$\overrightarrow{\boldsymbol{\Delta F'}} = -\overrightarrow{\boldsymbol{\Delta F}}.$$

Dividing both members by the area and passing to the limit, the mutual stress vector of action and reaction are equal and opposite, which can read:

$$\boldsymbol{T}(\boldsymbol{P}, -\boldsymbol{n}) = -\boldsymbol{T}(\boldsymbol{P}, \boldsymbol{n}). \qquad [8.1]$$

Along these lines, we consider that the external forces can be modeled by the elementary force vector $\overrightarrow{\boldsymbol{dF}_v}$ acting at $\boldsymbol{P}$ upon the volume element $d\mathcal{V}$ around $\boldsymbol{P}$.

DEFINITION 8.2.– The *volume force* acting at the point $\boldsymbol{P}$ upon the body $\mathcal{V}$ is:

$$\overrightarrow{f_v} = \frac{\overrightarrow{dF_v}}{d\mathcal{V}}.$$

Below, we prove *Cauchy's tetrahedron theorem.*

THEOREM 8.1.– If the maps $(\boldsymbol{P}, \boldsymbol{n}) \mapsto \boldsymbol{T}(\boldsymbol{P}, \boldsymbol{n})$ and $\boldsymbol{P} \mapsto \overrightarrow{f_v}(\boldsymbol{P})$ are continuous, balance equation [2.12] implies that there exists a 2-contravariant tensor field $\boldsymbol{P} \mapsto \boldsymbol{\sigma}(\boldsymbol{P})$, called *Cauchy's stress tensor*, such that:

$$\overrightarrow{t} = \boldsymbol{\sigma}(\boldsymbol{P}) \cdot \boldsymbol{n}. \qquad [8.2]$$

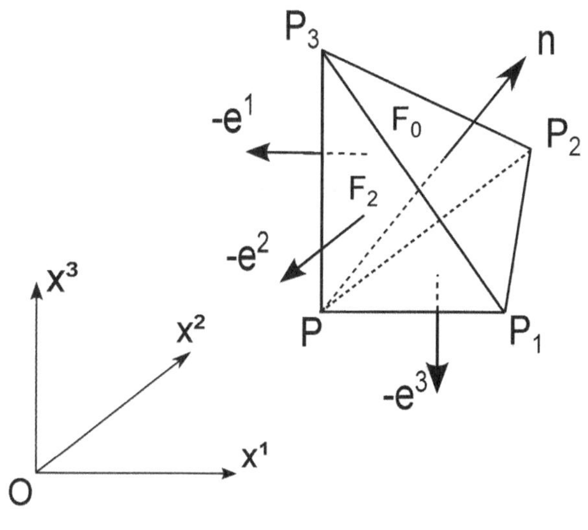

Figure 8.2. *Stress vector*

PROOF.– Because the set $\mathcal{V}$ is open, we can find, as a particular subdomain of $\mathcal{V}$, a tetrahedron $\mathcal{V}_t$ with vertex $\boldsymbol{P}$, three faces parallel to the coordinate planes and a face F_0 normal to $\boldsymbol{n}$ of area $\mathcal{S}_0$ and at the distance h of $\boldsymbol{P}$. As indicated in Figure 8.2, let F_i be the face opposite to another vertex $\boldsymbol{P}_j$, so its area is $\mathcal{S}_j = n_j \mathcal{S}_0$. The force balance equation [2.12] of the tetrahedron reads:

$$\sum_{k=0}^{3} \int_{F_k} \boldsymbol{T}(\boldsymbol{P}', \boldsymbol{n}) \, d\mathcal{S}(\boldsymbol{P}') + \int_{\mathcal{V}_t} \overrightarrow{f_v}(\boldsymbol{P}') \, d\mathcal{V}(\boldsymbol{P}') = \vec{0}.$$

and considering the projection on the axis Ox^i:

$$\sum_{k=0}^{3} \int_{F_k} T^i(P', n)\, dS(P') + \int_{\mathcal{V}_t} f_v^i(P')\, d\mathcal{V}(P') = 0.$$

Taking into account the continuity hypothesis, we can apply the mean value theorem for integrals. Hence, there exist points $\bar{P}_k \in F_k$ and $\bar{P} \in \mathcal{V}_t$ such that:

$$T^i(\bar{P}_0, n)\, S_0 + \sum_{j=1}^{3} T^i(\bar{P}_j, -e^j)\, S_j + f_v^i(\bar{P})\, \frac{h\, S_0}{3} = 0.$$

Thus, it holds:

$$T^i(\bar{P}_0, n) + \sum_{j=1}^{3} T^i(\bar{P}_j, -e^j)\, n_j + f_v^i(\bar{P})\, \frac{h}{3} = 0.$$

Keeping the covector n fixed when h approaches 0, the points $\bar{P}_0, \bar{P}_j$ and $\bar{P}$ coalesce into P. Using again the continuity hypothesis, we have:

$$T^i(P, n) = -\sum_{j=1}^{3} T^i(P, -e^j)\, n_j.$$

The coordinate system being arbitrary, we have:

$$T(P, n) = -\sum_{j=1}^{3} T(P, -e^j)\, n_j. \qquad [8.3]$$

Using next the continuity with respect to the second variable, let n approach a particular cobasis vector e^j in the previous relation, we obtain:

$$T(P, e^j) = -T(P, -e^j),$$

which is nothing else but Newton's third law in the form [8.1]. Taking into account of this, relation [8.3] leads to:

$$T(P, \sum_{j=1}^{3} n_j e^j) = T(P, n) = \sum_{j=1}^{3} n_j T(P, e^j).$$

that proves $\boldsymbol{T}$ is linear with respect to the second variable $\boldsymbol{n}$. As the value of $\boldsymbol{T}$ is a 1-contravariant tensor and $\boldsymbol{n}$ is a 1-covariant tensor, there exists a 2-contravariant tensor $\boldsymbol{\sigma}(\boldsymbol{P})$ such that $\boldsymbol{T}(\boldsymbol{P}, \boldsymbol{n})$ is obtained by contracting the product $\boldsymbol{\sigma}(\boldsymbol{P}) \otimes \boldsymbol{n}$. ■

In a given basis $(\vec{e}_i)$ and using the convention of summation, relation [8.2] reads:

$$t^i = \sigma^{ij} n_j.$$

If the stress vector $\overrightarrow{t}$ is represented by the column t, the stress tensor $\boldsymbol{\sigma}$ by the symmetric matrix σ and the components n_i are gathered into the column n, these relations read in matrix notations:

$$t = \sigma\, n. \tag{8.4}$$

In a new basis $(\vec{e}'_s)$ obtained from the previous one through a transformation matrix $\breve{P}$, the new components are given by the transformation law of 2-contravariant tensors:

$$\sigma'^{st} = (\breve{P}^{-1})^s_i (\breve{P}^{-1})^t_j \sigma^{ij}, \tag{8.5}$$

which, according to [14.3], reads in matrix notations:

$$\sigma' = \breve{P}^{-1} \sigma\, \breve{P}^{-T}.$$

In particular, when working in Galilean coordinate systems, the transformation matrix is an orthogonal transformation and we obtain the transformation law of Euclidean stress tensors:

$$\sigma' = R^T \sigma\, R. \tag{8.6}$$

The diagonal elements σ^{ii} are called *tensile stresses,* while the off diagonal ones are known as *shear stresses.*

To model realistic situations, it is often useful to consider piecewise continuous distributions of forces. Let us consider two regions $\mathcal{V}_-$ and $\mathcal{V}_+$ where the stress field is continuous and separated by a discontinuity surface $\mathcal{S}$. According to Newton's third law in the form [8.1], the only continuity requirement crossing $\mathcal{S}$ is the continuity of the stress vector:

$$\left[\overrightarrow{t}\right] = \vec{0}.$$

8.1.2. *Local equilibrium equations*

Now, we prove:

THEOREM 8.2.– If the map $x \mapsto \sigma(x)$ is continuously differentiable, a body is in internal equilibrium if and only if the *local equilibrium equations of 3D continua*:

$$\boxed{(div\,\sigma)^T + f_v = 0} \qquad [8.7]$$

$$\boxed{\sigma^T = \sigma} \qquad [8.8]$$

are satisfied.

PROOF.– Let us suppose that the body is in static equilibrium. Working in a given Galilean coordinate system, let dF be the 3-column gathering the components dF^i of the elementary force $\overrightarrow{\boldsymbol{dF}}$ acting at $\boldsymbol{P}$ of coordinates x^i. Its torsor [2.11] about the origin $\boldsymbol{O}$ is, owing to [7.11]:

$$d\breve{\tau} = \begin{pmatrix} 0 & dF^T \\ -dF & -j(x \times dF) \end{pmatrix} = \begin{pmatrix} 0 & dF^T \\ -dF & x\,dF^T - dF\,x^T \end{pmatrix}.$$

For an arbitrary subdomain $\mathcal{V}$ of the considered body, the resultant torsor is the integral of the elementary torsor of every infinitesimal parts corresponding to forces $\overrightarrow{\boldsymbol{dF}}_v$ on $\mathcal{V}$ and $\overrightarrow{\boldsymbol{dF}}_s$ on its boundary $\partial\mathcal{V}$:

$$\breve{\tau}(\mathcal{V}) = \int_{\partial\mathcal{V}} d\breve{\tau}_s(x) + \int_{\mathcal{V}} d\breve{\tau}_v(x) = \begin{pmatrix} 0 & F_{\mathcal{V}}^T \\ -F_{\mathcal{V}} & J_{\mathcal{V}} \end{pmatrix}, \qquad [8.9]$$

where:

$$F_{\mathcal{V}} = \int_{\partial\mathcal{V}} dF_s(x) + \int_{\mathcal{V}} dF_v(x) = \int_{\partial\mathcal{V}} t(x)\,dS(x) + \int_{\mathcal{V}} f_v(x)\,d\mathcal{V}(x), \qquad [8.10]$$

$$J_{\mathcal{V}} = \int_{\partial\mathcal{V}} (x\,(t(x))^T - t(x)\,x^T)\,dS(x) + \int_{\mathcal{V}} (x\,(f_v(x))^T - f_v(x)\,x^T)\,d\mathcal{V}(x). \qquad [8.11]$$

Working now with tensor components and simplified notations, taking into account theorem 8.1, the force resultant is represented by:

$$F_{\mathcal{V}}^i = \int_{\partial\mathcal{V}} t^i\,dS + \int_{\mathcal{V}} f_v^i\,d\mathcal{V} = \int_{\partial\mathcal{V}} \sigma^{ij} n_j\,dS + \int_{\mathcal{V}} f_v^i\,d\mathcal{V}.$$

According to definition 2.5, the body is in static equilibrium if the resultant torsor of each of its parts is null. Transforming the first term by the divergence theorem for tensor fields and owing to the balance equation [2.12], we have:

$$F_{\mathcal{V}}^i = \int_{\mathcal{V}} \left(\frac{\partial \sigma^{ij}}{\partial x^j} + f_v^i \right) d\mathcal{V} = 0,$$

for every subdomain $\mathcal{V}$. Therefore, we have at any point x:

$$\frac{\partial \sigma^{ij}}{\partial x^j} + f_v^i = 0. \qquad [8.12]$$

As the choice of the Galilean coordinate system is arbitrary, this proves the force equilibrium equation [8.7]. In a similar way, the moment resultant matrix $J_{\mathcal{V}}$ is represented by:

$$J_{\mathcal{V}}^{ik} = \int_{\partial\mathcal{V}} (x^i t^k - t^i x^k)\, d\mathcal{S} + \int_{\mathcal{V}} (x^i f_v^k - f_v^i x^k)\, d\mathcal{V}.$$

Transforming the first integral by the divergence theorem, we have:

$$\int_{\partial\mathcal{V}} (x^i t^k - t^i x^k)\, d\mathcal{S} = \int_{\partial\mathcal{V}} (x^i \sigma^{kj} - x^k \sigma^{ij}) n_j\, d\mathcal{S} = \int_{\mathcal{V}} \frac{\partial}{\partial x^j} (x^i \sigma^{kj} - x^k \sigma^{ij})\, d\mathcal{V},$$

$$\int_{\partial\mathcal{V}} (x^i t^k - t^i x^k)\, d\mathcal{S} = \int_{\mathcal{V}} \left(\delta_j^i \sigma^{kj} - \delta_j^k \sigma^{ij} + x^i \frac{\partial \sigma^{kj}}{\partial x^j} - x^k \frac{\partial \sigma^{ij}}{\partial x^j} \right) d\mathcal{V}.$$

Taking into account the force equilibrium [8.12] and owing to the balance equation [2.12] leads to:

$$J_{\mathcal{V}}^{ik} = \int_{\mathcal{V}} (\sigma^{ki} - \sigma^{ik}) d\mathcal{V} = 0,$$

for every subdomain $\mathcal{V}$, therefore, we have at any point x:

$$\sigma^{ki} = \sigma^{ik}. \qquad [8.13]$$

As the choice of the Galilean coordinate system is arbitrary, this proves the moment equilibrium equation [8.8]. Conversely, if the local equilibrium equations are satisfied, we have:

$$F_{\mathcal{V}}^i = \int_{\mathcal{V}} \left(\frac{\partial \sigma^{ij}}{\partial x^j} + f_v^i \right) d\mathcal{V} = 0, \qquad J_{\mathcal{V}}^{ik} = \int_{\mathcal{V}} (\sigma^{ki} - \sigma^{ik}) d\mathcal{V} = 0$$

for any part $\mathcal{V}$ and the body is in equilibrium, which achieves the proof. ■

Taking into account [8.10], [8.11] and [7.11], it is worth noting that the resultant torsor:

$$\breve{\tau}(\mathcal{V}) = \begin{pmatrix} 0 & F_{\mathcal{V}}^T \\ -F_{\mathcal{V}} & -j(M_{\mathcal{V}}) \end{pmatrix}, \qquad [8.14]$$

has components:

$$F_{\mathcal{V}} = \int_{\partial\mathcal{V}} t\, d\mathcal{S} + \int_{\mathcal{V}} f_v\, d\mathcal{V}, \qquad M_{\mathcal{V}} = \int_{\partial\mathcal{V}} x \times t\, d\mathcal{S} + \int_{\mathcal{V}} x \times f_v\, d\mathcal{V}. \quad [8.15]$$

8.2. Torsors

8.2.1. *Continuum torsor*

Due to Cauchy's tetrahedron theorem 8.1, we built a linear tensor, the stress tensor. In the previous section, we used the resultant torsor of the body but without discussing its tensor status. In fact, the torsors are affine tensors. For instance, the torsor of a force $\vec{\boldsymbol{F}}$ acting at point $\boldsymbol{P}$ is the skew-symmetric 2-contravariant affine tensor:

$$\breve{\boldsymbol{\tau}} = \boldsymbol{P} \otimes \vec{\boldsymbol{F}} - \vec{\boldsymbol{F}} \otimes \boldsymbol{P}. \qquad [8.16]$$

Up to now, we were working only with its local representation in a given affine frame f where, if x is the position of $\boldsymbol{P}$ and $\vec{\boldsymbol{F}}$ is represented by the column F, the torsor $\boldsymbol{\tau}$ is represented by the 4×4 skew-symmetric matrix:

$$\breve{\tau} = \begin{pmatrix} 1 \\ x \end{pmatrix} \begin{pmatrix} 0 \\ F \end{pmatrix}^T - \begin{pmatrix} 0 \\ F \end{pmatrix} \begin{pmatrix} 1 \\ x \end{pmatrix}^T,$$

hence, using [7.11], we recover [2.4] with $M = x \times F$ while the transformation law [2.5] of torsors is nothing else but the one [14.9] of the 2-contravariant affine tensors. Also, taking into account the decomposition in the affine frame of the point $\boldsymbol{P}$ and the force vector:

$$\boldsymbol{P} = \boldsymbol{O} + x^k \vec{e}_k, \qquad \vec{\boldsymbol{F}} = F^i \vec{e}_i,$$

the relation [8.16] leads to the decomposition of the force torsor:

$$\breve{\boldsymbol{\tau}} = F^i\,(\boldsymbol{O} \otimes \vec{e}_i - \vec{e}_i \otimes \boldsymbol{O}) + (x^k F^i - x^i F^k)\,\vec{e}_k \otimes \vec{e}_i.$$

Similarly, let us construct the torsor of the stress vector:

$$\frac{d\breve{\tau}_s}{dS}(\boldsymbol{P},\boldsymbol{n}) = \boldsymbol{P} \otimes \vec{t}(\boldsymbol{P},\boldsymbol{n}) - \vec{t}(\boldsymbol{P},\boldsymbol{n}) \otimes \boldsymbol{P}. \quad [8.17]$$

$$\frac{d\breve{\tau}_s}{dS}(\boldsymbol{P},\boldsymbol{n}) = t^i\left(\boldsymbol{O} \otimes \vec{e}_i - \vec{e}_i \otimes \boldsymbol{O}\right) + \left(x^k t^i - x^i t^k\right) \vec{e}_k \otimes \vec{e}_i. \quad [8.18]$$

Clearly, it is linearly depending on the covector $\boldsymbol{n}$, which leads to consider a vector valued torsor $\breve{\tau}_\sigma$, called the *stress torsor*, such that:

$$\boldsymbol{n}(\breve{\tau}_\sigma(\boldsymbol{P})) = \frac{d\breve{\tau}_s}{dS}(\boldsymbol{P},\boldsymbol{n}).$$

Hence,

$$\boldsymbol{n}(\breve{\tau}_\sigma) = \boldsymbol{P} \otimes (\boldsymbol{\sigma} \cdot \boldsymbol{n}) - (\boldsymbol{\sigma} \cdot \boldsymbol{n}) \otimes \boldsymbol{P}. \quad [8.19]$$

On this ground, it is worth to generalize the torsors in the following way:

DEFINITION 8.3.– A *continuum torsor* is a skew-symmetric 2-contravariant affine tensor $\boldsymbol{\tau}$ with vector value:

$$\boldsymbol{\tau}(\boldsymbol{\Psi}, \bar{\boldsymbol{\Psi}}) = -\boldsymbol{\tau}(\bar{\boldsymbol{\Psi}}, \boldsymbol{\Psi}).$$

For what we are concerned now, the continuum is 3D and we denote the torsors by $\breve{\tau}$. In the affine frame $(\boldsymbol{O}, (\vec{e}_i))$ its value is:

$$\vec{U} = \breve{\tau}(\boldsymbol{\Psi}, \bar{\boldsymbol{\Psi}}) = \breve{\tau}^j((\chi, \Phi), (\bar{\chi}, \bar{\Phi}))\, \vec{e}_j,$$

$$\vec{U} = \left[T^{ij}(\chi\bar{\Phi}_i - \bar{\chi}\Phi_i) + J^{kij}\, \Phi_k \bar{\Phi}_i\right] \vec{e}_j,$$

where T^{ij} and $J^{kij} = -J^{ikj}$ are the components of the torsor. Taking into account $\chi = \hat{\boldsymbol{O}}(\boldsymbol{\Psi})$ and $\Phi_i = \hat{\mathrm{e}}_i(\boldsymbol{\Psi})$, it reads:

$$\breve{\tau} = \breve{\tau}^j \vec{e}_j, \qquad \breve{\tau}^j = T^{ij}\left(\boldsymbol{O} \otimes \vec{e}_i - \vec{e}_i \otimes \boldsymbol{O}\right) + J^{kij}\, \vec{e}_k \otimes \vec{e}_i.$$

Let $(\boldsymbol{O}', (\vec{e}'_i))$ be a new affine frame obtained from the old one through an affine transformation $\breve{a} = (k, \breve{P})$. Hence, the transformation law of the torsor is:

$$T'^{st} = (\breve{P}^{-1})^s_i (\breve{P}^{-1})^t_j T^{ij}, \quad [8.20]$$

$$J'^{rst} = \left[(\breve{P}^{-1})^r_l (\breve{P}^{-1})^s_i J^{lij} + k'^r \left\{(\breve{P}^{-1})^s_i T^{ij}\right\}\right. \quad [8.21]$$

$$\left. - \left\{(\breve{P}^{-1})^r_i T^{ij}\right\} k'^s\right] (\breve{P}^{-1})^t_j, \quad [8.22]$$

with $k' = -\breve{P}^{-1}k$.

Now, let us return to the stress torsor. Taking into account the decomposition in the affine frame of the point $\boldsymbol{P}$ and the stress vector:

$$\boldsymbol{P} = \boldsymbol{O} + x^k \vec{e}_k, \qquad \boldsymbol{\sigma} \cdot \boldsymbol{n} = \sigma^{ij} \vec{e}_i n_j,$$

the relation [8.19] leads to the decomposition of the stress torsor:

$$\breve{\tau}_{\sigma} = \breve{\tau}^j_{\sigma} \vec{e}_j, \qquad \breve{\tau}^j_{\sigma} = \sigma^{ij} \left(\boldsymbol{O} \otimes \vec{e}_i - \vec{e}_i \otimes \boldsymbol{O}\right) + \left(x^k \sigma^{ij} - x^i \sigma^{kj}\right) \vec{e}_k \otimes \vec{e}_i,$$

hence, the components of the stress tensor are:

$$T^{ij} = \sigma^{ij}, \qquad J^{kij} = x^k \sigma^{ij} - x^i \sigma^{kj}. \qquad [8.23]$$

For the stress tensor, it is worth observing that [8.20] is nothing else but the transformation law [8.5] of the stress tensors. Also, the J^{lij} components of $\breve{\tau}_{\sigma}$ in the affine frame $(\boldsymbol{P}, (\vec{e}_i))$ vanish because $x = 0$ if $\boldsymbol{P}$ is the origin. In fact, this condition characterizes a stress torsor, according to the next theorem.

THEOREM 8.3.– A continuum torsor $\breve{\tau}$ is a stress torsor if and only if the components J^{lij} of $\breve{\tau}(\boldsymbol{P})$ in the affine frame $(\boldsymbol{P}, (\vec{e}_i))$ vanish.

PROOF.– We just showed the condition is necessary. To prove it is also sufficient, let us suppose $J^{lij} = 0$ for the representation of $\breve{\tau}(\boldsymbol{P})$ in $(\boldsymbol{P}, (\vec{e}_i))$. According to the transformation laws [8.20] and [8.22], the components of the continuum torsor $\breve{\tau}(\boldsymbol{P})$ in $(\boldsymbol{O}, (\vec{e}_i))$ are deduced from the ones in $(\boldsymbol{P}, (\vec{e}_i))$ through a translation $k = S^{-1}(\overrightarrow{\boldsymbol{OP}}) = x$ (thus, the transformation matrix P is the identity):

$$T'^{ij} = T^{ij}, \qquad J'^{lij} = x^l T^{ij} - x^i T^{lj}.$$

We can put:

$$\breve{\tau} = \breve{\tau}^j \vec{e}_j, \qquad \breve{\tau}^j = T^{ij} \left(\boldsymbol{O} \otimes \vec{e}_i - \vec{e}_i \otimes \boldsymbol{O}\right) + \left(x^l T^{ij} - x^i T^{lj}\right) \vec{e}_l \otimes \vec{e}_i.$$

By straightforward calculations, we obtain:

$$\boldsymbol{n}(\breve{\tau}) = \boldsymbol{P} \otimes (\boldsymbol{T} \cdot \boldsymbol{n}) - (\boldsymbol{T} \cdot \boldsymbol{n}) \otimes \boldsymbol{P} = \boldsymbol{n}(\breve{\tau}_T).$$

As the linear form $\boldsymbol{n}$ is arbitrary, the continuum torsor $\breve{\tau}$ is the stress torsor $\breve{\tau}_T$. ■

8.2.2. *Cauchy's continuum*

At first glance, the local equilibrium equations [8.12] and [8.13] are rather dissimilar. Considering the concept of torsor, is it possible to reveal an underlying structure? Considering the components of a continuum torsor $\breve{\tau}$ in the affine frame $(\boldsymbol{O}, (\vec{e}_i))$, we claim that its *divergence* is the scalar valued torsor:

$$\tilde{div}\, \breve{\tau}_\sigma = \frac{\partial T^{ij}}{\partial x^j} (\boldsymbol{O} \otimes \vec{e}_i - \vec{e}_i \otimes \boldsymbol{O}) + \frac{\partial J^{kij}}{\partial x^j} \vec{e}_k \otimes \vec{e}_i. \quad [8.24]$$

Also, let us define the torsor of the volume force:

$$\tau_{\vec{f}_v} = \frac{d\breve{\tau}_v}{d\mathcal{V}} (\boldsymbol{P}) = \boldsymbol{P} \otimes \vec{f}_v(\boldsymbol{P}) - \vec{f}_v(\boldsymbol{P}) \otimes \boldsymbol{P}, \quad [8.25]$$

$$\breve{\tau}_{\vec{f}_v} = f_v^i (\boldsymbol{O} \otimes \vec{e}_i - \vec{e}_i \otimes \boldsymbol{O}) + (x^k f_v^i - x^i f_v^k)\, \vec{e}_k \otimes \vec{e}_i. \quad [8.26]$$

Thus, the equation:

$$\boxed{\tilde{div}\, \breve{\tau}_\sigma + \breve{\tau}_{\vec{f}_v} = 0} \quad [8.27]$$

allows us to recover [8.12] and:

$$x^k \left(\frac{\partial \sigma^{ij}}{\partial x^j} + f_v^i \right) - x^i \left(\frac{\partial \sigma^{kj}}{\partial x^j} + f_v^k \right) + \sigma^{ik} - \sigma^{ki} = 0,$$

which, owing to [8.12], is nothing else but [8.13].

The drawback of definition [8.24] of the torsor divergence is that it is not general with respect to its status of affine tensor. As in Chapter 4, we introduce a *covariant divergence* of a continuum torsor $\breve{\tau}$, considering the components of $\breve{\tau}(\boldsymbol{P})$ in the current affine frame $(\boldsymbol{P}, (\vec{e}_i))$ (and not in $(\boldsymbol{O}, (\vec{e}_i))$!). We work in steps:

– We calculate the divergence at $\boldsymbol{P}'$ of the component system $\breve{\tau}$ of $\breve{\tau}\,(\boldsymbol{P}')$ in the affine frame $(\boldsymbol{P}, (\vec{e}_i))$ where $\boldsymbol{P}$ is a neighbor point of $\boldsymbol{P}'$.

– We consider its limit as $\boldsymbol{P}'$ approaches $\boldsymbol{P}$.

First, we express $\breve{\tau}$ with respect to the component system $\breve{\tau}'$ of $\breve{\tau}\,(\boldsymbol{P}')$ in the affine frame $(\boldsymbol{P}', (\vec{e}_i))$ by considering a translation $k' = x' - x$ (hence, the transformation matrix P is the identity). Transformation laws [8.20] and [8.22] give:

$$T^{ij} = T'^{ij}, \qquad J^{lij} = J'^{lij} + (x^l - x'^l)T^{ij} - (x^i - x'^i)T^{lj}.$$

Hence, the components of the divergence of $\breve{\tau}$ are:

$$\frac{\partial T^{ij}}{\partial x'^j} = \frac{\partial T'^{ij}}{\partial x'^j}, \qquad \frac{\partial J^{lij}}{\partial x'^j} = \frac{\partial J'^{lij}}{\partial x'^j} + T^{li} - T^{il}.$$

Considering their limits as $\boldsymbol{P}'$ approaches $\boldsymbol{P}$:

$$\tilde{\nabla}_j T^{ij} = \lim_{\boldsymbol{P}' \to \boldsymbol{P}} \frac{\partial T^{ij}}{\partial x'^j}, \qquad \tilde{\nabla}_j J^{lij} = \lim_{\boldsymbol{P}' \to \boldsymbol{P}} \frac{\partial J^{lij}}{\partial x'^j},$$

x' approaches x, T'^{ij} approaches T^{ij}, J'^{lij} approaches J^{lij} and it holds:

$$\tilde{\nabla}_j T^{ij} = \frac{\partial T^{ij}}{\partial x^j}, \qquad \tilde{\nabla}_j J^{lij} = \frac{\partial J^{lij}}{\partial x^j} + T^{li} - T^{il}. \qquad [8.28]$$

Then, the *covariant affine divergence* of the torsor field $\breve{\tau}$ is:

$$\tilde{\boldsymbol{div}}\, \breve{\tau} = \tilde{\nabla}_j T^{ij} \left(\boldsymbol{P} \otimes \vec{e}_i - \vec{e}_i \otimes \boldsymbol{P}\right) + \tilde{\nabla}_j J^{lij}\, \vec{e}_l \otimes \vec{e}_i. \qquad [8.29]$$

With this new definition, we can verify that the stress torsor and the volume force torsor verify the local equilibrium equations in the form [8.27]. Indeed, owing to [8.26], we have in the current affine frame $(\boldsymbol{P}, (\vec{e}_i))$:

$$\breve{\tau}_{\vec{f}_v} = f_v^i \left(\boldsymbol{P} \otimes \vec{e}_i - \vec{e}_i \otimes \boldsymbol{P}\right).$$

Moreover, according to theorem 8.3:

$$\tilde{\boldsymbol{div}}\, \breve{\tau}_\sigma = \frac{\partial \sigma^{ij}}{\partial x^j} \left(\boldsymbol{P} \otimes \vec{e}_i - \vec{e}_i \otimes \boldsymbol{P}\right) + \left(\sigma^{li} - \sigma^{il}\right) \vec{e}_l \otimes \vec{e}_i.$$

In short, we introduce a general definition of a continuum torsor and its divergence as affine tensor. Particularizing it to the stress torsors, we recover the local equilibrium equation [8.7] and [8.8] from the more compact formula [8.27].

DEFINITION 8.4.– Continua of which the torsor is a stress torsor are called *Cauchy's continua*.

As a consequence of equilibrium equations, the stress tensors of Cauchy's media are symmetric.

8.3. Invariants of the stress tensor

As the stress tensor $\boldsymbol{\sigma}$ is represented in an orthonormal basis by a symmetric matrix σ, it is diagonalizable and its eigenvalues $\sigma_1, \sigma_2, \sigma_3$ are real numbers called *principal stresses*. They are obtained by solving the characteristic equation:

$$\det(\sigma - \lambda\, 1_{\mathbb{R}^3}) = 0.$$

Owing to the transformation law [8.6] of Euclidean stress tensors, this equation is invariant under any orthogonal transformation R:

$$\det(\sigma' - \lambda\, 1_{\mathbb{R}^3}) = \det(R^T \sigma R - \lambda\, R^T 1_{\mathbb{R}^3} R) = \det\left[R^T(\sigma - \lambda\, 1_{\mathbb{R}^3})R\right],$$

$$\det(\sigma' - \lambda\, 1_{\mathbb{R}^3}) = (det(R))^2 \det(\sigma - \lambda\, 1_{\mathbb{R}^3}) = \det(\sigma - \lambda\, 1_{\mathbb{R}^3}).$$

As σ is a 3×3 matrix, the characteristic equation reads:

$$\det(\sigma - \lambda\, 1_{\mathbb{R}^3}) = -\lambda^3 + \iota_1(\sigma)\lambda^2 - \iota_2(\varepsilon)\lambda + \iota_3(\varepsilon) = 0,$$

where:

$$\iota_1(\sigma) = Tr(\sigma) = \sigma_1 + \sigma_2 + \sigma_3,$$

$$\iota_2(\sigma) = \frac{1}{2}\left[(Tr(\sigma))^2 - Tr(\sigma^2)\right] = \sigma_1\sigma_2 + \sigma_2\sigma_3 + \sigma_3\sigma_1,$$

$$\iota_3(\sigma) = \det(\sigma) = \sigma_1\sigma_2\sigma_3.$$

are called the *principal invariants* of σ. We denote:

$$\iota(\sigma) = (\iota_1(\sigma), \iota_2(\sigma), \iota_3(\sigma)).$$

As any function of invariants is also invariant, they allow us to generate other systems of invariants, for instance the one of the principal stresses.

9

Elasticity and Elementary Theory of Beams

9.1. Strains

In section 4.3.1, we consider commonly encountered situations where the deformations of materials are small. For a bulky body $\mathcal{V}$, we would like to generalize the definition [4.26] of the truss extension. A particle of the body, initially at $\boldsymbol{P}$, has position $\boldsymbol{P}'$ when the body is subjected to given external forces. We claim that the vector $\overrightarrow{\boldsymbol{PP}'}$, called *displacement*, is a smooth vector field $\boldsymbol{P} \mapsto \boldsymbol{u}(\overrightarrow{\boldsymbol{P}})$.

A body which does not undergo deformations is a rigid body (definition 3.4). Its motion preserves material length and angles or, in other words, the metric tensor $\breve{\boldsymbol{G}}$ of which Gram's matrix $\breve{G}$ is the identity in any Galilean coordinate system. To characterize the small deformation, we hope to build a tensor field depending on the displacement field and vanishing for every rigid motion. This suggests to study how the covariant metric tensor is perturbated by a rigid motion. We are working in two steps:

– Construct the curve $t \mapsto \boldsymbol{P}' = \varphi_t(\boldsymbol{P})$ solution of the ordinary differential equation:

$$\frac{d}{dt}(\varphi_t(\boldsymbol{P})) = \overrightarrow{\boldsymbol{u}}(\varphi_t(\boldsymbol{P})), \qquad [9.1]$$

with the initial condition $\varphi_0(\boldsymbol{P}) = \boldsymbol{P}$.

– Consider the pull-back of the metric tensor at $\boldsymbol{P}' = \varphi_t(\boldsymbol{P})$ and compare it to the metric at $\boldsymbol{P}$, next divide by t and calculate half the limit when t approaches zero:

$$\boldsymbol{\varepsilon} = \frac{1}{2}\, lim_{t\to 0}\, \frac{1}{t}\left[\varphi_t^*\breve{\boldsymbol{G}} - \breve{\boldsymbol{G}}\right]. \qquad [9.2]$$

As the metric tensor, this quantity is a symmetric 2-covariant tensor called the *strain tensor* (but it is not a metric). Now, let us determine it explicitly with respect to the displacement field. The question being local, we can work with a Galilean coordinate system in which the solution of [9.1] is expanded as:

$$x' = \varphi_t(x) = x + t\,u(x) + O(t^2),$$

where u, x and x' are the columns gathering, respectively, the contravariant components of the displacement $\overrightarrow{u}$, the coordinates of $\boldsymbol{P}$ and $\boldsymbol{P}' = \varphi_t(\boldsymbol{P})$. By differentiation, the tangent map to φ_t is represented by the Jacobean matrix:

$$\frac{\partial x'}{\partial x} = 1_{\mathbb{R}^3} + t\,\frac{\partial u}{\partial x} + O(t^2).$$

The strain tensor ε is represented by the symmetric 3×3 matrix:

$$\varepsilon = \frac{1}{2}\,lim_{t\to 0}\,\frac{1}{t}\left[\left(\frac{\partial x'}{\partial x}\right)^T \breve{G}'\,\frac{\partial x'}{\partial x} - \breve{G}\right],$$

where $\breve{G}'$ is Gram's matrix at x', that gives:

$$\varepsilon = \frac{1}{2}\,lim_{t\to 0}\frac{1}{t}\left[\breve{G}' + t\,\breve{G}'\,\frac{\partial u}{\partial x} + t\left(\frac{\partial u}{\partial x}\right)^T \breve{G}' - \breve{G} + O(t^2)\right].$$

In a Galilean coordinate system, Gram's matrix is constant, hence:

$$\varepsilon = \frac{1}{2}\left[\breve{G}\,\frac{\partial u}{\partial x} + \left(\frac{\partial u}{\partial x}\right)^T \breve{G}\right].$$

Going back to indicial notations gives the *internal compatibility equations*:

$$\varepsilon_{ij} = \frac{1}{2}\left(\breve{G}_{ik}\,\frac{\partial u^k}{\partial x^j} + \frac{\partial u^k}{\partial x^i}\breve{G}_{kj}\right) = \frac{1}{2}\left(\frac{\partial u_i}{\partial x^j} + \frac{\partial u_j}{\partial x^i}\right).$$

As the contravariant and covariant components of the displacement are identical, this relation reads in matrix notation:

$$\varepsilon = \frac{1}{2}\left[\frac{\partial u}{\partial x} + \left(\frac{\partial u}{\partial x}\right)^T\right] = grad_s\,u. \qquad [9.3]$$

We have to check this quantity vanishes for every rigid motion. Indeed, such a motion [3.26] is compound of a translation and a rotation. Under the hypothesis of small deformations, the rotation can be considered as infinitesimal hence of the form [3.24], which leads to the following modeling of the small *rigid displacement* fields:

$$u(x) = j(d\psi)\, x + dc = d\psi \times x + dc,$$

where $d\psi, dc \in \mathbb{R}^3$ are independent of x. For such a field, the derivative is skew-symmetric:

$$\frac{\partial u}{\partial x} = j(d\psi),$$

hence, its symmetric gradient vanishes. In other words, the condition of vanishing strain tensor is necessary for the motion to be rigid. It is also sufficient as claimed by *Kirchhoff's theorem*:

THEOREM 9.1.– If the domain $\mathcal{V}$ occupied by the body is connected, the displacement field is rigid if and only if the strain field vanishes.

PROOF.– We just showed the condition is necessary. To prove it is also sufficient, let us assume that the strain is null, then the derivative of the displacement is skew-symmetric. There exists a field $x \mapsto \omega(x) \in \mathbb{R}^3$ such that:

$$\omega = j^{-1}\left(\frac{\partial u}{\partial x}\right).$$

Let us prove it is independent of x. Indeed, we have:

$$\frac{1}{2}\, curl\, u = \frac{1}{2}\, j^{-1}\left(\frac{\partial u}{\partial x} - \left(\frac{\partial u}{\partial x}\right)^T\right) = \omega,$$

hence:

$$div\, \omega = \frac{1}{2}\, div\,(curl\, u) = 0.$$

On the other hand, we have:

$$curl\,(j(\omega)) = -curl\,((j(\omega))^T) = -curl\,(grad\, u) = 0$$

thus, owing to [14.26]:

$$\frac{\partial \omega}{\partial x} = curl\,(j(\omega)) + div\,\omega \cdot 1_{\mathbb{R}^3} = 0.$$

The field is uniform. The body being connected, we obtain by integration:

$$u(x) = j(\omega)\,(x - x_0) + u(x_0),$$

that achieves the proof. ■

Next, we would like to determine under which condition, a strain field being given, exists a displacement field satisfying the internal compatibility equations [9.3]. The answer is given by *Saint–Venant's theorem* below.

THEOREM 9.2.– For a given strain field, there exists a displacement field which fulfills the internal compatibility equations [9.3] if and only if *Saint–Venant compatibility conditions* are satisfied:

$$curl\,(curl\,\varepsilon)^T = 0. \qquad [9.4]$$

If the domain $\mathcal{V}$ occupied by the body is simply connected, it is given by:

$$u(x) = \int_{x_0}^{x} \left[\varepsilon(x'') + j\left(\int_{x_0}^{x''} (curl\,\varepsilon)(x')\,dx' + \omega(x_0) \right) \right] dx'' + u(x_0). [9.5]$$

PROOF.– The condition is necessary. Indeed, taking into account [14.28] and [14.27]:

$$curl\,\varepsilon = \frac{1}{2}\left[curl\left(\frac{\partial u}{\partial x}\right) + curl\,(grad\,u) \right] = \frac{1}{2}\,curl\left(\frac{\partial u}{\partial x}\right) = \frac{1}{2}\frac{\partial}{\partial x}\,(curl\,u), [9.6]$$

thus:

$$curl\,(curl\,\varepsilon)^T = \frac{1}{2}\,curl\,(grad\,(curl\,u)) = 0.$$

The condition is also sufficient. Indeed, as $\mathcal{V}$ is simply connected, using [14.28], there exists a vector field $x \longmapsto \omega(x) \in \mathbb{R}^3$ such that:

$$(curl\,\varepsilon)^T = grad\,\omega. \qquad [9.7]$$

The domain $\mathcal{V}$ being connected, we obtain by integration:

$$\omega(x) = \int_{x_0}^{x} (curl\,\varepsilon)(x')\,dx' + \omega(x_0). \qquad [9.8]$$

Moreover, taking into account [14.26] and [9.7], we have:

$$curl\,(\varepsilon + j(\omega))^T = curl\,\varepsilon - \frac{\partial \omega}{\partial x} + (div\,\omega)\,1_{\mathbb{R}^3} = (div\,\omega)\,1_{\mathbb{R}^3},$$

and, taking into account [14.29]:

$$div\,\omega = Tr\left(\frac{\partial \omega}{\partial x}\right) = Tr\,(curl\,\varepsilon) = div\,(j^{-1}(\varepsilon - \varepsilon^T)) = 0,$$

since ε is symmetric. Then, we proved that:

$$curl\,(\varepsilon + j(\omega))^T = 0.$$

As $\mathcal{V}$ is simply connected, using [14.28], there exists a vector field $x \longmapsto u(x) \in \mathbb{R}^3$ such that:

$$\varepsilon + j(\omega) = (grad\,u)^T = \frac{\partial u}{\partial x},$$

thus satisfying [9.3], and we obtain by integration:

$$u(x) = \int_{x_0}^{x} (\varepsilon(x'') + j(\omega(x''))\,dx'' + u(x_0).$$

Combining with [9.8] leads to [9.5] and achieves the proof. ■

Taking into account [9.6] and [9.8], we see that:

$$\omega = \frac{1}{2}\,curl\,u,$$

which is the *infinitesimal rotation vector of the elementary volume* around x.

The transformation law of 2-contravariant tensors gives:

$$\varepsilon'_{st} = (\breve{P}^T)^i_s \breve{P}^j_t\,\varepsilon_{ij}, \qquad [9.9]$$

which, according to [14.2], reads in matrix notations:

$$\varepsilon' = \breve{P}^T \varepsilon\, \breve{P}.$$

In particular, when working in Galilean coordinate systems, the transformation matrix is an orthogonal transformation and we obtain the transformation law of the Euclidean strain tensors:

$$\varepsilon' = R^T \varepsilon R. \qquad [9.10]$$

The diagonal elements ε_{ii} are called *normal strains*, while the off diagonal elements are known as *shear strains*. As the strain tensor $\boldsymbol{\varepsilon}$ is represented in an orthonormal basis by a symmetric matrix ε, it is diagonalizable and its eigenvalues $\varepsilon_1, \varepsilon_2, \varepsilon_3$ are real numbers called *principal strains*. They can be deduced from the corresponding system $\iota(\varepsilon)$ of principal invariants.

9.2. Internal work and power

Let us consider an arbitrary subdomain $\mathcal{V}$ of the body. According to definition 8.1, the elementary surface force acting at a point $\boldsymbol{P}$ of the boundary $\partial\mathcal{V}$ upon $\mathcal{V}$ through $d\mathcal{S}$ is:

$$\overrightarrow{d\boldsymbol{F}}_s = \overrightarrow{t}\, d\mathcal{S}.$$

The elementary work provided by the force to produce a displacement $\overrightarrow{u}$ of point $\boldsymbol{P}$ is:

$$d\mathcal{W}_s = \overrightarrow{d\boldsymbol{F}}_s \cdot \overrightarrow{u} = \overrightarrow{t} \cdot \overrightarrow{u}\, d\mathcal{S}.$$

Using matrix notations and taking into account [8.2] and [8.8], we have:

$$d\mathcal{W}_s = t^T u\, d\mathcal{S} = (\sigma\, n)^T u\, d\mathcal{S} = n^T \sigma\, u\, d\mathcal{S}.$$

According to definition 8.2, the elementary volume force acting at a point $\boldsymbol{P}$ of $\mathcal{V}$ upon the volume element $d\mathcal{V}$ around $\boldsymbol{P}$ is:

$$\overrightarrow{d\boldsymbol{F}}_v = \overrightarrow{f_v}\, d\mathcal{V}.$$

The elementary work provided by the force to produce a displacement $\overrightarrow{u}$ of point $\boldsymbol{P}$ is:

$$d\mathcal{W}_v = \overrightarrow{d\boldsymbol{F}}_v \cdot \overrightarrow{u} = \overrightarrow{f_v} \cdot \overrightarrow{u}\, d\mathcal{V} = f_v^T u\, d\mathcal{V}.$$

The total work provided by external forces is:

$$\mathcal{W} = \int_{\partial\mathcal{V}} n^T \sigma\, u\, d\mathcal{S} + \int_{\mathcal{V}} f_v^T u\, d\mathcal{V}. \qquad [9.11]$$

Owing to the internal equilibrium equations [8.7] and Green formula [14.20], we have:

$$\mathcal{W} = \int_{\mathcal{V}} (div\,(\sigma\, u) - (div\, \sigma)\, u)\; d\mathcal{V}.$$

Taking into account [14.15] and once again the symmetry [8.8] of the stress tensor, it holds:

$$\mathcal{W} = \int_{\mathcal{V}} Tr\,\left(\sigma\,\frac{\partial u}{\partial x}\right)\, d\mathcal{V} = \int_{\mathcal{V}} Tr\,\left(\sigma\,\varepsilon\right)\, d\mathcal{V}. \qquad [9.12]$$

Taking into account the continuity hypothesis, we can apply the mean value theorem for integrals. Hence, there exists a point $\bar{\boldsymbol{P}} \in \mathcal{V}$ such that:

$$\mathcal{W} = Tr\,\left(\sigma(\bar{\boldsymbol{P}})\,\varepsilon(\bar{\boldsymbol{P}})\right)\, \mathcal{V}.$$

Considering a subdomain $\mathcal{V}$ around a point $\boldsymbol{P}$ and approaching the limit as the volume of $\mathcal{V}$ approaches zero, $\bar{\boldsymbol{P}}$ coalesces into $\boldsymbol{P}$ and the *internal work by volume unit* is:

$$\mathcal{T}_{int} = \frac{d\mathcal{W}}{d\mathcal{V}} = \lim_{\mathcal{V}\to 0} \frac{\mathcal{W}}{\mathcal{V}} = Tr\,(\sigma\epsilon) = \boldsymbol{\sigma} : \epsilon$$

The elementary internal work by unit volume provided by $\boldsymbol{\sigma}$ to increment the strains of a quantity $d\varepsilon$ is:

$$d\mathcal{T}_{int} = \boldsymbol{\sigma} : d\varepsilon.$$

In a similar way, if $\vec{v}$ denotes the velocity of the point $\boldsymbol{P}$, the elementary power of the surface force acting at a point $\boldsymbol{P}$ of the boundary is:

$$d\mathcal{P}_s = \overrightarrow{\boldsymbol{dF}}_s \cdot \vec{v} = \vec{\boldsymbol{t}} \cdot \vec{v}\, dS.$$

the elementary power of the volume force acting at a point $\boldsymbol{P}$ of $\mathcal{V}$ is:

$$d\mathcal{P}_v = \overrightarrow{\boldsymbol{dF}}_v \cdot \vec{v} = \vec{\boldsymbol{f}}_v \cdot \vec{v}\, d\mathcal{V}.$$

By arguments analogous to the ones used for the work, we conclude that the *internal power by volume unit* is:

$$\mathcal{P}_{int} = Tr\,\left(\sigma\, D\right), \qquad [9.13]$$

where we introduced the *strain velocity*:

$$D = \frac{1}{2}\left[\frac{\partial v}{\partial x} + \left(\frac{\partial v}{\partial x}\right)^T\right] = grad_s\, v. \qquad [9.14]$$

9.3. Linear elasticity

9.3.1. *Hooke's law*

If the strains are small, many materials such as metals are elastic. We would like to generalize to bulky bodies Hooke's law 4.1 previously introduced for slender ones. Thus, we claim that the elastic behavior of the material is modeled by a linear map $\boldsymbol{E}$ mapping the 2-covariant strain tensor $\boldsymbol{\varepsilon}$ onto the 2-contravariant stress tensor $\boldsymbol{\sigma}$:

$$\boldsymbol{\sigma} = \boldsymbol{E}(\boldsymbol{\varepsilon}). \qquad [9.15]$$

Hence, there exists a 4-contravariant *elastic tensor* $\boldsymbol{C}$ such that:

$$\boldsymbol{\sigma} = \boldsymbol{C} : \boldsymbol{\varepsilon}. \qquad [9.16]$$

that reads with indicial notations:

$$\sigma^{ij} = C^{ijkl} \varepsilon_{kl}, \qquad [9.17]$$

or, alternatively:

$$\sigma = E(\varepsilon), \qquad [9.18]$$

where E is a linear map from the space $\mathbb{M}_{33}^{symm}$ of 3×3 matrices into itself. As the stress and strain tensors are both symmetric, the components of the elasticity tensor are subjected to the *minor symmetries*:

$$C^{ijkl} = C^{jikl} = C^{ijlk}.$$

Moreover, it is expected that, according to the experiments, the elastic behavior is *reversible* in the following sense. For any loop $\mathcal{C}$ in the space of strain tensors:

$$\oint_{\mathcal{C}} \boldsymbol{\sigma} : d\boldsymbol{\varepsilon} = 0, \qquad [9.19]$$

where $\boldsymbol{\sigma}$ depends on $\boldsymbol{\varepsilon}$ through [9.17]. In practice, this global condition is difficult to verify for every loop. In order to find an equivalent local condition, we consider a parallelogram of which the size approaches zero. Reasoning, as in section 3.3.2, we obtain:

$$\forall d\boldsymbol{\varepsilon}, \delta\boldsymbol{\varepsilon}, \qquad d\boldsymbol{\sigma} : \delta\boldsymbol{\varepsilon} - \delta\boldsymbol{\sigma} : d\boldsymbol{\varepsilon} = 0, \qquad [9.20]$$

where $\boldsymbol{\sigma}$ is given by [9.17] and this condition is necessary and sufficient for [9.19]. Using dummy indices, it reads:

$$d\sigma^{ij}\delta\varepsilon_{ij} - \delta\sigma^{kl}d\varepsilon_{kl} = (C^{ijkl} - C^{klij})d\varepsilon_{kl}\delta\varepsilon_{ij} = 0.$$

Because the infinitesimal variations $d\varepsilon$ and $\delta\varepsilon$ are arbitrary, the components of the elasticity tensor are subjected to the *major symmetries*:

$$C^{ijkl} = C^{klij}.$$

Taking into account the minor and major symmetries, among the $3^4 = 81$ components C^{ijkl}, there are only $6\,(6+1)/2 = 21$ independent ones.

One of the interesting features of the reversible behavior is the existence of a scalar *reversible energy potential* W generating the constitutive law, such that:

$$W(\boldsymbol{\varepsilon}) = W(\boldsymbol{\varepsilon}_0) + \int_{\boldsymbol{\varepsilon}_0}^{\boldsymbol{\varepsilon}} \boldsymbol{\sigma} : d\boldsymbol{\varepsilon}, \qquad [9.21]$$

where $\boldsymbol{\sigma}$ is given by [9.15], $\boldsymbol{\varepsilon}_0$ is any reference strain and the integration path from $\boldsymbol{\varepsilon}_0$ to $\boldsymbol{\varepsilon}$ can be chosen arbitrarily. Indeed, $\mathcal{C}_1$ and $\mathcal{C}_2$ being to such paths, let us consider the loop $\mathcal{C}$ obtained by concatenation of $\mathcal{C}_1$ and $\mathcal{C}_2$. If the sense of $\mathcal{C}$ is the same as $\mathcal{C}_1$ and opposite to the one of $\mathcal{C}_2$, we have:

$$\oint_{\mathcal{C}} \boldsymbol{\sigma} : d\boldsymbol{\varepsilon} = \int_{\mathcal{C}_1} \boldsymbol{\sigma} : d\boldsymbol{\varepsilon} - \int_{\mathcal{C}_2} \boldsymbol{\sigma} : d\boldsymbol{\varepsilon} = 0.$$

By differentiating [9.21], we recover the constitutive law:

$$\boldsymbol{\sigma} = \frac{\partial W}{\partial \boldsymbol{\varepsilon}},$$

which reads in indicial notation:

$$\sigma^{ij} = \frac{\partial W}{\partial \varepsilon_{ij}}.$$

Of course, the potential is definite modulo an arbitrary constant $W(\boldsymbol{\varepsilon}_0)$ which is not relevant for the constitutive law. Choosing $\boldsymbol{\varepsilon}_0 = \mathbf{0}$, $W(\mathbf{0}) = 0$ and the straight path from $\mathbf{0}$ to $\boldsymbol{\varepsilon}$, the integration is straightforward for Hooke's law [9.16] and gives:

$$W(\boldsymbol{\varepsilon}) = \frac{1}{2}\boldsymbol{\varepsilon} : (\boldsymbol{C} : \boldsymbol{\varepsilon}),$$

or, in indicial notations:

$$W(\varepsilon) = \frac{1}{2} C^{ijkl} \varepsilon_{ij} \varepsilon_{kl} = \frac{1}{2} \sigma^{ij} \varepsilon_{ij}. \qquad [9.22]$$

It is reasonable to claim that the energy to deform elastically the elementary volume must be provided by the environment:

$$\forall \varepsilon \neq \mathbf{0}, \qquad W(\varepsilon) > 0, \qquad [9.23]$$

then W is a strictly convex function and the constitutive law [9.16] can be inversed:

$$\varepsilon = \boldsymbol{S} : \boldsymbol{\sigma}, \qquad [9.24]$$

where $\boldsymbol{S}$ is a 4-covariant tensor with the minor and major symmetries:

$$S_{ijkl} = S_{jikl} = S_{ijlk}. \qquad [9.25]$$

9.3.2. *Isotropic materials*

It is worth to noting that the law [9.18] is the local representation in a given basis S of the intrinsic elasticity law [9.15]. Later on, we only consider the basis associated with reference coordinate systems in which the distances and other mechanical quantities are measured, the Galilean coordinate ones. Restricting us to the 3D space, we are now working with Euclidean tensors. Let us now introduce the class of materials with strongest symmetries.

DEFINITION 9.1.– A material is *isotropic* if the behavior of the material is the same in every Galilean coordinate system.

In other words, we have in any orthonormal basis S':

$$\sigma' = E(\varepsilon'),$$

As an orthonormal basis S is transformed into another one S' through an orthogonal transformation $R = S^{-1}S' \in \mathbb{O}(3)$, according to the transformation laws [9.10] and [8.6], the law is isotropic if:

$$\forall \varepsilon \in \mathbb{M}_{33}^{symm}, \qquad \forall R \in \mathbb{O}(3), \qquad R^T E(\varepsilon)\, R = E(R^T \varepsilon\, R). \qquad [9.26]$$

The general form of isotropic constitutive law of materials is given by *Rivlin–Ericksen representation theorem* below.

THEOREM 9.3.– The elasticity law is isotropic if and only if the map E is of the form:

$$E(\varepsilon) = \lambda_0(\iota(\varepsilon))\, 1_{\mathbb{R}^3} + \lambda_1(\iota(\varepsilon))\, \varepsilon + \lambda_2(\iota(\varepsilon))\, \varepsilon^2. \quad [9.27]$$

PROOF.– The demonstration is decomposed into three steps:

– Step 1: *We prove that any matrix R which diagonalizes ε also diagonalizes $E(\varepsilon)$*. As the matrix ε is symmetric, it is diagonalizable and its eigenvalues ε_i are real numbers. Hence, there exists an orthogonal transformation $R = (V_1, V_2, V_3)$ such that:

$$R^T \varepsilon\, R = diag(\varepsilon_1, \varepsilon_2, \varepsilon_2). \quad [9.28]$$

Without loss of generality, we may assume that R is a rotation (otherwise, replace one of its column by its opposite). In addition, let us consider the *mirror symmetry* with respect to the plan Ox_2x_3:

$$M_1 = diag(-1, 1, 1), \quad [9.29]$$

which is clearly an orthogonal transformation. The relation [9.28] means that for $j = 1, 2, 3$ the column V_i of R is an eigenvector of ε corresponding to the eigenvalue ε_i. Hence, we likewise have:

$$(RM_1)^T \varepsilon\, (RM_1) = diag(\varepsilon_1, \varepsilon_2, \varepsilon_2) = R^T \varepsilon\, R,$$

since the effect of multiplying R by M_1 on the right is to replace its first column by its opposite. Owing to [9.26], we have:

$$M_1^T (R^T E(\varepsilon)\, R)\, M_1 = (RM_1)^T E(\varepsilon)\, (RM_1) = E((RM_1)^T \varepsilon\, (RM_1))$$
$$M_1^T (R^T E(\varepsilon)\, R)\, M_1 = E(R^T \varepsilon\, R) = R^T E(\varepsilon)\, R.$$

As a straightforward calculation shows, the effect to multiply the matrix $R^T E(\varepsilon)\, R$ by M_1^T on the left and M_1 on the right is to change the sign of the elements of its first column (and first row) which are not on the diagonal so they can can be canceled. Next, we repeat this reasoning with the mirror symmetry M_2 with respect to the plan Ox_1x_3 that has the effect of canceling the elements of the second column (and second row) of $R^T E(\varepsilon)\, R$ which are not on the diagonal. Then, the matrix $R^T E(\varepsilon)\, R$ is diagonal. We just proved that any matrix R which diagonalizes ε also diagonalizes $E(\varepsilon)$.

– Step 2: *We prove that the map E is necessarily of the form [9.27] where λ_i are real-valued functions*. Three cases must be distinguished. Assume first that the matrix ε has three distinct eigenvalues ε_i, with associated orthonomalized eigenvectors V_i.

Then, the two sets $\{1_{\mathbb{R}^3}, \varepsilon, \varepsilon^2\}$ and $\{V_1V_1^T, V_2V_2^T, V_3V_3^T\}$ span the same subspace of the linear space $\mathbb{M}_{33}^{symm}$. Indeed, we observe that:

$$1_{\mathbb{R}^3} = V_1V_1^T + V_2V_2^T + V_3V_3^T,$$
$$\varepsilon = \varepsilon_1 V_1V_1^T + \varepsilon_2 V_2V_2^T + \varepsilon_3 V_3V_3^T,$$
$$\varepsilon^2 = \varepsilon_1^2 V_1V_1^T + \varepsilon_2^2 V_2V_2^T + \varepsilon_3^2 V_3V_3^T,$$

and that the van der Monde determinant:

$$det \begin{pmatrix} 1 & 1 & 1 \\ \varepsilon_1 & \varepsilon_2 & \varepsilon_3 \\ \varepsilon_1^2 & \varepsilon_2^2 & \varepsilon_3^2 \end{pmatrix},$$

does not vanish, since the three eigenvalues are assumed to be distinct. Denoting by σ_i the eigenvalues of $\sigma = E(\varepsilon)$, the result of Step 1 shows that we can expand $E(\varepsilon)$ as:

$$E(\varepsilon) = \sigma_1 V_1V_1^T + \sigma_2 V_2V_2^T + \sigma_3 V_3V_3^T,$$

and consequently also as:

$$E(\varepsilon) = \lambda_0(\varepsilon)\, 1_{\mathbb{R}^3} + \lambda_1(\varepsilon)\, \varepsilon + \lambda_2(\varepsilon)\, \varepsilon^2, \qquad [9.30]$$

where the components $\lambda_i(\varepsilon)$ are uniquely determined because the matrices $1_{\mathbb{R}^3}$, ε and ε^2 are linearly independent in this case.

Assume next the matrix ε has a double eigenvalue, let us say $\varepsilon_2 = \varepsilon_3 \neq \varepsilon_1$. Then, the two sets $\{1_{\mathbb{R}^3}, \varepsilon\}$ and $\{V_1V_1^T, V_2V_2^T + V_3V_3^T\}$ span the same subspace of the linear space $\mathbb{M}_{33}^{symm}$ since in this case:

$$1_{\mathbb{R}^3} = V_1V_1^T + (V_2V_2^T + V_3V_3^T), \qquad \varepsilon = \varepsilon_1 V_1V_1^T + \varepsilon_2 (V_2V_2^T + V_3V_3^T).$$

By a reasoning similar to the one of the first case, we prove that $E(\varepsilon)$ also has a double eigenvalue and we have:

$$E(\varepsilon) = \lambda_0(\varepsilon)\, 1_{\mathbb{R}^3} + \lambda_1(\varepsilon)\, \varepsilon. \qquad [9.31]$$

Finally, assume that the matrix ε has a triple eigenvalue. We likewise show that $E(\varepsilon)$ has also a triple eigenvalue and we have:

$$E(\varepsilon) = \lambda_0(\varepsilon)\, 1_{\mathbb{R}^3}. \qquad [9.32]$$

It is worth noticing that [9.31] and [9.32] are particular cases of [9.30] and that we can use the latter as a general expression.

– Step 3: *We observe that the* λ_i *are invariant functions of* ε. Indeed, combining the isotropy condition [9.26] and the expansion formula [9.30] gives:

$$
\begin{aligned}
E(R^T\varepsilon\, R) &= R^T\left[\lambda_0(R^T\varepsilon\, R)\, 1_{\mathbb{R}^3} + \lambda_1(R^T\varepsilon\, R)\,\varepsilon + \lambda_2(R^T\varepsilon\, R)\,\varepsilon^2\right] R \\
&= R^T E(\varepsilon)\, R = R^T\left[\lambda_0(\varepsilon)\, 1_{\mathbb{R}^3} + \lambda_1(\varepsilon)\,\varepsilon + \lambda_2(\varepsilon)\,\varepsilon^2\right] R.
\end{aligned}
$$

Hence, $\lambda_i(R^T\varepsilon\, R) = \lambda_i(\varepsilon)$ because of the uniqueness of the expansion of $E(\varepsilon)$ in the spaces spanned by the sets $\{1_{\mathbb{R}^3}, \varepsilon, \varepsilon^2\}$, $\{1_{\mathbb{R}^3}, \varepsilon\}$ or $\{1_{\mathbb{R}^3}\}$, according to which case is considered. Then, the invariance of the λ_i is satisfied if they depend on ε through the system $\iota(\varepsilon)$ of principal invariants. ■

We are in conditions to generalize Hooke's law (Law 4.1 in section 4.3.1) in the sense that stresses are proportional to corresponding deformations and the material is isotropic. According to the previous theorem, the elastic law has the form [9.27]. The last term containing the nonlinear factor ε^2, we cancel it by assuming $\lambda_2 = 0$. For the first term, being linear, λ_0 must be linear, depending on ε through the only linear principal invariant $\iota_1(\varepsilon) = Tr(\varepsilon)$. Finally, for the second term, containing the factor ε, λ_1 must be constant. A straightforward calculation shows that the reversibility condition [9.20] is fulfilled. Hence, we must assume the following.

LAW 9.1.– There exist two material constants λ, μ called *Lamé coefficients* such that the *isotropic linear elastic law* or *generalized Hooke's law* is reversible and given by:

$$\sigma = \lambda\, Tr(\varepsilon)\, 1_{\mathbb{R}^3} + 2\,\mu\,\varepsilon. \qquad [9.33]$$

It is easy to calculate by [9.21] that the corresponding potential vanishing at $\varepsilon = 0$ is:

$$W(\varepsilon) = \frac{1}{2}\,\lambda\,(Tr(\varepsilon))^2 + \mu\, Tr(\varepsilon^2).$$

Of course, the relation [9.33] is only valid in Galilean coordinate systems. To find the free coordinate version of this law, let us remark that it reads in tensorial notations preserving the covariant and contravariant indices of ε and $\boldsymbol{\sigma}$:

$$\sigma^{ij} = \lambda\,(\delta^{kl}\varepsilon_{kl})\,\delta^{ij} + 2\,\mu\,\delta^{ik}\delta^{jl}\varepsilon_{kl}.$$

As the contravariant metric tensors $\breve{\boldsymbol{G}}^{-1}$ is represented by the identity matrix in Galilean coordinate systems, it is easy to guess the free coordinate form of Hooke's law using contracted products:

$$\boldsymbol{\sigma} = \lambda\,(\breve{\boldsymbol{G}}^{-1} : \boldsymbol{\varepsilon})\,\breve{\boldsymbol{G}}^{-1} + 2\,\mu\,\breve{\boldsymbol{G}}^{-1}\cdot\boldsymbol{\varepsilon}\cdot\breve{\boldsymbol{G}}^{-1}.$$

A standard method to measure the properties of an isotropic elastic material is to use a truss specimen, let us say of axis x^1, loaded only at each extremity by the same uniform stress distribution σ^{11}. If the cross-section is constant, it is easy to be convinced that the stress field is uniform inside with a unique non-vanishing component σ^{11} because, in the absence of volume forces, the local equilibrium equation [8.7] is obviously satisfied and the free stress condition $t = \sigma\, n = 0$ so is on the lateral surface. From Hooke's law [9.33], we see that shear components vanish. The only non-zero components are extensions ε_{ii}. Because of the isotropy of the material, the transversal components are equal: $\varepsilon_{22} = \varepsilon_{33}$. Measuring $\sigma^{11}, \varepsilon_{11}, \varepsilon_{22}$, we determine the *engineering constants* (i.e. which can be physically measured):

– *Young's modulus* $E = \sigma^{11} / \varepsilon_{11}$;

– *Poisson's coefficient*: $\nu = -\varepsilon_{22} / \varepsilon_{11}$.

The conditions $E = \sigma^{11} / \varepsilon_{11}$ and $\sigma^{22} = 0$, combined with Hooke's law [9.33], lead to:

$$(1 - 2\,\nu)\,\lambda + 2\,\mu = E, \qquad (1 - 2\,\nu)\,\lambda - 2\,\mu\,\nu = 0.$$

If ν is different from -1 and $1/2$, we obtain Lamé's coefficients in terms of engineer constants:

$$\lambda = \frac{\nu\, E}{(1 - 2\,\nu)\,(1 + \nu)}, \qquad \mu = \frac{E}{2\,(1 + \nu)}. \qquad [9.34]$$

We leave to the readers to verify that [9.23] is satisfied if $E > 0$ and $-1 < \nu < 1/2$, which justifies *a posteriori* to exclude the limit values.

9.3.3. *Elasticity problems*

Let $\mathcal{V}$ be an elastic body subjected to external actions, volume forces f on $\mathcal{V}$, imposed displacement $\bar{u}$ on a part $\mathcal{S}_u$ of the boundary $\partial\mathcal{V}$ and imposed stress vector $\bar{t}$ on the remaining part $\mathcal{S}_t$ of the boundary. Recording [8.4], [8.7], [9.3] and [9.33], we can state the following boundary value problem.

DEFINITION 9.2.– *Elasticity problem.* Find fields of displacement u, strain ε and stresses σ such that:

– in $\mathcal{V}$: $\varepsilon = grad_s\, u, \qquad div\, \sigma + f_v = 0, \qquad \sigma = \lambda\, Tr(\varepsilon)\, 1_{\mathbb{R}^3} + 2\ \mu\, \varepsilon$;

– on $\mathcal{S}_u$: $u = \bar{u}$;

– on $\mathcal{S}_t$: $t = \sigma\, n = \bar{t}$.

The set of equations to satisfy in the volume is a partial derivative system of order one which is fortunately linear. The difficulty lays in the large number of scalar equations, 15 (for 15 scalar unknowns). Before solving the problem, it is convenient to reduce its size. There are two dual methods:

– The simplest method consists of eliminating the stress and strain fields to conserve uniquely the displacement field as unknown. For this reason, it is called *displacement method*. In the sequel, we consider only problems with no volume forces (because they are often negligeable) and homogeneous material properties, then Lamé's coefficients are uniform on $\mathcal{V}$.

First, let us remark that $Tr(\varepsilon) = div\, u$. Combining [9.3] and [9.33] leads to:

$$\sigma = \lambda\,(div\, u)\, 1_{\mathbb{R}^3} + \mu\,\frac{\partial u}{\partial x} + \mu\, grad\, u,$$

hence, taking into account [14.16] and [14.18]:

$$div\,\sigma = \lambda\,\frac{\partial}{\partial x}(div\, u) + \mu\, div\left(\frac{\partial u}{\partial x}\right) + \mu\, div\,(grad\, u),$$

$$div\,\sigma = (\lambda + \mu)\,\frac{\partial}{\partial x}(div\, u) + \mu\, div\,(grad\, u) = 0.$$

By transposition and owing to [14.23], we obtain *Navier–Lamé equation*:

$$(\lambda + \mu)\, grad\,(div\, u) + \mu\,\Delta u = 0, \qquad [9.35]$$

a system of three scalar linear partial derivative equations.

– On the other hand, the *stress method* consists of eliminating the strain field by introduction of the expression of the stresses given by the inverse of Hooke's law into Saint–Venant compatibility conditions [9.4]. The resulting equations are solved with respect to the stress field together with the local equilibrium equation [8.7], which leads to a system of nine scalar linear partial derivative equations. The existence of the corresponding displacement field is ensured by theorem 9.2.

9.4. Elementary theory of elastic trusses and beams

9.4.1. *Multiscale analysis: from the beam to the elementary volume*

On this basis, we would like to develop a simplified theory of slender bodies. In section 4.3, we defined the internal forces at the beam scale, the normal force N, the shear force T, the torque M_t and the bending moment M_b due to the external actions. But, at the local scale of the elementary volume, what are the corresponding stresses and strains?

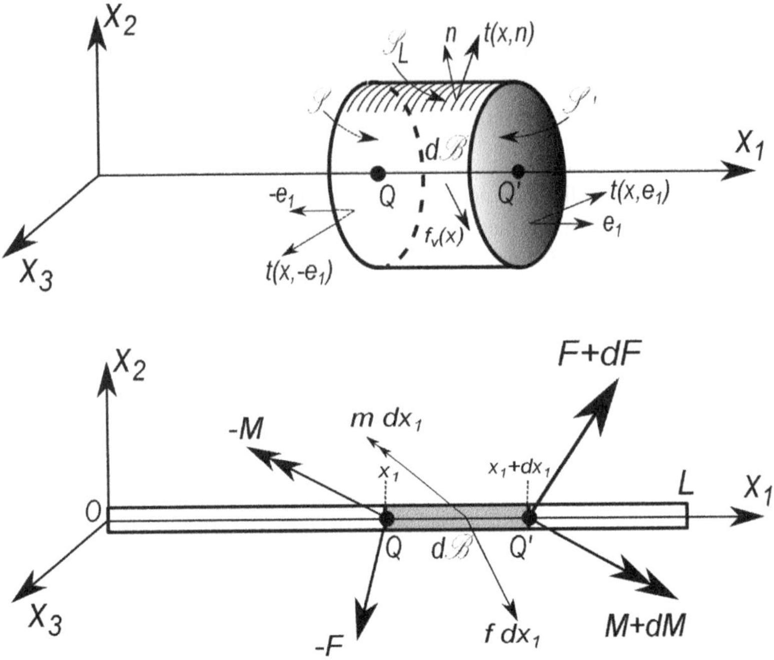

Figure 9.1. *Above: free body diagram of the slice $\mathcal{B}$ as 3D body. Below: the same slice as 1D body (as it was seen from a long way off)*

To perform this scale change, we consider a straight slender body of constant cross-section subjected to axial and transversal loads, hence it works as both truss and beam. For convenience, we call it a beam, we choose an Euclidean coordinate system $\mathbf{O}x_1x_2x_3$, then all the indices may be lowered, and the mean line of the beam is x_1's axis then the cross-sections are parallel to the plane Ox_1x_2. As in section 4.1.2, let us consider a slice $d\mathcal{B}$ of infinitesimal length dx_1 between the cross-sections $\mathcal{S}$ and $\mathcal{S}'$ corresponding, respectively, to the points Q and Q' on the mean line (Figure 9.1). Let $\mathcal{V}$ be the domain occupied by the slice. Its boundary is decomposed into $\mathcal{S}$ and $\mathcal{S}'$ and the lateral boundary $\mathcal{S}_L$, a cylindrical surface generated by a segment of length dx_1 parallel to x_1's axis and passing through a piecewise smooth guiding curve $\mathcal{C}$ parametrized by the arclength s.

If the beam is in static equilibrium, the resultant torsor of the slice vanishes:

$$\check{\tau}(d\mathcal{B}) = \int_{\mathcal{S}} d\check{\tau}_s + \int_{\mathcal{S}'} d\check{\tau}_s + \int_{\mathcal{S}_L} d\check{\tau}_s + \int_{\mathcal{V}} d\check{\tau}_v.$$

When drawing the free body diagram, the slice is isolated by making a cut along the cross-sections $\mathcal{S}$ and $\mathcal{S}'$. If the beam is idealized by a 1D material body as in Chapter 4.1, the two former terms must be considered as torsors of internal forces acting upon $\mathcal{S}$ and $\mathcal{S}'$, while the remaining terms define the torsor of external forces:

$$d\breve{\mathcal{T}}^{ext} = \int_{\mathcal{S}_L} d\breve{\mathcal{T}}_s + \int_{\mathcal{V}} d\breve{\mathcal{T}}_v.$$

Taking into account [8.17] and [8.25], we have:

$$\frac{d\breve{\mathcal{T}}^{ext}}{dx_1} = \int_{\mathcal{C}} (\boldsymbol{P} \otimes \vec{\boldsymbol{t}}(\boldsymbol{P},\boldsymbol{n}) - \vec{\boldsymbol{t}}(\boldsymbol{P},\boldsymbol{n}) \otimes \boldsymbol{P})\, ds + \int_{\mathcal{S}} (\boldsymbol{P} \otimes \vec{\boldsymbol{f}}_v(\boldsymbol{P}) - \vec{\boldsymbol{f}}_v(\boldsymbol{P}) \otimes \boldsymbol{P})\, d\mathcal{S},$$

where $\boldsymbol{n}$ is the unit normal to $\mathcal{S}_L$ at $\boldsymbol{P}$. Using the decompositions [8.18] and [8.26] in the affine frame, and presenting the result in a matrix form as in [8.14] and [8.15], we have:

$$\frac{d\breve{\mathcal{T}}^{ext}}{dx_1} = \begin{pmatrix} 0 & f^T \\ -f & -j(m) \end{pmatrix},$$

where:

$$f = \int_{\mathcal{C}} t(x,n)\, ds(x) + \int_{\mathcal{S}} f_v(x)\, d\mathcal{S}(x), \qquad [9.36]$$

$$m = \int_{\mathcal{C}} x \times t(x,n)\, ds(x) + \int_{\mathcal{S}} x \times f_v(x)\, d\mathcal{S}(x).$$

Likewise, the torsor of the internal forces acting upon the slice through the cross-section $\mathcal{S}$ is:

$$\breve{\mathcal{T}}^{int} = \int_{\mathcal{S}} d\breve{\mathcal{T}}_s = \int_{\mathcal{S}} (\boldsymbol{P} \otimes \vec{\boldsymbol{t}}(\boldsymbol{P}, -\boldsymbol{e}_1) - \vec{\boldsymbol{t}}(\boldsymbol{P}, -\boldsymbol{e}_1) \otimes \boldsymbol{P})\, d\mathcal{S}.$$

According to the sign convention of Figure 9.1, its decomposition in the affine frame leads to:

$$\breve{\mathcal{T}}^{int} = -\begin{pmatrix} 0 & F^T \\ -F & -j(M) \end{pmatrix},$$

where, owing to [8.1]:

$$F = \int_{\mathcal{S}} t(x,e_1)\, d\mathcal{S}(x), \qquad M = \int_{\mathcal{S}} x \times t(x,e_1)\, d\mathcal{S}(x). \qquad [9.37]$$

According to the decomposition of the force into the normal force N and shear forces T_i, the one of the moments into torque M_t and bending moments M_{bi} (see section 4.1.1) and the fact that the beam is parallel to x_1's axis, we have:

$$F = \begin{pmatrix} N \\ T_2 \\ T_3 \end{pmatrix}, \qquad M = \begin{pmatrix} M_t \\ M_{b2} \\ M_{b3} \end{pmatrix}, \qquad U = \begin{pmatrix} 1 \\ 0 \\ 0 \end{pmatrix}.$$

The local equilibrium equations of arches [4.2] and [4.3] read:

$$\frac{dN}{dx_1} + f_1 = 0, \qquad \frac{dT_2}{dx_1} + f_2 = 0, \qquad \frac{dT_3}{dx_1} + f_3 = 0, \tag{9.38}$$

$$\frac{dM_t}{dx_1} + m_1 = 0, \qquad \frac{dM_{b2}}{dx_1} - T_3 + m_2 = 0, \qquad \frac{dM_{b3}}{dx_1} + T_2 + m_3 = 0. \tag{9.39}$$

Eliminating the shear forces between these equations leads to:

$$-\frac{dM_{b3}^2}{dx_1^2} - \frac{dm_3}{dx_1} + f_2 = 0 \qquad \frac{dM_{b2}^2}{dx_1^2} + \frac{dm_2}{dx_1} + f_3 = 0.$$

According to section 4.1.2, the column x represents the vector $\overrightarrow{\boldsymbol{QP}}$ joining, within the cross-section $\mathcal{S}$, the point $\boldsymbol{Q}$ on the mean line to the current point $\boldsymbol{P}$:

$$x = \begin{pmatrix} 0 \\ x_2 \\ x_3 \end{pmatrix}.$$

The beam force and moment are related to the stress field through [9.37], which gives:

$$N = \int_{\mathcal{S}} \sigma_{11}\, dx_2 dx_3, \qquad T_2 = \int_{\mathcal{S}} \sigma_{12}\, dx_2 dx_3, \qquad T_3 = \int_{\mathcal{S}} \sigma_{13}\, dx_2 dx_3, \tag{9.40}$$

$$M_t = \int_{\mathcal{S}} (x_2\sigma_{13} - x_3\sigma_{12})\, dx_2 dx_3, \tag{9.41}$$

$$M_{b2} = \int_{\mathcal{S}} x_3\sigma_{11}\, dx_2 dx_3, \qquad M_{b3} = -\int_{\mathcal{S}} x_2\sigma_{11}\, dx_2 dx_3.$$

Likewise, the external forces and moments are given by [9.37]:

$$f_i = \int_S f_{vi}\, dx_2 dx_3 + \int_C t_i\, ds, \qquad [9.42]$$

$$m_1 = \int_S (x_2 f_{v3} - x_3 f_{v2})\, dx_2 dx_3 + \int_C (x_2 t_3 - x_3 t_2)\, ds, \qquad [9.43]$$

$$m_2 = \int_S x_3 f_{v1}\, dx_2 dx_3 + \int_C x_3 t_1\, ds, \qquad [9.44]$$

$$m_3 = -\int_S x_2 f_{v1}\, dx_2 dx_3 - \int_C x_2 t_1\, ds.$$

To determine the stress distribution in the cross-section, we need the elasticity theory of bulky bodies. Exact solutions are difficult to find except for very simple geometries and external actions in which we are looking for reasonable approximations. To develop some intuition of a simplified modeling, we wish to spot the quantities that can be neglected in the equations. Let us consider characteristic lengths, L for the beam, h for the cross-section. As the body is slender, the adimensional parameter $\epsilon = h/L$ is small with respect to the unity, which will allow us to simplify the equations by neglecting small terms of order ϵ or higher ones. For this aim, let us introduce adimensional reduced variables:

$$x_1' = \frac{x_1}{L}, \qquad x_2' = \frac{x_2}{h}, \qquad x_3' = \frac{x_3}{h}, \qquad s' = \frac{s}{h}. \qquad [9.45]$$

For a reference force value F_0, the reference stress and volume force are, respectively, $\sigma_0 = F_0/h^2$ and $f_{v0} = F_0/h^3$, hence we introduce the adimensional reduced variables:

$$\sigma_{ij}' = \frac{\sigma_{ij}}{\sigma_0}, \qquad t_i' = \frac{t_i}{\sigma_0}, \qquad f_{vi}' = \frac{f_{vi}}{f_{v0}}$$

Introducing the expressions N and f_1 given by [9.40] and [9.43] into the former equation in [9.38], eliminating old variables with respect to the reduced ones, next leaving out the primes, we obtain:

$$\epsilon \frac{dN}{dx_1} + f_1 = 0.$$

Likewise, we obtain:

$$\epsilon \frac{dM_t}{dx_1} + m_1 = 0, \quad -\epsilon^2 \frac{d^2 M_{b3}}{dx_1^2} - \epsilon \frac{dm_3}{dx_1} + f_2 = 0$$

$$\epsilon^2 \frac{d^2 M_{b2}}{dx_1^2} + \epsilon \frac{dm_2}{dx_1} + f_3 = 0.$$

These relations show us that:

– The bending effects are dominant.

– The external forces f and moments m are at the most of order ϵ and can be neglected in first approximation. Owing to [9.43] and [9.50], the stress vector on the lateral surface S_L and the body forces will be neglected.

9.4.2. *Transversely rigid body model*

After identifying the negligeable quantities, we want to determine the transversal strain and stress fields:

$$\bar{\varepsilon} = \begin{pmatrix} \varepsilon_{22} & \varepsilon_{23} \\ \varepsilon_{23} & \varepsilon_{33} \end{pmatrix}, \qquad \bar{\sigma} = \begin{pmatrix} \sigma_{22} & \sigma_{23} \\ \sigma_{23} & \sigma_{33} \end{pmatrix},$$

by the displacement method. As the body is slender, it is worth distinguishing the quantities x_1 and u_1 related to the *fibers* parallel to x_1's axis (out of plane quantities) and the ones related to the cross-section (in-plane quantities):

$$\bar{u} = \begin{pmatrix} u_2 \\ u_3 \end{pmatrix}, \qquad \overline{div}\,\bar{u} = \frac{\partial u_2}{\partial x_2} + \frac{\partial u_3}{\partial x_3}, \qquad \bar{\Delta}\,\bar{u} = \frac{\partial^2 \bar{u}}{\partial x_2^2} + \frac{\partial^2 \bar{u}}{\partial x_3^2},$$

thus, the in-plane equations of system [9.35] in the absence of volume forces read:

$$(\lambda + \mu)\,\overline{grad}\left(\overline{div}\,\bar{u} + \frac{\partial u_1}{\partial x_1}\right) + \mu\left(\bar{\Delta}\,\bar{u} + \frac{\partial^2 \bar{u}}{\partial x_1^2}\right) = 0.$$

For a reference displacement U_0, we eliminate the old variables with respect to the reduced displacements $u_i' = u_i/U_0$ and coordinates [9.45], next leaving out the primes, we obtain:

$$(\lambda + \mu)\,\overline{grad}\,(\overline{div}\,\bar{u}) + \mu\,\bar{\Delta}\,\bar{u} + \epsilon\,(\lambda + \mu)\,\overline{grad}\left(\frac{\partial u_1}{\partial x_1}\right) + \epsilon^2 \mu\,\frac{\partial^2 \bar{u}}{\partial x_1^2} = 0,$$

and neglecting the small quantities, the in-plane equations are reduced to:

$$(\lambda + \mu)\,\overline{grad}\,(\overline{div}\,\bar{u}) + \mu\,\bar{\Delta}\,\bar{u} = 0,$$

to be satisfied at constant x_1 in S together with the free stress boundary condition $\bar{t} = \bar{\sigma}\,\bar{n} = 0$ on C:

$$\begin{pmatrix} t_2 \\ t_3 \end{pmatrix} = \begin{pmatrix} \sigma_{22} & \sigma_{23} \\ \sigma_{23} & \sigma_{33} \end{pmatrix} \begin{pmatrix} n_2 \\ n_3 \end{pmatrix} = \begin{pmatrix} 0 \\ 0 \end{pmatrix}.$$

As we transformed [9.11] into [9.12], we obtain (a change of dimension):

$$\int_{\mathcal{C}} \bar{n}^T \bar{\sigma}\, \bar{u}\, ds + \int_{\mathcal{S}} \bar{f}_v^T \bar{u}\, d\mathcal{V} = \int_{\mathcal{S}} Tr\ (\bar{\sigma}\, \bar{\varepsilon})\ d\mathcal{S}.$$

Because the volume forces and stress vector vanish, so does the left-hand member and, owing to [9.22], we have:

$$2 \int_{\mathcal{S}} W(\bar{\varepsilon})\, d\mathcal{S} = 0.$$

Because of [9.23]:

$$\bar{\varepsilon} = 0.$$

On this ground, we take a radical but strategic decision to modify the constitutive law. We claim that:

– the material is linear elastic and reversible;

– it is *transversely rigid*:

$$\varepsilon_{22} = \varepsilon_{33} = \varepsilon_{23} = 0, ; \qquad [9.46]$$

– the behavior is isotropic in the cross-section. In other words, its symmetry group is composed of the orthogonal transformations in the Ox_2x_3 plane (forming the group $\mathbb{O}(2)$ seen as a subgroup of $\mathbb{O}(3)$), the mirror symmetry with respect to this plane and all their products.

It is convenient to introduce the columns:

$$\sigma_n = \begin{pmatrix} \sigma_{11} \\ \sigma_{22} \\ \sigma_{33} \end{pmatrix}, \qquad \sigma_s = \begin{pmatrix} \sigma_{23} \\ \sigma_{13} \\ \sigma_{12} \end{pmatrix}, \qquad \varepsilon_n = \begin{pmatrix} \varepsilon_{11} \\ \varepsilon_{22} \\ \varepsilon_{33} \end{pmatrix}, \qquad \varepsilon_s = \begin{pmatrix} \varepsilon_{23} \\ \varepsilon_{13} \\ \varepsilon_{12} \end{pmatrix},$$

the compliance matrices, respectively, relative to the normal and shear components:

$$S_{nn} = \begin{pmatrix} S_{1111} & S_{1122} & S_{1133} \\ S_{1122} & S_{2222} & S_{2233} \\ S_{1133} & S_{2233} & S_{3333} \end{pmatrix}, \qquad S_{ss} = \begin{pmatrix} S_{2323} & S_{2313} & S_{2312} \\ S_{2313} & S_{1313} & S_{1312} \\ S_{2312} & S_{1312} & S_{1212} \end{pmatrix},$$

and the one relative to the coupling between normal and shear components:

$$S_{ns} = \begin{pmatrix} S_{1123} & S_{1113} & S_{1112} \\ S_{1113} & S_{2213} & S_{2212} \\ S_{1112} & S_{2212} & S_{3312} \end{pmatrix}.$$

Taking into account the minor and major symmetries [9.25], the constitutive law takes the matrix form $\sigma = S\,\varepsilon$ which is decomposed according to:

$$\begin{pmatrix} \varepsilon_n \\ \sqrt{2}\,\varepsilon_s \end{pmatrix} = \begin{pmatrix} S_{nn} & \sqrt{2}\,S_{ns} \\ \sqrt{2}\,S_{ns}^T & 2\,S_{ss} \end{pmatrix} \begin{pmatrix} \sigma_n \\ \sqrt{2}\,\sigma_s \end{pmatrix},$$

in *Kelvin representation* [THO 56, THO 90], rediscovered later on by Walpole [WAL 84] and Rychlewski [RYC 84]. To satisfy [9.46] for every value of the stress components, the second, third and later rows of the compliance matrix vanish in the previous relation. Because of its symmetry, so are the second, third and later columns. Thus, the constitutive law reduces to [9.46] and:

$$\begin{pmatrix} \varepsilon_{11} \\ \sqrt{2}\,\varepsilon_{12} \\ \sqrt{2}\,\varepsilon_{13} \end{pmatrix} = \begin{pmatrix} S_{11} & S_{12} & S_{13} \\ S_{12} & S_{22} & S_{23} \\ S_{13} & S_{23} & S_{33} \end{pmatrix} \begin{pmatrix} \sigma_{11} \\ \sqrt{2}\,\sigma_{12} \\ \sqrt{2}\,\sigma_{13} \end{pmatrix}, \qquad [9.47]$$

where:

$$S_{11} = S_{1111}, \quad S_{12} = \sqrt{2}\,S_{1112}, \quad S_{13} = \sqrt{2}\,S_{1113},$$

$$S_{22} = 2\,S_{1212}, \quad S_{23} = 2\,S_{1213}, \quad S_{33} = 2\,S_{1313}.$$

In short, relation [9.47] reads:

$$e = S\,s.$$

If the orthogonal transformations of $\mathbb{O}(2)$ are symmetries, they preserve the compliance matrix, then we are going to show that it is diagonal and $S_{22} = S_{33}$ in two steps:

– First, we claim that the compliance matrix is preserved by the mirror symmetry [9.29] with respect to the plane Ox_2x_3. According to the transformation laws [8.6] of the stresses and [9.10] of the strains, the components ε_{11} and σ_{11} are preserved while the sign of the shear components is changed, hence:

$$\begin{pmatrix} \varepsilon_{11} \\ -\sqrt{2}\,\varepsilon_{12} \\ -\sqrt{2}\,\varepsilon_{13} \end{pmatrix} = \begin{pmatrix} S_{11} & S_{12} & S_{13} \\ S_{12} & S_{22} & S_{23} \\ S_{13} & S_{23} & S_{33} \end{pmatrix} \begin{pmatrix} \sigma_{11} \\ -\sqrt{2}\,\sigma_{12} \\ -\sqrt{2}\,\sigma_{13} \end{pmatrix}.$$

Comparing to [9.47], we see that $S_{12} = S_{13} = 0$.

– Moreover, we consider as other symmetries, the rotations around the x_1's axis $x' = R_\vartheta^T x$ where:

$$R_\vartheta = \begin{pmatrix} 1 & 0 & 0 \\ 0 & \cos\vartheta & -\sin\vartheta \\ 0 & \sin\vartheta & \cos\vartheta \end{pmatrix}$$

Using [8.6] and [9.10] once again, it is easy to verify that the transformation laws of the strains and stresses read in matrix notations:

$$e' = R_\vartheta^T e, \qquad s' = R_\vartheta^T s,$$

hence, the transformation law of the compliance matrix takes the form:

$$S' = R_\vartheta^T S\, R_\vartheta,$$

that leads to $S'_{11} = S_{11}$ which is an obvious invariant and:

$$S'_{22} = S_{22}\cos^2\vartheta + S_{23}\sin(2\vartheta) + S_{33}\sin^2\vartheta,$$

$$S'_{33} = S_{33}\cos^2\vartheta - S_{23}\sin(2\vartheta) + S_{22}\sin^2\vartheta,$$

$$S'_{23} = S_{23}\cos(2\vartheta) + \frac{1}{2}(S_{33} - S_{22})\sin(2\vartheta).$$

It is easy to verify that the compliances are invariant if and only if:

$$S_{23} = 0, \qquad S_{22} = S_{33}.$$

In short, the constitutive law of the transversely rigid body has the form:

$$\varepsilon_{22} = \varepsilon_{33} = \varepsilon_{23} = 0, \quad \varepsilon_{11} = \frac{\sigma_{11}}{E}, \quad \varepsilon_{12} = \frac{\sigma_{12}}{2\mu}, \quad \varepsilon_{13} = \frac{\sigma_{13}}{2\mu}, \qquad [9.48]$$

where it is natural to choose μ given by [9.34].

9.4.3. *Calculating the local fields*

Until now, we used the displacement method by solving the Navier–Lamé equation. At this stage, we change track by solving the elasticity problem by the stress method, integrating Saint–Venant compatibility conditions [9.4] together with the local equilibrium equations. Using [14.25] and owing to [9.46] gives:

$$curl\,\varepsilon = \begin{pmatrix} \dfrac{\partial \varepsilon_{13}}{\partial x_2} - \dfrac{\partial \varepsilon_{12}}{\partial x_3} & 0 & 0 \\ \dfrac{\partial \varepsilon_{11}}{\partial x_3} - \dfrac{\partial \varepsilon_{13}}{\partial x_1} & \dfrac{\partial \varepsilon_{12}}{\partial x_3} & \dfrac{\partial \varepsilon_{13}}{\partial x_3} \\ \dfrac{\partial \varepsilon_{12}}{\partial x_1} - \dfrac{\partial \varepsilon_{11}}{\partial x_2} & -\dfrac{\partial \varepsilon_{12}}{\partial x_2} & -\dfrac{\partial \varepsilon_{13}}{\partial x_2} \end{pmatrix}. \qquad [9.49]$$

Then, the compatibility conditions [9.4] are reduced to:

$$\frac{\partial^2 \varepsilon_{11}}{\partial x_3^2} = 2\,\frac{\partial^2 \varepsilon_{13}}{\partial x_1 \partial x_3}, \quad \frac{\partial^2 \varepsilon_{11}}{\partial x_2^2} = 2\,\frac{\partial^2 \varepsilon_{12}}{\partial x_1 \partial x_2}, \qquad [9.50]$$

$$\frac{\partial^2 \varepsilon_{11}}{\partial x_2 \partial x_3} = \frac{\partial^2 \varepsilon_{12}}{\partial x_1 \partial x_3} + \frac{\partial^2 \varepsilon_{13}}{\partial x_1 \partial x_2},$$

$$\frac{\partial}{\partial x_3}\left(\frac{\partial \varepsilon_{13}}{\partial x_2} - \frac{\partial \varepsilon_{12}}{\partial x_3}\right) = 0, \quad \frac{\partial}{\partial x_2}\left(\frac{\partial \varepsilon_{13}}{\partial x_2} - \frac{\partial \varepsilon_{12}}{\partial x_3}\right) = 0. \qquad [9.51]$$

In reduced variables, the three former equations read:

$$\frac{\partial^2 \varepsilon_{11}}{\partial x_3^2} = 2\,\epsilon\,\frac{\partial^2 \varepsilon_{13}}{\partial x_1 \partial x_3}, \quad \frac{\partial^2 \varepsilon_{11}}{\partial x_2^2} = 2\,\epsilon\,\frac{\partial^2 \varepsilon_{12}}{\partial x_1 \partial x_2}, \quad \frac{\partial^2 \varepsilon_{11}}{\partial x_2 \partial x_3} = \epsilon\left(\frac{\partial^2 \varepsilon_{12}}{\partial x_1 \partial x_3} + \frac{\partial^2 \varepsilon_{13}}{\partial x_1 \partial x_2}\right).$$

Neglecting the right-hand members, we see that ε_{11} is affine with respect to x_2 and x_3 hence:

$$\varepsilon_{11} = u'_{10}(x_1) - \theta'_2(x_1)\,x_2 - \theta'_3(x_1)\,x_3, \qquad [9.52]$$

where $u_{10}, \theta_2, \theta_3$ are arbitrary functions and the prime denotes the derivative with respect to x_1. Moreover, conditions [9.51] show that:

$$\frac{\partial \varepsilon_{13}}{\partial x_2} - \frac{\partial \varepsilon_{12}}{\partial x_3} = \theta'_1(x_1) \qquad [9.53]$$

where θ_1 is an arbitrary function. Once the dominant terms are determined, we intend to solve the compatibility conditions exactly. As the left-hand members vanish in [9.50], so must the right-hand members. The condition:

$$\frac{\partial}{\partial x_3}\left(\frac{\partial \varepsilon_{12}}{\partial x_1}\right) + \frac{\partial}{\partial x_2}\left(\frac{\partial \varepsilon_{13}}{\partial x_1}\right) = 0,$$

is satisfied provided that there exists a function χ such that:

$$\frac{\partial \varepsilon_{12}}{\partial x_1} = \frac{\partial \chi}{\partial x_2}, \qquad \frac{\partial \varepsilon_{13}}{\partial x_1} = -\frac{\partial \chi}{\partial x_3}. \qquad [9.54]$$

Introducing these expressions into the two former conditions gives:

$$\frac{\partial^2 \chi}{\partial x_2^2} = \frac{\partial^2 \chi}{\partial x_3^2} = 0,$$

hence, χ is separately affine with respect to x_2 and x_3. There exist arbitrary functions $\chi_0, \gamma_1, \gamma_2, \chi_3$ such that:

$$2\chi = \chi_0'(x_1) + \gamma_2'(x_1)\, x_2 - \gamma_3'(x_1)\, x_3 - \chi_3'(x_1)\, x_2\, x_3,$$

where χ_0' must be canceled without loss of generality because ε_{12} and ε_{13} depend on χ through its partial derivative with respect to x_2 and x_3. Introducing these expressions into [9.54] and integrating lead to:

$$2\,\varepsilon_{12} = \gamma_2(x_1) - \chi_3(x_1)\, x_3 + \varphi_2(x_2, x_3),$$

$$2\,\varepsilon_{13} = \gamma_3(x_1) + \chi_3(x_1)\, x_2 + \varphi_3(x_2, x_3),$$

where φ_2, φ_3 are arbitrary functions. Introducing these expressions into [9.53] gives:

$$\frac{\partial \varphi_3}{\partial x_2}(x_2, x_3) - \frac{\partial \varphi_2}{\partial x_3}(x_2, x_3) = 2\,(\theta_1'(x_1) - \chi_3(x_1)),$$

which is satisfied for any value of the coordinates if each member of the equation is constant. Without lost of generality, we can suppose that this constant is null, hence:

$$\chi_3 = \theta_1', \qquad \frac{\partial \varphi_3}{\partial x_2} = \frac{\partial \varphi_2}{\partial x_3}.$$

There exists a function φ of x_2 and x_3 such that:

$$\varphi_2 = \frac{\partial \varphi}{\partial x_2}, \qquad \varphi_3 = \frac{\partial \varphi}{\partial x_3}.$$

Taking into account the previous two relations, we have:

$$2\,\varepsilon_{12} = \gamma_2(x_1) + \frac{\partial \varphi}{\partial x_2}(x_2, x_3) - \theta_1'(x_1)\, x_3, \qquad [9.55]$$

$$2\,\varepsilon_{13} = \gamma_3(x_1) + \frac{\partial \varphi}{\partial x_3}(x_2, x_3) + \theta_1'(x_1)\, x_2. \qquad [9.56]$$

Taking into account these two expressions, [9.52] and [9.46], the strain field reads:

$$\varepsilon = \begin{pmatrix} u'_{10} - \theta'_2 x_2 - \theta'_3 x_3 & \frac{1}{2}\left(\gamma_2 + \frac{\partial \varphi}{\partial x_2} - \theta'_1 x_3\right) & \frac{1}{2}\left(\gamma_3 + \frac{\partial \varphi}{\partial x_3} + \theta'_1 x_2\right) \\ \frac{1}{2}\left(\gamma_2 + \frac{\partial \varphi}{\partial x_2} - \theta'_1 x_3\right) & 0 & 0 \\ \frac{1}{2}\left(\gamma_3 + \frac{\partial \varphi}{\partial x_3} + \theta'_1 x_2\right) & 0 & 0 \end{pmatrix}.$$

Introducing it into [9.49] and integrating by [9.8] leads to the expression of the rotation vector of the elementary volume:

$$\omega = \begin{pmatrix} \theta_1 \\ -\theta_3 + \frac{1}{2}\left(-\gamma_3 + \dfrac{\partial \varphi}{\partial x_3} - \theta'_1 x_2\right) \\ \theta_2 + \frac{1}{2}\left(\gamma_2 - \dfrac{\partial \varphi}{\partial x_2} - \theta'_1 x_3\right) \end{pmatrix}.$$

Introducing functions u_{20} and u_{30} of x_1 such that:

$$\gamma_2 + \theta_2 = u'_{20}, \qquad \gamma_3 + \theta_3 = u'_{30},$$

and integrating by [9.5], we obtain the displacement field u of components:

$$u_1 = u_{10}(x_1) - \theta_2(x_1)\, x_2 - \theta_3(x_1)\, x_3 + \varphi(x_2, x_3), \tag{9.57}$$

$$u_2 = u_{20}(x_1) - \theta_1(x_1)\, x_3, \qquad u_3 = u_{30}(x_1) + \theta_1(x_1)\, x_2. \tag{9.58}$$

In addition, let us remark that in the stress method we also have to consider the local equilibrium equations. Combining [9.55], [9.56] and the constitutive law of the transversely rigid body [9.48], and introducing them into the axial equilibrium equation:

$$\frac{\partial \sigma_{11}}{\partial x_1} + \frac{\partial \sigma_{12}}{\partial x_2} + \frac{\partial \sigma_{13}}{\partial x_3} = 0,$$

where the volume force is neglected, we have:

$$E\,(u''_{10} - \theta''_2 x_2 - \theta''_3 x_3) + 2\,\mu\,\bar{\Delta}\,\varphi = 0,$$

which reads in reduced variables:

$$\epsilon^2\, E\,(u''_{10} - \theta''_2 x_2 - \theta''_3 x_3) + 2\,\mu\,\bar{\Delta}\,\varphi = 0,$$

neglecting the terms of order ϵ^2 gives:

$$\bar{\Delta}\,\varphi = 0,$$

hence, φ is a plane harmonic function called the *warping function*. It is worth noting that under the previous approximation the axial equilibrium equation reduces to:

$$\frac{\partial \sigma_{12}}{\partial x_2} + \frac{\partial \sigma_{13}}{\partial x_3} = 0. \qquad [9.59]$$

An alternative way to satisfy it is to introduce *Prandtl's stress function* ψ such that:

$$\sigma_{12} = \frac{\partial \psi}{\partial x_3}, \qquad \sigma_{13} = -\frac{\partial \psi}{\partial x_2}. \qquad [9.60]$$

Taking into account once again [9.48], the compatibility condition [9.53] becomes:

$$\bar{\Delta}\,\psi = -2\,\mu\,\theta_1'. \qquad [9.61]$$

Moreover, neglecting the surface force on the boundary $\mathcal{C}$ of the cross-section, we have to satisfy the free stress condition:

$$n_2\sigma_{12} + n_3\sigma_{13} = 0. \qquad [9.62]$$

It is worth noting that [9.60] means that the vector of components σ_{12} and σ_{13} is perpendicular to $\overline{grad}\,\psi$ then tangential to the curves $\psi = C^{te}$. Hence, the previous relations are satisfied provided that Prandtl's function is constant on $\mathcal{C}$ and, if the cross-section is simply connected, we can choose this constant equal to zero because the stresses depend only on its partial derivatives.

9.4.4. *Multiscale analysis: from the elementary volume to the beam*

We are now able to express the beam variables N, T, M_t and M_b in terms of the local ones:

– *Shear forces*. Introducing:

$$\bar{x} = \begin{pmatrix} x_2 \\ x_3 \end{pmatrix}, \qquad \bar{\tau} = \begin{pmatrix} \sigma_{12} \\ \sigma_{13} \end{pmatrix}, \qquad \bar{T} = \begin{pmatrix} T_2 \\ T_3 \end{pmatrix},$$

the equilibrium equations [9.59] and [9.62] read:

$$\overline{div}\,\bar{\tau} = 0, \qquad \bar{n}\cdot\bar{\tau} = 0.$$

Owing to [14.17], we have:

$$\overline{div}\,(\bar{\tau}\,\bar{x}^T) = (\overline{div}\,\bar{\tau})\,\bar{x}^T + \bar{\tau}^T,$$

where the first term of the right-hand member vanishes because of the internal equilibrium equation. Taking into account Green's formula [14.20] and [9.40], we have:

$$\bar{T} = \int_S \bar{\tau}\, dx_2 dx_3 = \int_C (\bar{n} \cdot \bar{\tau})\, \bar{x}\, ds = 0, \qquad [9.63]$$

because of the free stress condition. Hence, the expected shear forces vanish because the latter two equations [9.39] become reduced in variables:

$$\epsilon \frac{dM_{b2}}{dx_1} - T_3 + m_2 = 0, \qquad \epsilon \frac{dM_{b3}}{dx_1} + T_2 + m_3 = 0,$$

that shows the shear forces are negligeable. In particular, if φ and θ_1' are null, combining [9.63] with [9.48], [9.55] and [9.56] shows that:

$$\gamma_2 = u_{20}' - \theta_2 = 0, \qquad \gamma_3 = u_{30}' - \theta_3 = 0,$$

conditions which are generally considered as valid anyway and known as *Bernoulli's hypothesis*. The geometric interpretation is that, during the deformation, the cross-section remains orthogonal to the mean line. The displacement field [9.57], [9.58] becomes:

$$u_1 = u_{10}(x_1) - u_{20}'(x_1)\, x_2 - u_{30}'(x_1)\, x_3 + \varphi(x_2, x_3),$$
$$u_2 = u_{20}(x_1) - \theta_1(x_1)\, x_3, \qquad u_3 = u_{30}(x_1) + \theta_1(x_1)\, x_2.$$

– *Normal force*. Owing to [9.48] and [9.52], the stress fields read:

$$\sigma_{11} = E\, \varepsilon_{11} = E\, (u_{10}'(x_1) - u_{20}''(x_1)\, x_2 - u_{30}''(x_1)\, x_3). \qquad [9.64]$$

For convenience, we choose the origin in order that the *static moments* vanish:

$$\int_S x_2\, dx_2 dx_3 = \int_S x_3\, dx_2 dx_3 = 0.$$

Integrating σ_{11} on the cross-section and denoting its area by A, the relation between the normal force and the extension u_{10}' is given by [9.40]:

$$\boxed{N = E\, A\, u_{10}',} \qquad [9.65]$$

that shows the truss stiffness in [4.27] is $K_t = E\, A$.

– *Bending moments*. We define the *moment of inertia matrix* of the cross-section in its plane as:

$$\mathcal{I} = \int_S \bar{x}\,\bar{x}^T \, dx_2 dx_3.$$

As it is symmetric, it is possible to choose the axis in such a way that it is diagonal:

$$\int_S x_2 x_3 \, dx_2 dx_3 = 0, \qquad [9.66]$$

and the *principal moments of inertia* are denoted by:

$$I_2 = \int_S x_3^2 \, dx_2 dx_3, \qquad I_3 = \int_S x_2^2 \, dx_2 dx_3. \qquad [9.67]$$

Taking into account [9.42], [9.64], [9.66] and [9.67], we have:

$$\boxed{M_{b3} = E\, I_3 u_{20}'', \qquad M_{b2} = -E\, I_2 u_{30}'',} \qquad [9.68]$$

that shows the flexural stiffness in [4.31] is $K_b = E\, I_2$. Taking into account [9.65] and [9.68], the axial tensile stress is given by *Navier formula*:

$$\boxed{\sigma_{11} = \frac{N}{A} - \frac{M_{b3}}{I_3} x_2 + \frac{M_{b2}}{I_2} x_3.} \qquad [9.69]$$

– *Torque*. Owing to [9.41] and [9.60], we have:

$$M_t = -\int_S \left(x_2 \frac{\partial \psi}{\partial x_2} + x_3 \frac{\partial \psi}{\partial x_2} \right) dx_2 dx_3 = -\int_S (\overline{grad}\, \psi) \cdot \bar{x} \, dx_2 dx_3.$$

Taking into account [7.41], it holds:

$$M_t = \int_S (\psi\, \overline{div}\, \bar{x} - \overline{div}(\psi\, \bar{x})) \, dx_2 dx_3.$$

Using Green's formula [14.19], we obtain:

$$M_t = 2 \int_S \psi \, dx_2 dx_3 - \int_C (\bar{n} \cdot \bar{x})\, \psi \, ds.$$

As ψ is null on C, we see that the torque is twice the integral of Prandtl's function on the cross-section:

$$\boxed{M_t = 2 \int_S \psi \, dx_2 dx_3.} \qquad [9.70]$$

In particular, it is easy to determine Prandtl's function for a *shaft* of radius R, for example in a steam turbine. As the problem is axisymmetric and ψ must be zero on the circle C of radius R, we can try the function:

$$\psi = \frac{\mu \, \theta_1'}{2} \, (R^2 - x_2^2 - x_3^2),$$

which satisfies [9.61]. The torque carried by the shaft is:

$$M_t = \mu \, \frac{\pi \, R^4}{2} \, \theta_1'.$$

In regions near supports and concentrated forces, the previous simplified solution is perturbated by local effects. To take them into account, the characteristic length h must be considered in all the directions when building the reduced variables, the small parameter ϵ disappears and no term can be neglected. Therefore, we claim *Saint–Venant principle* below.

PRINCIPLE 9.1.– The elementary theory of beams is valid except for the regions of small size h at the vicinity of supports and concentrated forces.

10

Dynamics of 3D Continua and Elementary Mechanics of Fluids

10.1. Deformation and motion

Once again, we recover the extra dimension of time to work in the space–time with convention 1.1 on indices.

DEFINITION 10.1.– A body which is not rigid is called a *deformed body*. The material lengths and angles are in general not preserved by its motion. The motion of a deformed body is called a *deformation*.

We model the deformations into two steps:

♢ Construct the group of coordinate changes which preserve the uniform straight motions, the durations and the orientations of volumes (but not the distances, angles and volumes). We denote it by $\mathbb{GD}$ and we call it the *deformation group*.

♡ Determine the coordinate changes which are admissible with $\mathbb{GD}$ in the sense that their Jacobian matrix belongs to $\mathbb{GD}$.

♢ In the first step, we wish to characterize the most simple deformations, those which are homogeneous, which is uniform on the body. It lies in the following theorem.

THEOREM 10.1.– The coordinate changes for which are invariant.

– the uniform straight motions;

– the durations;

– the orientations of volumes;

are regular affine maps of the following form:

$$dX = P\,dX' + C, \quad C = \begin{pmatrix} \tau \\ k \end{pmatrix}, \quad P = \begin{pmatrix} 1 & 0 \\ v & F \end{pmatrix}, \qquad [10.1]$$

where $\tau \in \mathbb{R}$, $k \in \mathbb{R}^3$, $v \in \mathbb{R}^3$ and F is a matrix of $\mathbb{GL}(3)$ such that:

$$\det(F) > 0, \qquad [10.2]$$

Their set is a group containing Galileo's one.

PROOF.– It is the same as theorem 1.1 up to formula [16.47]. Taking into account that oriented volumes are transformed as:

$$V' = \det(F)V,$$

their sign is preserved because of [10.2]. The verification of the group structure of $\mathbb{GD}$ is straightforward. The Galilean transformations are particular deformations for which F is a rotation. ■

♡ Next, we wish to determine the coordinate change $X \mapsto X'$ such that:

$$\frac{\partial X}{\partial X'} = P = \begin{pmatrix} 1 & 0 \\ v & F \end{pmatrix} \in \mathbb{GD}, \qquad [10.3]$$

This partial derivative system involves $(4 \times 4 = 16)$ equations for 4 unknowns (X^α). It is overdetermined and has generally no solutions, except if the equations satisfy suitable compatibility conditions. We are faced with the same kind of problem than in section 3.3.2 (excepted that the gravitation Γ is replaced by [10.3]). In order to find the compatibility conditions, we consider a parallelogram of which the size approaches zero. Reasoning likewise, we obtain:

$$\forall dX', \delta X', \qquad \delta P\, dX' - dP\, \delta X' = 0, \qquad [10.4]$$

where P is given by [10.3]. In the following theorem, we introduce the column:

$$X' = \begin{pmatrix} t' \\ s' \end{pmatrix}. \qquad [10.5]$$

THEOREM 10.2.– Any coordinate change $X' \mapsto X$ of which the Jacobian matrix belongs to $\mathbb{GD}$ is compound of a smooth map:

$$x = \varphi(t', s'), \qquad [10.6]$$

and a clock change:

$$t = t' + \tau_0. \qquad [10.7]$$

PROOF.– Differentiating [10.3] provides:

$$\delta P\, dX' - dP\, \delta X' = \begin{pmatrix} 0 \\ \delta v\, dt' - dv\, \delta t' + \delta F\, ds' - dF\, \delta s' \end{pmatrix}.$$

The compatibility condition $\delta(dt') - d(\delta t') = 0$ is automatically fulfilled. Moreover, denoting by $d_{s'}F$ the infinitesimal variation of F resulting from the variation ds', we obtain:

$$\delta F\, ds' - dF\, \delta s' = \frac{\partial F}{\partial t'}(ds'\, \delta t' - \delta s'\, dt') + \delta_{s'} F\, ds' - d_{s'} F\, \delta s'.$$

Owing to [14.24] and after simplification, the second condition of compatibility reads:

$$\left(\frac{\partial v}{\partial s'} - \frac{\partial F}{\partial t'}\right)(\delta s'\, dt' - ds'\, \delta t') + (curl\, F^T)^T(\delta s' \times ds') = 0.$$

The infinitesimal perturbations dX' and $\delta X'$ being arbitrary, this condition is satisfied if and only if:

$$\frac{\partial v}{\partial s'} = \frac{\partial F}{\partial t'}, \qquad curl\, F^T = 0. \qquad [10.8]$$

Under these conditions, the equation system [10.3] can be integrated, let:

$$\frac{\partial t}{\partial t'} = 1, \qquad \frac{\partial t}{\partial s'} = 0, \qquad \frac{\partial x}{\partial s'} = F, \qquad \frac{\partial x}{\partial t'} = v. \qquad [10.9]$$

The integration of the two former equations leads to the clock change [10.7]. The third equation is satisfied if and only if there exists an arbitrary column field $\varphi(t', s') \in \mathbb{R}^3$ such that:

$$F = \frac{\partial \varphi}{\partial s'}. \qquad [10.10]$$

Introducing this last relation into the first condition [10.8], we obtain:

$$\frac{\partial}{\partial s'}\left(v - \frac{\partial \varphi}{\partial t'}\right) = 0.$$

There exists an arbitrary column $v_0(t') \in \mathbb{R}^3$ such that:

$$v = \frac{\partial \varphi}{\partial t'}(t', s') + v_0(t').$$

Taking into account this last relation [10.10] and the two latter equations of [10.9], we obtain, after integration:

$$x = \varphi(t', s') + \varphi_0(t'),$$

where φ_0 is a primitive of v_0. Without loss of generality, φ_0 may be absorbed in φ, leading to [10.6]. ■

Of course, the rigid motions [5.3] are particular cases:

$$x = \varphi(t', s') = R(t')s' + x_0(t').$$

For this reason, considering arbitrary deformations φ, s' will be called *Lagrangian* or *material coordinates*, by opposition to *Eulerian* or *spatial coordinates* x. People often choose φ such that:

$$s' = \varphi(0, s'),$$

that allows us to identify the Lagrangian coordinates with the initial position at $t = 0$ but this choice is not compulsory. Without lost of generality, we can forget in the sequel the change clock, putting $\tau_0 = 0$ and identifying t and t'.

On this ground, we wish to model the motion of three-dimension (3D) continua. Fixing s' in [10.6], we claim the map $t \mapsto \varphi(t, s')$ is the trajectory of the particle identified by s' and its velocity in the Eulerian representation is given by:

$$v = \frac{dx}{dt}\,|_{s'=C^{te}} = \frac{\partial \varphi}{\partial t}.$$

According to theorem 10.1, the Jacobian matrix [10.3] is composed of a boost v and a linear transformation F. Because of the implicit function theorem and taking into account [10.2] and [10.10], there exists a map $(t, x) \mapsto s' = \kappa(t, x)$ such that

$x = \varphi(t, \kappa(t, x))$, at least locally. As s' is constant along the trajectory, $s' = \kappa(t, x)$ is an integral of the motion, thus in the Lagrangian representation:

$$v' = \frac{ds'}{dt} = \frac{\partial \kappa}{\partial t} + \frac{\partial \kappa}{\partial x}\frac{dx}{dt} = 0, \qquad [10.11]$$

or with simplified notations:

$$\frac{ds'}{dt} = \frac{\partial s'}{\partial t} + \frac{\partial s'}{\partial x} v = 0,$$

Introducing the 4-velocity vector represented by [1.12], it is worth to observe that it reads:

$$\frac{\partial s'}{\partial X} U = 0. \qquad [10.12]$$

More generally, we introduce:

DEFINITION 10.2.– For any (scalar or vector) field $(t, x) \mapsto q(t, x)$, its *material derivative* or *total derivative* is:

$$\frac{dq}{dt} = \frac{\partial q}{\partial t} + \frac{\partial q}{\partial x} v = \frac{\partial q}{\partial X} U.$$

For that matter, it is interesting to work directly in the space–time to model transport phenomena such as flows and heat transfer. For any scalar field q, the corresponding flux is represented by the 4-column $q\,U$. If it is free divergence:

$$div_X\,(q\,U) = \frac{\partial q}{\partial t} + div\,(q\,v) = 0, \qquad [10.13]$$

q is conserved in the following sense. Integrating this relation on the volume $\mathcal{V}$ between the dates t_0 and t_1 and using Green formula [14.19] leads to:

$$\int_{\mathcal{V}} q\,d\mathcal{V}\mid_{t=t_1} = \int_{\mathcal{V}} q\,d\mathcal{V}\mid_{t=t_0} - \int_{t_0}^{t_1}\int_{\partial\mathcal{V}} q\,(v \cdot n)\,d\mathcal{V}\,dt.$$

Hence, the total quantity of q on $\mathcal{V}$ at the final time t_1 is equal to the corresponding quantity at the initial time t_0 minus the quantity of q lost by transport at velocity v through the boundary of $\mathcal{V}$ between these two dates.

DEFINITION 10.3.– The 3-column $q\,v$ measuring the rate of flow of a scalar q by unit area is called the *flux* of q. The corresponding space–time quantity is the *4-flux* $q\,\vec{U}$ of q. Condition [10.13] is the local expression of the *balance of* q.

10.2. Flash-back: Galilean tensors

Recalling some remarks already presented in section 1.3.4, the set of Galilean transformations is a group (even if the word "group" has not yet been used). It is called *Galileo's group* and is denoted by $\mathbb{GAL}$ in the following. As subgroup of the affine group $\mathbb{A}ff(4)$, Galileo's group naturally acts onto the tensors by restriction of their transformation laws. The $\mathbb{GAL}$-tensors are called *Galilean tensors*. In fact, we already know such tensors, for instance the 4-*velocity* $\vec{\mathbf{U}}$ of which the components in Galilean coordinate systems are modified according to the transformation law [1.11] of a vector:

$$U = \begin{pmatrix} 1 \\ v \end{pmatrix} = \begin{pmatrix} 1 & 0 \\ u & R \end{pmatrix} \begin{pmatrix} 1 \\ v' \end{pmatrix},$$

which provides the velocity addition formula $v = u + R\,v'$. The 4-velocites are Galilean vectors. Incidentally, it is worth noting the following result:

THEOREM 10.3.– Any non-vanishing Galilean vector $\vec{\boldsymbol{V}}$ is the 4-flux of $q = V^0$ in any Galilean coordinate system.

PROOF.– Let $\vec{\boldsymbol{V}}$ be represented in a Galilean coordinate system X by the 4-column:

$$V = \begin{pmatrix} q \\ w \end{pmatrix},$$

where $q \in \mathbb{R}$ and $w \in \mathbb{R}^3$. According to [1.15], its transformation law [7.24] gives:

$$q' = q, \qquad w' = R^T(w - q u). \qquad [10.14]$$

Galilean transformations leave q invariant and there is hence no problem in using q instead of q' in the following. By a method similar to the one in section 3.1.1, we annihilate w' by choosing the Galilean boost $u = w/q$. Conversely, let us consider a Galilean coordinate system X' in which the vector $\vec{\boldsymbol{V}}$ has a reduced form:

$$V' = \begin{pmatrix} q \\ 0 \end{pmatrix},$$

Along the lines of the boost method initiated in section 3.1.2, let X be another Galilean coordinate system obtained from X' through a Galilean boost v. Then, $V = q\,U$, which proves $\vec{\boldsymbol{V}}$ is the 4-flux of q. ■

If the 4-*flux* $q\,\vec{\mathbf{U}}$ is free covariant divergence:

$$\boldsymbol{Div}\,(q\,\vec{U}) = 0,$$

owing to [14.40] and [14.37], the readers can verify that the balance [10.13] of q is satisfied for a Galilean gravitation [3.38].

Let us pick up a Galilean coordinate system and consider a linear form represented by the key-row:

$$e^0 = \begin{pmatrix} 1 & 0 & 0 & 0 \end{pmatrix}. \qquad [10.15]$$

According to the transformation law [7.28], it is represented in any other Galilean coordinate system by the same key-row. We denote by $\boldsymbol{e}^0$ this Galilean linear form and we call it the *time arrow*.

Likewise, let us consider a 2-contravariant linear tensor represented in a given Galilean coordinate system by the matrix:

$$\gamma = \begin{pmatrix} 0 & 0 \\ 0 & 1_{\mathbb{R}^3} \end{pmatrix}. \qquad [10.16]$$

According to the transformation law [14.3], it is represented in any other Galilean coordinate system by the same matrix and is denoted by $\boldsymbol{\gamma}$. It is worth noting that the contracted product of the two previous tensors, represented in a Galilean coordinate system by $e^0_\alpha \gamma^{\alpha\beta}$, vanishes:

$$\boldsymbol{e}^0 \cdot \boldsymbol{\gamma} = \boldsymbol{0}.$$

Also, let us consider an object $\boldsymbol{G}$ represented in any given Galilean coordinate system by the matrix:

$$G = \begin{pmatrix} -2\phi & A^T \\ A & 1_{\mathbb{R}^3} \end{pmatrix}, \qquad [10.17]$$

where $\phi \in \mathbb{R}$ and $A \in \mathbb{R}^3$ are the potentials of the Galilean gravitation. According to their transformation law [6.22], $\boldsymbol{G}$ is a Galilean symmetric 2-covariant linear tensor because the transformation law [14.2] is satisfied.

Nevertheless, it is worth observing that $\boldsymbol{G}$ generally *is not a metric* – particularly in the simplest situation where there is no gravitation – because it is not necessarily non-degenerate.

The gravitation itself, although represented by Christoffel's symbols $\Gamma^\alpha_{\mu\beta}$, *is not a tensor* but is a covariant differential of which the transformation law is [14.35].

The study of the dynamics of particles (Chapter 3) revealed an object called the linear 4-momentum, represented in a Galilean coordinate system by the column T gathering the components T^α. T being modified in a coordinate change according to [3.30], it is a vector. Now, have a look at the definition [3.37]. We recognize the definition [14.34] of the covariant differential of a vector field. Hence, $\boldsymbol{d}T$ is nothing else but $\nabla_{dX}T$. It is not a simple change of notations insofar as the covariant differential was introduced in Chapter 3 using heuristic arguments, while the definition of Chapter 14 is more rigorous and general. The advantage of the new viewpoint offered by the tensor analysis is to write the physical laws in an intrinsic form. For instance, the general equation of the motion provided by the law 3.4 is recast in a coordinate-free style as:

$$\nabla_{\overrightarrow{dX}}\vec{\boldsymbol{T}} = \vec{\boldsymbol{H}},$$

where $\vec{\boldsymbol{T}}$ is the linear 4-momentum vector field and $\vec{\boldsymbol{H}}$ is the vector field representing the resultant of other forces.

As the force torsor, the dynamical torsor $\boldsymbol{\tau}$ of a particle or a rigid body is a skew-symmetric 2-contravariant affine tensor, the transformation law [3.4] being nothing else than [15.9]. In the affine frame $(\boldsymbol{X}_0, (\vec{\boldsymbol{e}}_\alpha))$, it is decomposed as:

$$\boldsymbol{\tau} = T^\alpha\,(\boldsymbol{X}_0 \otimes \vec{\boldsymbol{e}}_\alpha - \vec{\boldsymbol{e}}_\alpha \otimes \boldsymbol{X}_0) + J^{\alpha\beta}\,\vec{\boldsymbol{e}}_\alpha \otimes \vec{\boldsymbol{e}}_\beta.$$

When the transformation law is restricted to the Galilean transformations, it is a Galilean affine tensor. We call it *Galilean torsor*. Before going further, let us calculate the covariant differential of a torsor considered as affine tensor. Using the rule [14.51], we have:

$$\begin{aligned}\tilde{\nabla}_{\overrightarrow{dX}}\boldsymbol{\tau} = {} & dT^\alpha\,(\boldsymbol{X}_0 \otimes \vec{\boldsymbol{e}}_\alpha - \vec{\boldsymbol{e}}_\alpha \otimes \boldsymbol{X}_0) + T^\alpha\,(\boldsymbol{X}_0 \otimes \tilde{\nabla}_{\overrightarrow{dX}}\vec{\boldsymbol{e}}_\alpha - \tilde{\nabla}_{\overrightarrow{dX}}\vec{\boldsymbol{e}}_\alpha \otimes \boldsymbol{X}_0)\\ & + T^\alpha\,(\tilde{\nabla}_{\overrightarrow{dX}}\boldsymbol{X}_0 \otimes \vec{\boldsymbol{e}}_\alpha - \vec{\boldsymbol{e}}_\alpha \otimes \tilde{\nabla}_{\overrightarrow{dX}}\boldsymbol{X}_0) + dJ^{\alpha\beta}\,\vec{\boldsymbol{e}}_\alpha \otimes \vec{\boldsymbol{e}}_\beta\\ & + J^{\alpha\beta}\,(\tilde{\nabla}_{\overrightarrow{dX}}\vec{\boldsymbol{e}}_\alpha \otimes \vec{\boldsymbol{e}}_\beta + \vec{\boldsymbol{e}}_\alpha \otimes \tilde{\nabla}_{\overrightarrow{dX}}\vec{\boldsymbol{e}}_\beta).\end{aligned} \quad [10.18]$$

Taking into account the infinitesimal motion of the basis vectors [14.36] and of the origin [14.51], it holds:

$$\begin{aligned}\tilde{\nabla}_{\overrightarrow{dX}}\boldsymbol{\tau} = {} & dT^\alpha\,(\boldsymbol{X}_0 \otimes \vec{\boldsymbol{e}}_\alpha - \vec{\boldsymbol{e}}_\alpha \otimes \boldsymbol{X}_0) + \Gamma^\rho_\alpha T^\alpha\,(\boldsymbol{X}_0 \otimes \vec{\boldsymbol{e}}_\rho - \vec{\boldsymbol{e}}_\rho \otimes \boldsymbol{X}_0)\\ & + \Gamma^\beta_A T^\alpha\,(\vec{\boldsymbol{e}}_\beta \otimes \vec{\boldsymbol{e}}_\alpha - \vec{\boldsymbol{e}}_\alpha \otimes \vec{\boldsymbol{e}}_\beta) + dJ^{\alpha\beta}\,\vec{\boldsymbol{e}}_\alpha \otimes \vec{\boldsymbol{e}}_\beta\\ & + \Gamma^\rho_\alpha J^{\alpha\beta}\,\vec{\boldsymbol{e}}_\rho \otimes \vec{\boldsymbol{e}}_\beta + \Gamma^\rho_\beta J^{\alpha\beta}\,\vec{\boldsymbol{e}}_\alpha \otimes \vec{\boldsymbol{e}}_\rho,\end{aligned} \quad [10.19]$$

and, by renaming the dummy indices, we obtain:

$$\tilde{\nabla}_{\overrightarrow{dX}}\,\boldsymbol{\tau} = \tilde{\nabla}_{dX} T^{\beta}\,(\boldsymbol{X}_0 \otimes \vec{e}_{\beta} - \vec{e}_{\beta} \otimes \boldsymbol{X}_0) + \tilde{\nabla}_{dX} J^{\alpha\beta}\,\vec{e}_{\alpha} \otimes \vec{e}_{\beta}, \qquad [10.20]$$

with:

$$\tilde{\nabla}_{dX} T^{\beta} = dT^{\beta} + \Gamma^{\beta}_{\rho} T^{\rho}, \qquad [10.21]$$

$$\tilde{\nabla}_{dX} J^{\alpha\beta} = dJ^{\alpha\beta} + \Gamma^{\alpha}_{\rho} J^{\rho\beta} + \Gamma^{\beta}_{\rho} J^{\alpha\rho} + \Gamma^{\alpha}_{A} T^{\beta} - T^{\alpha}\Gamma^{\beta}_{A}. \qquad [10.22]$$

In matrix form, the covariant differential of a torsor field is given by:

$$\tilde{\nabla}_{dX} T = dT + \Gamma(dX)\,T,$$

$$\tilde{\nabla}_{dX} J = dJ + \Gamma(dX)\,J + J\,(\Gamma(dX))^T + \Gamma_A(dX)\,T^T - T\,(\Gamma_A(dX))^T,$$

We recover, in a more rigorous framework, the covariant differential [5.43] of the dynamical torsor of a rigid body introduced in a heuristic way. Introducing the resultant torsor of the other forces (i.e. different from the gravitation):

$$\boldsymbol{\tau}^{\star} = H^{\alpha}\,(\boldsymbol{X}_0 \otimes \vec{e}_{\alpha} - \vec{e}_{\alpha} \otimes \boldsymbol{X}_0) + G^{\alpha\beta}\,\vec{e}_{\alpha} \otimes \vec{e}_{\beta},$$

the generalized equation of rigid body motion given by the law 5.3 can be recast in a coordinate-free style as:

$$\tilde{\nabla}_{\overrightarrow{dX}}\,\boldsymbol{\tau} = \boldsymbol{\tau}^{\star}.$$

In this regard, it is worth to recalling that, according to section 5.3.2, the map $\boldsymbol{A}$ of section 14.5.3 was choosen as the identity. Thus, putting $A = 1_{\mathbb{R}^4}$ in [14.47], we recover [5.6] in the form:

$$\Gamma_A(dX) = dX - \nabla_{dX} C.$$

Returning to tensor notations, [14.50] reads:

$$\Gamma^{\alpha}_{A\beta} = \delta^{\alpha}_{\beta} - \nabla_{\beta} C^{\alpha}. \qquad [10.23]$$

10.3. Dynamical torsor of a 3D continuum

We would like to define the dynamical torsor of a continuum. Imitating the model of the statics developed in Chapter 9, we extend the notion of torsor in a space–time framework under the form of a vector-valued torsor, according to definition 8.3. In an affine frame $(\boldsymbol{X}_0, (\vec{e}_\alpha))$, it is decomposed as:

$$\boldsymbol{\tau} = \boldsymbol{\tau}^\gamma \vec{e}_\gamma, \qquad \boldsymbol{\tau}^\gamma = T^{\beta\gamma}\,(\boldsymbol{X}_0 \otimes \vec{e}_\beta - \vec{e}_\beta \otimes \boldsymbol{X}_0) + J^{\alpha\beta\gamma}\,\vec{e}_\alpha \otimes \vec{e}_\beta.$$

Let $(\boldsymbol{X}'_0, (\vec{e}\,'_\alpha))$ be a new affine frame obtained from the old one through an affine transformation $a = (C, P)$. Hence, the transformation law of the torsor is:

$$T'^{\rho\sigma} = (P^{-1})^\rho_\beta (P^{-1})^\sigma_\gamma T^{\beta\gamma}, \qquad [10.24]$$

$$J'^{\rho\sigma\tau} = \Big[(P^{-1})^\rho_\alpha (P^{-1})^\sigma_\beta J^{\alpha\beta\gamma} + C'^\rho\,\{(P^{-1})^\sigma_\beta T^{\beta\gamma}\} - \Big\{(P^{-1})^\rho_\beta T^{\beta\gamma}\Big\}\,C'^\sigma\Big]\,(P^{-1})^\tau_\gamma. \qquad [10.25]$$

with: $C' = -P^{-1}C$.

Next, let us calculate the covariant differential of the continuum torsor:

$$\tilde{\nabla}_{\overrightarrow{dX}}\,\boldsymbol{\tau} = \tilde{\nabla}_{\overrightarrow{dX}}\,(\boldsymbol{\tau}^\gamma \vec{e}_\gamma) = (\tilde{\nabla}_{\overrightarrow{dX}}\,\boldsymbol{\tau}^\gamma)\,\vec{e}_\gamma + \boldsymbol{\tau}^\gamma\,(\tilde{\nabla}_{\overrightarrow{dX}}\,\vec{e}_\gamma).$$

Taking into account [14.36], we have:

$$\tilde{\nabla}_{\overrightarrow{dX}}\,\boldsymbol{\tau} = (\tilde{\nabla}_{\overrightarrow{dX}}\,\boldsymbol{\tau}^\gamma + \Gamma^\gamma_\rho \boldsymbol{\tau}^\rho)\,\vec{e}_\gamma.$$

The calculation of the first term of the right-hand member is similar to the one [10.20] of the scalar valued dynamical torsor (add indices γ in [10.20], [10.21] and [10.22]). Finally, we obtain:

$$\tilde{\nabla}_{\overrightarrow{dX}}\,\boldsymbol{\tau} = \Big[\tilde{\nabla}_{dX} T^{\beta\gamma}\,(\boldsymbol{X}_0 \otimes \vec{e}_\beta - \vec{e}_\beta \otimes \boldsymbol{X}_0) + \tilde{\nabla}_{dX} J^{\alpha\beta\gamma}\,\vec{e}_\alpha \otimes \vec{e}_\beta\Big]\,\vec{e}_\gamma,$$

with:

$$\tilde{\nabla}_{dX} T^{\beta\gamma} = dT^{\beta\gamma} + \Gamma^\beta_\rho T^{\rho\gamma} + \Gamma^\gamma_\rho T^{\beta\rho},$$

$$\tilde{\nabla}_{dX} J^{\alpha\beta\gamma} = dJ^{\alpha\beta\gamma} + \Gamma^\alpha_\rho J^{\rho\beta\gamma} + \Gamma^\beta_\rho J^{\alpha\rho\gamma} + \Gamma^\gamma_\rho J^{\alpha\beta\rho} + \Gamma^\alpha_A T^{\beta\gamma} - T^{\alpha\gamma}\Gamma^\beta_A.$$

It is worth recalling that coefficients Γ^α_ρ represent the gravitation, as discussed at length in Chapter 3 devoted to the dynamics. Hence, there exists a field $\tilde{\nabla}\,\boldsymbol{\tau}$ of 1-covariant and 3-contravariant affine tensors such that:

$$\tilde{\nabla}_{\overrightarrow{dX}}\,\boldsymbol{\tau} = (\tilde{\nabla}\,\boldsymbol{\tau}) \cdot \overrightarrow{dX}.$$

Using Christoffel's symbols [14.38] and additional symbols [14.49], we have:

$$\tilde{\nabla}\boldsymbol{\tau} = \left[\tilde{\nabla}_\sigma T^{\beta\gamma}\,(\boldsymbol{X}_0 \otimes \vec{\boldsymbol{e}}_\beta - \vec{\boldsymbol{e}}_\beta \otimes \boldsymbol{X}_0) + \tilde{\nabla}_\sigma J^{\alpha\beta\gamma}\,\vec{\boldsymbol{e}}_\alpha \otimes \vec{\boldsymbol{e}}_\beta\right]\vec{\boldsymbol{e}}_\gamma \otimes \boldsymbol{e}^\sigma,$$

with:

$$\tilde{\nabla}_\sigma T^{\beta\gamma} = \frac{\partial T^{\beta\gamma}}{\partial X^\sigma} + \Gamma^\beta_{\sigma\rho} T^{\rho\gamma} + \Gamma^\gamma_{\sigma\rho} T^{\beta\rho},$$

$$\tilde{\nabla}_\sigma J^{\alpha\beta\gamma} = \frac{\partial J^{\alpha\beta\gamma}}{\partial X^\sigma} + \Gamma^\alpha_{\sigma\rho} J^{\rho\beta\gamma} + \Gamma^\beta_{\sigma\rho} J^{\alpha\rho\gamma} + \Gamma^\gamma_{\sigma\rho} J^{\alpha\beta\rho} + \Gamma^\alpha_{A\sigma} T^{\beta\gamma} - T^{\alpha\gamma}\Gamma^\beta_{A\sigma}.$$

By contraction, we define the *covariant divergence* of the dynamical tensor of the continuum:

$$\tilde{\boldsymbol{Div}}\,\boldsymbol{\tau} = \tilde{\nabla}_\gamma T^{\beta\gamma}\,(\boldsymbol{X}_0 \otimes \vec{\boldsymbol{e}}_\beta - \vec{\boldsymbol{e}}_\beta \otimes \boldsymbol{X}_0) + \tilde{\nabla}_\gamma J^{\alpha\beta\gamma}\,\vec{\boldsymbol{e}}_\alpha \otimes \vec{\boldsymbol{e}}_\beta. \qquad [10.26]$$

with:

$$\tilde{\nabla}_\gamma T^{\beta\gamma} = \frac{\partial T^{\beta\gamma}}{\partial X^\gamma} + \Gamma^\beta_{\gamma\rho} T^{\rho\gamma} + \Gamma^\gamma_{\gamma\rho} T^{\beta\rho}, \qquad [10.27]$$

$$\tilde{\nabla}_\gamma J^{\alpha\beta\gamma} = \frac{\partial J^{\alpha\beta\gamma}}{\partial X^\gamma} + \Gamma^\alpha_{\gamma\rho} J^{\rho\beta\gamma} + \Gamma^\beta_{\gamma\rho} J^{\alpha\rho\gamma}$$

$$+\Gamma^\gamma_{\gamma\rho} J^{\alpha\beta\rho} + \Gamma^\alpha_{A\gamma} T^{\beta\gamma} - T^{\alpha\gamma}\Gamma^\beta_{A\gamma}. \qquad [10.28]$$

At this stage, have a break to look back to the static of 3D continua (just a matter to erase the time dimension). In the absence of gravitation, Christoffel's symbols vanish and we recover the covariant affine divergence of the torsor field given by [8.28] and [8.29].

We are now able to generalize the local equation [8.27] to the dynamics of 3D continua. Although there is *a priori* a degree of arbitrariness in this choice, we decide to consider a *generalized Cauchy medium*, claiming that:

– There exists a field of vectors $\boldsymbol{X} \mapsto \vec{\boldsymbol{H}}(\boldsymbol{X})$ representing the resultant of other forces (i.e. different from the gravitation) and its torsor is:

$$\boldsymbol{\tau}_{\vec{\boldsymbol{H}}} = \boldsymbol{X} \otimes \vec{\boldsymbol{H}}(\boldsymbol{X}) - \vec{\boldsymbol{H}}(\boldsymbol{X}) \otimes \boldsymbol{X},$$

$$\boldsymbol{\tau}_{\vec{\boldsymbol{H}}} = H^\beta\,(\boldsymbol{X}_0 \otimes \vec{\boldsymbol{e}}_\beta - \vec{\boldsymbol{e}}_\beta \otimes \boldsymbol{X}_0) + G^{\alpha\beta}\,\vec{\boldsymbol{e}}_\alpha \otimes \vec{\boldsymbol{e}}_\beta, \qquad [10.29]$$

with:

$$G^{\alpha\beta} = X^\alpha H^\beta - H^\alpha X^\beta, \qquad [10.30]$$

– There exists a field of linear 2-contravariant tensors $\boldsymbol{X} \mapsto \boldsymbol{T}(\boldsymbol{X})$. The corresponding vector valued torsor τ_T is such that for any covector $\boldsymbol{N}$:

$$\boldsymbol{N}(\tau_T) = \boldsymbol{X} \otimes (\boldsymbol{T} \cdot \boldsymbol{N}) - (\boldsymbol{T} \cdot \boldsymbol{N}) \otimes \boldsymbol{X}. \qquad [10.31]$$

– The motion of the 3D continuum is governed by the free-coordinate equation:

$$\boxed{\tilde{\boldsymbol{Div}}\, \tau_T + \tau_{\vec{H}} = \boldsymbol{0}} \qquad [10.32]$$

These assumptions are justified insofar as the predictions agree with a very wide spectrum of experimental observations. We start with a straightforward result:

THEOREM 10.4.– A torsor satisfies [10.31] if and only if its components $J^{\alpha\beta\gamma}$ in the affine frame $(\boldsymbol{X}, (\vec{e}_\alpha))$ vanish.

PROOF.– Using the transformation laws [10.24] and [10.25], we argue by analogy with the proof of theorem 8.3. ■

Next, we consider the simple case where:

– we put $C^\alpha = 0$ in [10.23], which amounts to restrict ourself to proper coordinate systems (in the sense of section 3.1.2) where the elementary volume is at rest;

– we are working in the affine frame $(\boldsymbol{X}, (\vec{e}_\alpha))$, then $J^{\alpha\beta\gamma}$ components of τ_T and $G^{\alpha\beta}$ components of $\tau_{\vec{H}}$ vanish.

Taking into account [10.28], the law [10.32] leads to:

$$\boxed{T^{\beta\alpha} - T^{\alpha\beta} = 0.} \qquad [10.33]$$

We leave it to the readers to verify that it is true in every affine frame, owing to the transformation law [10.24].

10.4. The stress-mass tensor

10.4.1. *Transformation law and invariants*

Decomposing [10.31] in an affine frame, we verify that:

$$J^{\alpha\beta\gamma} = X^\alpha T^{\beta\gamma} - T^{\alpha\gamma} X^\beta.$$

Hence, the structure of the dynamical torsor of the continuum is fixed by the one of the $T^{\alpha\gamma}$ components of the 2-contravariant linear tensor $\boldsymbol{T}$. For the sake of easiness, they are gathered into a symmetric matrix T. Their transformation law [10.24] is given in matrix form by [14.3]:

$$T' = P^{-1} T \, P^{-T}. \qquad [10.34]$$

Taking into account the structure of the space–time, T is decomposed by blocks:

$$T = \begin{pmatrix} \rho & p^T \\ p & -\sigma_\star \end{pmatrix},$$

where ρ is a scalar, $p \in \mathbb{R}^3$ and $\sigma_\star \in \mathbb{M}_{33}^{symm}$. From now on, we consider $\boldsymbol{T}$ as a Galilean tensor by restriction of its transformation law to Galilean transformations. Owing to their decomposition by block [1.15], the transformation law [10.34] itemizes in:

$$\boxed{\rho' = \rho,} \qquad [10.35]$$

$$\boxed{p' = R^T(p - \rho\, u),} \qquad [10.36]$$

$$\boxed{\sigma'_\star = R^T\left(\sigma_\star + u\, p^T + p\, u^T - \rho\, u\, u^T\right) R.} \qquad [10.37]$$

It is worth noting that Galilean transformations preserve the component ρ and there is no trouble to put ρ instead of ρ' in the following. To find the other invariants of $\boldsymbol{T}$, we follow the method applied in section 3.1.1 to the dynamical torsor of a particle. We discuss only the case that the invariant component ρ does not vanish. Starting in any Galilean coordinate system X, we choose the Galilean boost:

$$u = \frac{p}{\rho},$$

which annihilates p' and reduces [10.37] to:

$$\sigma'_\star = R^T \left(\sigma_\star + \frac{1}{\rho}\, p\, p^T\right) R. \qquad [10.38]$$

This suggests to cast a glance to the matrix:

$$\sigma = \sigma_\star + \frac{1}{\rho}\, p\, p^T. \qquad [10.39]$$

Taking into account [10.35], [10.36] and [10.37], we obtain its transformation law:

$$\sigma' = R^T \sigma\, R. \qquad [10.40]$$

As the matrix σ is symmetric, it is diagonalizable then its eigenvalues $\sigma_1, \sigma_2, \sigma_3$ are real numbers and the missing invariants of $\boldsymbol{T}$.

10.4.2. *Boost method*

Conversely, let us consider a Galilean coordinate system in which the tensor field $\boldsymbol{T}$ at a given point of coordinates X' has a *reduced form*:

$$T' = \begin{pmatrix} \rho & 0 & 0 & 0 \\ 0 & -\sigma_1 & 0 & 0 \\ 0 & 0 & -\sigma_2 & 0 \\ 0 & 0 & 0 & -\sigma_3 \end{pmatrix} = \begin{pmatrix} \rho & 0 \\ 0 & -\sigma' \end{pmatrix},$$

where the diagonal matrix σ' containing the σ_i is the matrix [10.39] in the coordinates X'. In the spirit of the boost method initiated in section 3.1.2, we claim now the elementary volume around the point x' is at rest at time t'. Let X be another Galilean coordinate system obtained from X' through a Galilean boost v combined with a rotation R. Applying the inverse transformation law of [10.34]:

$$T = P\, T'\, P^T, \qquad [10.41]$$

we obtain:

$$p = \rho\, v, \qquad \sigma_\star = R\, \sigma'\, R^T - \rho\, v\, v^T.$$

The boost method turns out the physical meaning of the components:

– by analogy with section 3.1.2, the invariant quantity ρ is interpreted as the mass per unit volume or *density* (*of mass*);

– the quantity p, product of the density and velocity, is the *linear momentum* (per unit volume);

– eliminating the velocity v between the two previous relations and taking into account [10.40], we recover [10.39]. Besides, we recognize in [10.40] the transformation law 8.6 of Euclidean stress tensors. Thus, σ can be identified to the *statical stresses*, while:

$$\sigma_\star = \sigma - \rho\, v\, v^T,$$

are the *dynamical stresses*.

DEFINITION 10.4.– The *stress-mass tensor* $\boldsymbol{T}$ is structured into three components:

– the *density* ρ;

– the *linear momentum* p;

– the *dynamical stresses* $\sigma_\star$.

The invariants of the stress-mass tensor are:

– the *density* ρ;

– the *principal stresses* $\sigma_1, \sigma_2, \sigma_3$.

In matrix form, the stress-mass tensor reads:

$$\boxed{T = \begin{pmatrix} \rho & p^T \\ p & -\sigma_\star \end{pmatrix} = \begin{pmatrix} \rho & \rho\, v^T \\ \rho\, v & \rho\, v\, v^T - \sigma \end{pmatrix}.} \qquad [10.42]$$

Before going further, let us stop to adapt the boost method according to the motion of the particles modeled in section 10.1. Let us consider the coordinate system X' associated with the Lagrangian representation, given by [10.5]. It is not generally Galilean. Because of [10.11], the stress–mass tensor reads in the Lagrangian representation:

$$T' = \begin{pmatrix} \rho' & 0 \\ 0 & -\sigma' \end{pmatrix}, \qquad [10.43]$$

but the matrix σ' of *material stresses* is not in general diagonal. Applying the inverse transformation law [10.34] with the boost $u = v$, we obtain:

$$\rho = \rho', \qquad p = \rho\, v, \qquad \sigma_\star = \sigma - \rho\, v\, v^T, \qquad [10.44]$$

where *spatial stresses* σ in the Eulerian representation are related to the material stresses according to:

$$\sigma = F\, \sigma'\, F^T. \qquad [10.45]$$

10.5. Euler's equations of motion

We claimed that the motions of 3D continua are [10.32] which, taking into account [10.36], are structured into two groups:

$$\tilde{\nabla}_\gamma T^{\beta\gamma} + H^\beta = 0, \qquad \tilde{\nabla}_\gamma J^{\alpha\beta\gamma} + G^{\alpha\beta} = 0. \qquad [10.46]$$

The second group led to the symmetry conditions [10.33] of the stress–mass tensor. It remains to examine the consequences of the first group, owing to the structure [10.42] of the stress-mass tensor in Galilean coordinate systems which reads in tensor notations:

$$T^{00} = \rho, \qquad T^{0j} = T^{j0} = \rho v^j, \qquad T^{ij} = \rho v^i v^j - \sigma^{ij}. \qquad [10.47]$$

Excluding thrusts, we assume that the other forces (different from the gravitation) are modeled by 4-column H of the form [3.76]:

$$H = \begin{pmatrix} 0 \\ -f_v \end{pmatrix},$$

where f_v is the volume force introduced by definition [8.2]. In indicial notations, we have:

$$H^0 = 0, \qquad H^j = -f_v^j. \qquad [10.48]$$

Besides, the gravitation being represented in a Galilean coordinate system by [2.23], the non-vanishing Christoffel's symbols are:

$$\Gamma^j_{00} = -g^j, \qquad \Gamma^j_{0k} = \Gamma^j_{k0} = \Omega^j_k, \qquad [10.49]$$

where Ω^j_k is a simplified notation for the elements of the skew-symmetric matrix $j(\Omega)$ defined by [7.10]. Taking into account [10.27], the first group of [10.46] reads:

$$\tilde{\nabla}_\gamma T^{\beta\gamma} + H^\beta = \frac{\partial T^{\beta\gamma}}{\partial X^\gamma} + \Gamma^\beta_{\gamma\rho} T^{\rho\gamma} + \Gamma^\gamma_{\gamma\rho} T^{\beta\rho} + H^\beta = 0. \qquad [10.50]$$

By putting $\beta = 0$ which corresponds to the time coordinate, we obtain:

$$\tilde{\nabla}_\gamma T^{0\gamma} + H^0 = \frac{\partial T^{00}}{\partial t} + \frac{\partial T^{0j}}{\partial x^j} = \frac{\partial \rho}{\partial t} + \frac{\partial}{\partial x^j}(\rho v^j) = 0. \qquad [10.51]$$

For the spatial coordinates, we put $\beta = i$, that gives, taking into account the vanishing terms:

$$\tilde{\nabla}_\gamma T^{i\gamma} + H^i = \frac{\partial T^{i0}}{\partial t} + \frac{\partial T^{ij}}{\partial x^j} + \Gamma^i_{00} T^{00} + \Gamma^i_{0k} T^{k0} + \Gamma^i_{k0} T^{0k} - f^i_v = 0,$$

$$\tilde{\nabla}_\gamma T^{i\gamma} + H^i = \frac{\partial}{\partial t}(\rho\, v^i) + \frac{\partial}{\partial x^j}(\rho\, v^i v^j - \sigma^{ij}) - \rho\, g^i + 2\rho\, \Omega^i_k\, v^k - f^i_v = 0,$$

or, after differentiation and rearrangement:

$$\rho\, \frac{\partial v^i}{\partial t} + \left[\frac{\partial \rho}{\partial t} + \frac{\partial}{\partial x^j}(\rho\, v^j)\right] v^i + \rho\, v^j\, \frac{\partial v^i}{\partial x^j} = \frac{\partial \sigma^{ij}}{\partial x^j} + f^i_v + \rho\,(g^i - 2\Omega^i_k\, v^k).$$

Taking into account [10.51] leads to:

$$\rho\, \frac{\partial v^i}{\partial t} + \rho\, v^j\, \frac{\partial v^i}{\partial x^j} = \frac{\partial \sigma^{ij}}{\partial x^j} + f^i_v + \rho\,(g^i - 2\Omega^i_k\, v^k). \qquad [10.52]$$

In short, the first group of [10.46] is recast as [10.51] and [10.52] that can be formulated in matrix notation as follows.

LAW 10.1.– The motion of a 3D continuum obeys *Euler's equations*:

♢ *Balance of mass*:

$$\boxed{\frac{\partial \rho}{\partial t} + div\,(\rho\, v) = 0} \qquad [10.53]$$

♡ *Balance of linear momentum*:

$$\boxed{\rho\, \left[\frac{\partial v}{\partial t} + \frac{\partial v}{\partial x}\, v\right] = (div\, \sigma)^T + f_v + \rho(g - 2\,\Omega \times v)} \qquad [10.54]$$

This requires some comments:

– This formulation is consistent with Galileo's principle of relativity 1.1 in the sense that the form of these equations is the same in all the Galilean coordinate systems. It is ensured particularly to the last term in [10.54].

♢ Comparing [10.35], [10.36] and [10.14] shows that:

$$N = \begin{pmatrix} \rho \\ p \end{pmatrix},$$

represents a Galilean vector $\vec{\boldsymbol{N}}$ and, according to theorem 10.3, it is the 4-flux of mass density ρ:

$$\vec{\boldsymbol{N}} = \rho \vec{\boldsymbol{U}}.$$

Equation [11.50] means that this field is divergence free:

$$\boldsymbol{Div}\, \vec{\boldsymbol{N}} = 0,$$

Taking into account definition 10.2 of the material derivative and [7.41], it is worth noting that it can also read:

$$\frac{\partial \rho}{\partial t} + div\,(\rho\, v) = \frac{d\rho}{dt} + \rho\, div\, v = 0. \qquad [10.55]$$

We wish to show it traduces the balance of mass. Let ρ_0 be the value of the density at the date $t = 0$ of the particle identified by the Lagrangian coordinates s'. Then, ρ_0 is a function of s'. As ρ is the mass by unit volume, the actual value at the date t is:

$$\rho = \frac{\rho_0}{det\ F} = \rho_0\, det\left(\frac{\partial s'}{\partial x}\right). \qquad [10.56]$$

Next, we need the following proposition:

LEMMA 10.1.– For every motion, we have:

$$\frac{d}{dt}\left(\frac{\partial s'}{\partial x}\right) + \frac{\partial s'}{\partial x}\frac{\partial v}{\partial x} = 0. \qquad [10.57]$$

PROOF.– Let δX be any uniform vector field on $\mathcal{M}$ and U be the 4-velocity. Differentiating [10.12] gives:

$$\frac{\partial}{\partial X}\left(\frac{\partial s'}{\partial X}\, U\right)\,\delta X = \delta\left(\frac{\partial s'}{\partial X}\, U\right) = \delta\left(\frac{\partial s'}{\partial X}\right)\, U + \frac{\partial s'}{\partial X}\,\delta U = 0,$$

$$= \frac{\partial}{\partial X}\,(\delta s')\, U + \frac{\partial s'}{\partial X}\frac{\partial U}{\partial X}\,\delta X = \frac{\partial}{\partial X}\left(\frac{\partial s'}{\partial X}\,\delta X\right)\, U + \frac{\partial s'}{\partial X}\frac{\partial U}{\partial X}\,\delta X = 0.$$

Consequently:

$$\frac{d}{dt}\left(\frac{\partial s'}{\partial X}\,\delta X\right) = \frac{\partial}{\partial X}\left(\frac{\partial s'}{\partial X}\,\delta X\right)\, U = -\frac{\partial s'}{\partial X}\frac{\partial U}{\partial X}\,\delta X.$$

As δX is uniform, we have:

$$\frac{d}{dt}\left(\frac{\partial s'}{\partial X}\right)\,\delta X = -\frac{\partial s'}{\partial X}\frac{\partial U}{\partial X}\,\delta X.$$

As δX is arbitrary, it holds:

$$\frac{d}{dt}\left(\frac{\partial s'}{\partial X}\right) = -\frac{\partial s'}{\partial X}\frac{\partial U}{\partial X},$$

or, in detail:

$$\frac{d}{dt}\left(\frac{\partial s'}{\partial t}\ \frac{\partial s'}{\partial x}\right) = -\left(\frac{\partial s'}{\partial t}\ \frac{\partial s'}{\partial x}\right)\begin{pmatrix} 0 & 0 \\ \frac{\partial\, v}{\partial t} & \frac{\partial\, v}{\partial x} \end{pmatrix},$$

that proves [10.57]. ■

Then, we prove the following proposition:

THEOREM 10.5.– If ρ_0 is a smooth function of s', then the density of mass:

$$\rho = \rho_0\,(s')\,det\left(\frac{\partial s'}{\partial x}\right), \qquad [10.58]$$

verifies the balance of mass [11.50].

PROOF.– Differentiating [10.58], we have:

$$\frac{d\rho}{dt} = \frac{\partial \rho}{\partial s'}\frac{ds'}{dt} + Tr\left(\frac{\partial \rho}{\partial\left(\frac{\partial s'}{\partial x}\right)}\frac{d}{dt}\left(\frac{\partial s'}{\partial x}\right)\right),$$

or, owing to [10.11] and [14.32]:

$$\frac{d\rho}{dt} = \rho\, Tr\left(\frac{\partial x}{\partial s'}\frac{d}{dt}\left(\frac{\partial s'}{\partial x}\right)\right). \qquad [10.59]$$

Then, introducing [10.57] into [10.59], it holds:

$$\frac{d\rho}{dt} = -\rho\, Tr\ (grad\ v\,) = -\rho\ div\ v,$$

that proves the balance of mass [11.50]. ■

♡ Applying definition 10.2 of the material derivative, equation [10.54] is reduced to:

$$\boxed{\rho\,\frac{dv}{dt} = (div\ \sigma)^T + f_v + \rho(g - 2\,\Omega\times v).} \qquad [10.60]$$

It clearly appears as representing the balance of linear momentum for a unit volume because the product of the density of mass by the acceleration dv/dt is equal to the sum of the gravitational forces, the other external forces f_v and the internal forces given by the divergence of the static stresses.

Another relevant equation is the balance of energy, a simple consequence of one of the linear momentum. By taking the scalar product of both members of [11.51] by v, we have:

$$\rho \frac{dv}{dt} \cdot v = (div\ \sigma)\ v + (f_v + \rho g) \cdot v,$$

the scalar triple product in the last term disapearing because v occurs twice in it. Owing to [14.15] and the symmetry of σ, we obtain the *balance of energy*:

$$\boxed{\rho \frac{d}{dt}\left(\frac{1}{2} \parallel v \parallel^2\right) = div\,(\sigma v) - Tr\,(\sigma D) + (f_v + \rho g) \cdot v,} \qquad [10.61]$$

where the strain velocity [10.14] occurs. This equation shows us that the time rate of kinetic energy is balanced by the divergence of the *stress transport* σv, the opposite of the internal power by volume unit, the power of the volume forces and the gravity.

10.6. Constitutive laws in dynamics

A *constitutive law* is a relation describing the mechanical behavior of a material (solid or fluid). We already know the most simple of them, Hooke's law, but there is a very large spectrum of phenomenological behaviors that cannot be deduced from simple axioms of the statics or dynamics of continua. Their accurate description requires additional information resulting from experimental testing. Nevertheless, it is possible to determine general conditions satisfied by them in order to be consistent with Galileo's principle of relativity.

If the strains are small, we claimed that for elastic bodies in statics, the stress tensor representing the internal forces is proportional to the strain tensor associated with the motion of the continuum. In dynamics, the stress tensor is generalized in the form of the stress-mass tensor T while the motion of the continuum can be described by s' and $\partial s'/\partial X$ as explained in section 10.1. Provisionally forgetting the dependence with respect to s', we would like to specify the conditions of consistency of constitutive laws modeled by a map:

$$T = \mathcal{F}\left(\frac{\partial s'}{\partial X}\right).$$

They are given by the *principle of material indifference*. The key idea is to have a look at the Lagrangian representation.

THEOREM 10.6.– In the Lagrangian representation, the stress-mass tensor T' depends on $\partial s'/\partial X$ through *right Cauchy strains* $C = F^T F$.

PROOF.– Let X be any Galilean coordinate system and X' be the coordinate system 10.5 associated with the Lagrangian coordinates s'. According to theorem 10.1, the Jacobian matrix of the coordinate change $X' \mapsto X$ is a deformation:

$$\frac{\partial X}{\partial X'} = \begin{pmatrix} 1 & 0 \\ v & F \end{pmatrix} \in \mathbb{GD}. \qquad [10.62]$$

For another Galilean coordinate system $\bar{X}$, the Jacobian matrix of the coordinate change $X' \mapsto \bar{X}$ is also a deformation:

$$\frac{\partial \bar{X}}{\partial X'} = \begin{pmatrix} 1 & 0 \\ \bar{v} & \bar{F} \end{pmatrix} \in \mathbb{GD}.$$

Then, we have:

$$\frac{\partial \bar{X}}{\partial X'} = \frac{\partial \bar{X}}{\partial X} \frac{\partial X}{\partial X'},$$

where the Jacobian matrix of the coordinate change $X \mapsto \bar{X}$ is a linear Galilean transformation:

$$\frac{\partial \bar{X}}{\partial X'} = \begin{pmatrix} 1 & 0 \\ u & R \end{pmatrix},$$

that provides:

$$\bar{v} = u + R\,v, \qquad \bar{F} = R\,F. \qquad [10.63]$$

As T' are the components of the stress–mass tensor $\mathbf{T}$ in the coordinate system X' associated with the Lagrangian coordinates s', it is preserved by any change $X \mapsto \bar{X}$ of Galilean coordinate system. Hence, if there is some constitutive law:

$$T' = \mathcal{F}'\left(\frac{\partial s'}{\partial X}\right),$$

the value T' must depend on the variable $\partial s'/\partial X$ through a quantity:

$$C = \mathcal{C}\left(\frac{\partial s'}{\partial X}\right),$$

invariant by Galilean transformations. In other words, we claim that:

$$\left\{ \mathcal{C}\left(\frac{\partial s'}{\partial X}\right) = \mathcal{C}\left(\frac{\partial s'}{\partial \bar{X}}\right) \right\} \quad \Rightarrow \quad \left\{ \mathcal{F}'\left(\frac{\partial s'}{\partial X}\right) = \mathcal{F}'\left(\frac{\partial s'}{\partial \bar{X}}\right) \right\}.$$

There exists a unique map $\tilde{\mathcal{F}}' = \mathcal{F}'/\mathcal{C}$ from $\mathbb{M}_{33}^{symm}$ into $\mathbb{M}_{44}^{symm}$ such that:

$$\mathcal{F}' = \tilde{\mathcal{F}}' \mathcal{C}.$$

It reminds to guess the invariant C. As the variable $\partial s'/\partial X$ is obtained by erasing the first row in the inverse $\partial X'/\partial X$ of the Jacobian matrix [10.62]:

$$\frac{\partial s'}{\partial X} = \left(-F^{-1} v \; F^{-1} \right), \qquad [10.64]$$

we wish to identify invariants of v and F, variables which are transformed under Galilean transformations according to [10.63]. As this was done in section 3.1.1 for the dynamical torsor, we intend annihilating some of them by suitable Galilean transformations (because zeros are constant and then obvious invariants). Starting in any coordinate system X, we pick $u = -R\,v$ which annihilates $\bar{v}$. There is nothing more to do because the rotation R obviously cannot annihilate $\bar{F}$. Hence, the expected invariant depends only on F. We can verify that $C = F^T F$ are invariants since:

$$\bar{C} = \bar{F}^T \bar{F} = (R\,F)^T (R\,F) = F^T (R^T R)\, F = F^T F = C,$$

that achieves the proof. ■

It is worth noting that the matrix C of right Cauchy strains is symmetric and, owing to [10.2]:

$$J = \det(F) = \sqrt{\det(C)}.$$

Recovering the variable s', owing to [10.43], [10.56] and according to theorem 10.5, the constitutive law in the Lagrangian representation reads:

$$\rho' = \frac{\rho_0(s')}{\sqrt{\det(C)}}$$

$$\sigma' = \sigma'(s', C). \qquad [10.65]$$

If the continuum is *homogeneous*, i.e. the material properties are the same at each point, ρ' and σ' do not depend explicitly on s'. Otherwise, it is *heterogeneous*.

We can now return to the Eulerian representation of the constitutive law due to the transformation laws [10.44] and [10.45] of the stress-mass tensor:

$$T = \begin{pmatrix} \frac{\rho_0(s')}{\sqrt{\det(C)}} & \frac{\rho_0(s')}{\sqrt{\det(C)}}\, v^T \\ \frac{\rho_0(s')}{\sqrt{\det(C)}}\, v & \frac{\rho_0(s')}{\sqrt{\det(C)}}\, v\, v^T - F\,\sigma'(s', C)\, F^T \end{pmatrix}.$$

As an example of such constitutive laws, we can quote the *barotropic fluids* of which the material stresses read:

$$\sigma' = -q\,(s', \det(C))\, C^{-1}, \qquad [10.66]$$

where q is a given scalar valued function representing the *pressure*. The corresponding stress-mass tensor in Eulerian representation is:

$$T = \begin{pmatrix} \frac{\rho_0(s')}{\sqrt{\det(C)}} & \frac{\rho_0(s')}{\sqrt{\det(C)}}\, v^T \\ \frac{\rho_0(s')}{\sqrt{\det(C)}}\, v & \frac{\rho_0(s')}{\sqrt{\det(C)}}\, v\, v^T + q\,(s', \det(C))\, 1_{\mathbb{R}^3} \end{pmatrix}. \qquad [10.67]$$

Owing to [14.16], the balance of linear momentum reads for barotropic fluids:

$$\boxed{\rho\, \frac{dv}{dt} = -grad\, q + f_v + \rho(g - 2\,\Omega \times v).} \qquad [10.68]$$

If the time is fixed at a given date t, we can consider only *spatial tensors*. From the time, Galileo's group is reduced to the group of special Euclidean transformations preserving the components of the covariant metric tensor $\breve{\boldsymbol{G}}$. The material stresses σ'^{ij}, gathered in the matrix σ', are the components in the Lagrangian representation of a 2-contravariant tensor $\boldsymbol{\sigma}$. On the other hand, theorem 10.2 showed that the position of the body at time t is given by a map:

$$\varphi_t : s' \mapsto x = \varphi_t(s') = \varphi(t, s'),$$

which is the local expression in given charts of a map φ_t. Its tangent map $\boldsymbol{F}$ is represented in the coordinates s' and x by the matrix F. To find the free coordinate version of the definition $C = F^T F$, let us remark that it reads in tensorial notations:

$$C_{ij} = \delta_{kl} F^k_i F^l_j.$$

Thus, right Cauchy strains are the components of a 2-covariant spatial tensor $\boldsymbol{C}$ which is the pull-back by φ_t of the metric tensor $\breve{\boldsymbol{G}}$:

$$\boldsymbol{C} = \varphi_t^* \breve{\boldsymbol{G}}$$

Omitting the dependence with respect to s', we can, therefore, write the constitutive law [10.65] in a coordinate-free form:

$$\boldsymbol{\sigma} = \boldsymbol{H}(\boldsymbol{C}). \qquad [10.69]$$

Having a look at definition 9.2 of the strain tensor suggests to express as usual $\boldsymbol{\sigma}$ with respect to $\boldsymbol{C}$ through:

$$\boldsymbol{E} = \frac{1}{2}\left[\varphi_t^* \breve{\boldsymbol{G}} - \breve{\boldsymbol{G}}\right] = \frac{1}{2}\left[\boldsymbol{C} - \breve{\boldsymbol{G}}\right],$$

called *Euler-Lagrange strain tensor*. In every Galilean coordinate system, Gram's matrix being the identity, it is represented by:

$$E = \frac{1}{2}\left(C - 1_{\mathbb{R}^3}\right) = \frac{1}{2}\left(F^T F - 1_{\mathbb{R}^3}\right). \qquad [10.70]$$

It gives a new version of the constitutive law [10.69]:

$$\boldsymbol{\sigma} = \boldsymbol{H}(\boldsymbol{E}). \qquad [10.71]$$

10.7. Hyperelastic materials and barotropic fluids

Another example is the *hyperelastic materials* of which the behavior is reversible. In linear elasticity (section 9.3.1), we formulated the reversibility condition in terms of internal work by unit volume, which does not create difficulties because the strains are very small and the volume is almost unchanged. On the other hand, for fluids and solids exhibiting very large deformations, it is preferable to relate the internal power to the mass to model the constitutive law and to formulate the reversibility condition in terms of *specific internal work*, that is of internal work by mass unit $\mathcal{P}_{int}/\rho$. Hence, we claim that for any loop in the space of Euler–Lagrange strain tensors:

$$\oint Tr\left(\frac{\sigma'}{\rho}\, dE\right) = 0, \qquad [10.72]$$

where σ' is given by:

$$\sigma' = H'(E),$$

and ρ is also depending on E through C. Reasoning as in section 3.3.2, we obtain the equivalent local condition:

$$\forall dE, \delta E, \qquad d\left(\frac{\sigma'}{\rho}\right) : \delta E - \delta\left(\frac{\sigma'}{\rho}\right) : dE = 0,$$

As in section 9.3.1, there exists a *reversible energy potential* e_{int} generating the constitutive law in terms of material stresses:

$$\sigma' = \rho \frac{\partial e_{int}}{\partial E}. \qquad [10.73]$$

Moreover, differentiating [10.70], we have $dC = 2\,dE$ and taking into account the definition [14.30] of the derivative with respect to a matrix, we have:

$$\sigma' = 2\,\rho \frac{\partial e_{int}}{\partial C}. \qquad [10.74]$$

Combining [10.45] with [10.73] and [10.74] gives the constitutive law in terms of spatial stresses:

$$\sigma = \rho\, F \frac{\partial e_{int}}{\partial E} F^T = 2\rho\, F \frac{\partial e_{int}}{\partial C} F^T. \qquad [10.75]$$

Let us show that barotropic fluids are particular cases of hyperelastic materials. Indeed, let us choose a potential e_{int} depending on C through $J = \det(F) = \sqrt{\det(C)}$. Applying the chain rule to calculate [10.74]:

$$\sigma' = 2\,\rho \frac{de_{int}}{dJ} \frac{\partial J}{\partial(\det(C))} \frac{\partial}{\partial C}(\det(C)).$$

Taking into account [10.56] and [14.32] leads to:

$$\sigma' = \rho_0 \frac{de_{int}}{dJ} C^{-1},$$

which can be identified to [10.66] provided:

$$q = -\rho_0 \frac{de_{int}}{dJ}. \qquad [10.76]$$

Let us return now to the general case of hyperelastic materials. First, let us remark that [10.57] reads:

$$\frac{d}{dt}(F^{-1}) = -F^{-1} \frac{\partial v}{\partial x},$$

that, owing to [14.33], leads to:

$$\frac{\partial v}{\partial x} = \frac{dF}{dt} F^{-1}. \quad [10.77]$$

and:

$$\frac{dF}{dt} = \frac{\partial v}{\partial x} F.$$

Hence, differentiating [10.70] gives:

$$\frac{dE}{dt} = \frac{1}{2}\frac{dC}{dt} = \frac{1}{2}\left(\frac{dF}{dt}^T F + F^T \frac{dF}{dt}\right) = \frac{1}{2} F^T \left[\frac{\partial v}{\partial x} + \left(\frac{\partial v}{\partial x}\right)^T\right] F,$$

or, introducing the strain velocity [10.14]:

$$\frac{dE}{dt} = \frac{1}{2}\frac{dC}{dt} = F^T D\, F. \quad [10.78]$$

Remembering the dependence of e_{int} with respect to s' and differentiating it provides:

$$\rho \frac{de_{int}}{dt} = \rho \frac{\partial e_{int}}{\partial s'}\frac{ds'}{dt} + \rho Tr\left(\frac{\partial e_{int}}{\partial E}\frac{dE}{dt})\right),$$

or, owing to [10.11] and [10.78]:

$$\rho \frac{de_{int}}{dt} = \rho Tr\left(\frac{\partial e_{int}}{\partial E} F^T D\, F\right) = Tr\left(\rho F \frac{\partial e_{int}}{\partial E} F^T D\right),$$

that is, taking into account the expression [10.75] of the spatial stresses, the internal power by volume unit [9.13]:

$$\rho \frac{de_{int}}{dt} = Tr\,(\sigma D), \quad [10.79]$$

We are now able to transform the general expression of the balance of energy [10.61] for the particular case of hyperelastic materials with zero volume forces f_v, that gives:

$$\rho \frac{d}{dt}\left(\frac{1}{2} \parallel v \parallel^2\right) = div\,(\sigma v) - Tr\,(\sigma D) + \rho g \cdot v.$$

Taking into account [10.79] and expressing the gravity in terms of the Galilean gravitation potentials due to [6.14]:

$$\rho \frac{d}{dt}\left(\frac{1}{2} \parallel v \parallel^2 + e_{int}\right) = div\,(\sigma v) - \rho \left(grad\,\phi + \frac{\partial A}{\partial t}\right) \cdot v,$$

or:

$$\rho \frac{d}{dt}\left(\frac{1}{2} \parallel v \parallel^2 + \phi + e_{int}\right) = div\,(\sigma v) + \rho \left(\frac{\partial \phi}{\partial t} - \frac{\partial A}{\partial t} \cdot v\right). \qquad [10.80]$$

Inspired by the Hamiltonian [6.18] of a particle subjected to a Galilean gravitation and introducing the reversible energy potential, we define the *Hamiltonian density* as:

$$\mathcal{H} = \rho \left(\frac{1}{2} \parallel v \parallel^2 + \phi + e_{int}\right). \qquad [10.81]$$

or $\mathcal{H} = \rho\eta$ after introducing, for the sake of simplicity the *specific Hamiltonian*:

$$\eta = \frac{1}{2} \parallel v \parallel^2 + \phi + e_{int}. \qquad [10.82]$$

Taking into account the balance of mass [10.55], we have:

$$\frac{d\mathcal{H}}{dt} = \frac{d}{dt}(\rho\eta) = \rho \frac{d\eta}{dt} + \frac{d\rho}{dt}\eta = \rho\left(\frac{d\eta}{dt} - \eta\, div\, v\right),$$

which, owing to definition 10.2 of the material derivative, gives:

$$\rho \frac{d\eta}{dt} = \frac{d\mathcal{H}}{dt} + \rho\eta\, div\, v = \frac{\partial \mathcal{H}}{\partial t} + \frac{\partial \mathcal{H}}{\partial x} v + \mathcal{H}\, div\, v,$$

and, because of [7.41]:

$$\rho \frac{d\eta}{dt} = \frac{\partial \mathcal{H}}{\partial t} + div\,(\mathcal{H} v). \qquad [10.83]$$

Introducing this expression into [10.80], we obtain a new version of the balance of energy dedicated to the hyperelastic materials and in particular to the barotropic fluids:

$$\boxed{\frac{\partial \mathcal{H}}{\partial t} + div\ (\mathcal{H} v - \sigma v) = \rho\left(\frac{\partial \phi}{\partial t} - \frac{\partial A}{\partial t} \cdot v\right).} \qquad [10.84]$$

For a time independent gravitation field, the right hand member vanishes.

11

Dynamics of Continua of Arbitrary Dimensions

11.1. Modeling the motion of one-dimensional (1D) material bodies

The mathematical modeling developed in the present chapter can be given for the dynamics of media of other dimensions such as 4. For instance, we consider in Chapter 4 slender bodies called arches, modeled as a curve of the space, which is a continuum of dimension 1, submanifold of the physical space of dimension 3. We can extend this mathematical construction to the dynamics. The motion of a 1D material body – solid of fluid – can be described by a space–time submanifold parametrized by two coordinates, the arclength and the time, hence by a submanifold $\mathcal{N}$ of dimension 2. Let s_0 be the arclength with respect to a given reference point of the initial slender body (i.e. at $t = 0$). In the *Lagrangian coordinates* t and s_0 and the Galilean coordinates X, the position at time t of the material particle pinpointed by s_0 is represented by:

$$x = \varphi(t, s_0).$$

Its velocity is:

$$v = \frac{\partial \varphi}{\partial t}. \qquad [11.1]$$

Let s be the arclength with respect to a given reference point of the same slender body at time t. In the *Eulerian coordinates* t and s, the motion is represented by:

$$x = \psi(t, s).$$

We claim there exists a function:

$$s_0 = f(t, s), \qquad [11.2]$$

monotonically strictly increasing with respect to s:

$$\frac{\partial f}{\partial s} > 0, \qquad [11.3]$$

such that:

$$x = \psi(t, s) = \varphi(t, f(t, s)). \qquad [11.4]$$

In the Eulerian representation, each point $\boldsymbol{\xi}$ of the submanifold $\mathcal{N}$ has local coordinates:

$$\xi = \begin{pmatrix} \xi^0 \\ \xi^1 \end{pmatrix} = \begin{pmatrix} t \\ s \end{pmatrix},$$

according to the convention 1.1. Let us determine the tangent map $\boldsymbol{U}$ to the injection $i : \boldsymbol{\xi} \mapsto \boldsymbol{X}$ of $\mathcal{N}$ into the space–time $\mathcal{M}$. Differentiating [11.4] and taking into account [11.1], it is represented by:

$$\begin{pmatrix} dt \\ dx \end{pmatrix} = \begin{pmatrix} 1 & 0 \\ v + \dfrac{\partial f}{\partial t}\dfrac{\partial \varphi}{\partial s_0} & \dfrac{\partial f}{\partial s}\dfrac{\partial \varphi}{\partial s_0} \end{pmatrix} \begin{pmatrix} dt \\ ds \end{pmatrix}.$$

As s is the arclength, $ds = \| dx \|$ at constant t, hence:

$$n = \frac{\partial \psi}{\partial s} = \frac{\partial f}{\partial s}\frac{\partial \varphi}{\partial s_0},$$

is the unit tangent vector to the curve. Differentiating [11.2] at constant s_0 and owing to [11.3], we obtain the tangential component of the velocity:

$$v_t = v \cdot n = \frac{ds}{dt}\,|_{s_0 = C^{te}} = -\frac{\partial f}{\partial t} \Big/ \frac{\partial f}{\partial s}.$$

Thus, the tangent map $\boldsymbol{U}$ is represented by the 4×2 matrix:

$$U = \begin{pmatrix} 1 & 0 \\ v - v_t\, n & n \end{pmatrix}. \qquad [11.5]$$

11.2. Group of the 1D linear Galilean transformations

The group $\mathbb{GAL}_0$ of the linear Galilean transformations P acts on the components dX of the tangent vectors to the space–time $\overrightarrow{d\boldsymbol{X}} \in T_{\boldsymbol{X}}\,\mathcal{M}$ by:

$$dX = P\,dX'. \qquad [11.6]$$

It forwards an action on the components $d\xi$ of the tangent vectors to the submanifold $\overrightarrow{d\boldsymbol{\xi}} \in T_{\boldsymbol{\xi}}\,\mathcal{N}$:

$$d\xi = S\,d\xi'. \qquad [11.7]$$

The tangent map:

$$\overrightarrow{d\boldsymbol{X}} = \boldsymbol{U}\,\overrightarrow{d\boldsymbol{\xi}} \qquad [11.8]$$

is represented in the Galilean coordinate system X (respectively, X') by:

$$dX = U\,d\xi \quad (\text{resp. } dX' = U'\,d\xi'). \qquad [11.9]$$

Combining [11.6], [11.7] and [11.9] leads to:

$$\forall d\xi', \qquad U\,S\,d\xi' = P\,U'\,d\xi',$$

hence, for any linear Galilean transformation P, there exists a 2×2 matrix:

$$S = \begin{pmatrix} \alpha & \beta \\ \gamma & \delta \end{pmatrix},$$

such that:

$$U\,S = P\,U'.$$

Putting:

$$U' = \begin{pmatrix} 1 & 0 \\ v' - v'_t\,n' & n' \end{pmatrix},$$

and owing to [1.4], we have:

$$\begin{aligned} &\alpha = 1, \qquad \beta = 0, \qquad \beta\,(v - v_t n) + \delta\,n = R\,n', \\ &\alpha\,(v - v_t n) + \gamma\,n = u + R\,(v' - v'_t n'). \end{aligned} \qquad [11.10]$$

from which it results $\delta\, n = R\, n'$ then $\delta = 1$ since the unit tangent vector components change according to:

$$n = R\, n'. \qquad [11.11]$$

Condition [11.10] reads:

$$v - (u + R\, v') - (v_t - \gamma - v'_t)\, n = 0.$$

The transformation law [11.11] and the velocity addition formula [1.13] entail $v_t = u_t + v'_t$, hence the previous relation is fulfilled provided that $\gamma = u_t$. Finally, at each linear Galilean transformation, [1.4] is associated with a unique matrix:

$$S = \begin{pmatrix} 1 & 0 \\ u_t & 1 \end{pmatrix}, \qquad [11.12]$$

with $u_t = u \cdot n$. The set of such matrices is an abelian group for the matrix product. We call it the *group of 1D linear Galilean transformation*. We obtained a linear representation of Galileo's group into this group. Likewise, we could associate a group of two-dimensional (2D) Galilean transformations with a submanifold $\mathcal{N}$ of dimension 2 representing the motion of a material surface in dynamics.

The space–time $\mathcal{M}$ is equipped with a connection representing the gravitation and induces on the submanifold $\mathcal{N}$ a connection that we denote also by Γ but with left-hand indices, according to the above convention 11.1. We want to characterize the maps $d\xi \mapsto dS = \Gamma(d\xi)$ of which the values are infinitesimal 1D Galilean transformations and satisfying the condition [3.36]. Differentiating the expression of S in [11.12], a 1D infinitesimal Galilean transformation around the identity reads:

$$dS = \begin{pmatrix} 0 & 0 \\ du_t & 0 \end{pmatrix}.$$

where $du_t = B\, ds - g_t\, dt$. The condition [3.36] is fulfilled provided that $B\,(ds\, \delta t - \delta s\, dt) = 0$, hence B is null. The induced connection representing the *1D Galilean gravitation* reads:

$$\Gamma(d\xi) = \begin{pmatrix} 0 & 0 \\ -g_t dt & 0 \end{pmatrix}, \qquad [11.13]$$

where g_t is interpreted as the gravity in the direction tangent to the material line.

11.3. Torsor of a continuum of arbitrary dimension

Our aim is to develop a general approach for the dynamics of material bodies of dimension d represented by a submanifold of dimension $d+1$ of the space–time. It can be given for:

– $d = 0$: the dynamics of material particles and rigid bodies represented by a curve, already presented in Chapters 3 and 5;

– $d = 1$: the dynamics of 1D material bodies (arch if solid, flow in a pipe or jet if fluid) represented by a surface, sketched in the previous section;

– $d = 2$: the dynamics of 2D material bodies (plate or shell if solid, sheet of fluid) represented by a volume (we leave it to the readers to do);

– $d = 3$: the dynamics of 3D material bodies, represented by a submanifold of dimension 4, already presented in Chapter 10.

The readers interested in geometric approaches for shells can consult [VAL 95]. In order to distinguish the indices related to the submanifold and the space–time, we put the former ones at the left-hand side of the symbol, adopting the following convention:

CONVENTION 11.1.– Coordinate labels:

– Left-hand Greek indices run over the submanifold coordinate labels $0, \cdots, d$.

– Right-hand Greek indices run over the four space–time coordinate labels $0, \cdots, 3$.

while the convention 1.1 concerning the Latin indices is still valid.

According to definition 8.3, we can attach to a continuum of arbitrary dimension d a field of torsors $\boldsymbol{\xi} \mapsto \boldsymbol{\tau}(\boldsymbol{\xi})$ on a submanifold $\mathcal{N}$ of dimension $d+1$ of which the values are tangent vectors to $\mathcal{N}$ at $\boldsymbol{\xi}$. More precisely, if i is the injection of $\mathcal{N}$ into the space–time $\mathcal{M}$, the torsor $\boldsymbol{\tau}(\boldsymbol{\xi})$ is a bilinear skew-symmetric affine tensor from $A^*T_{\boldsymbol{X}}\mathcal{M} \times A^*T_{\boldsymbol{X}}\mathcal{M}$ into $T_{\boldsymbol{\xi}}\mathcal{N}$ with $\boldsymbol{X} = i(\boldsymbol{\xi})$. In an affine frame $(\boldsymbol{X}_0, (\vec{e}_\alpha))$ of $T_{\boldsymbol{X}}\mathcal{M}$ and a basis $({}_\gamma\vec{\eta})$ of $T_{\boldsymbol{\xi}}\mathcal{N}$, it is decomposed as:

$$\boldsymbol{\tau} = {}^\gamma\boldsymbol{\tau}\, {}_\gamma\vec{\eta}, \qquad {}^\gamma\boldsymbol{\tau} = {}^\gamma T^\beta \left(\boldsymbol{X}_0 \otimes \vec{e}_\beta - \vec{e}_\beta \otimes \boldsymbol{X}_0\right) + {}^\gamma J^{\alpha\beta}\, \vec{e}_\alpha \otimes \vec{e}_\beta .$$

Let $(\boldsymbol{X}'_0, (\vec{e}\,'_\alpha))$ be a new affine frame obtained from the old one through an affine transformation $a = (C, P)$ and $({}_\gamma\vec{\eta}\,')$ be a new basis obtained from the old one through a linear transformation S. Hence, the transformation law of the torsor is:

$${}^\sigma T'^\rho = (P^{-1})^\rho_\beta\, {}^\sigma_\gamma(S^{-1})\, {}^\gamma T^\beta, \qquad [11.14]$$

$${}^\tau J'^{\rho\sigma} = \Big[(P^{-1})^\rho_\alpha (P^{-1})^\sigma_\beta\, {}^\gamma J^{\alpha\beta} + C'^\rho \left\{(P^{-1})^\sigma_\beta\, {}^\gamma T^\beta\right\} - \left\{(P^{-1})^\rho_\beta\, {}^\gamma T^\beta\right\} C'^\sigma\Big]\, {}^\tau_\gamma(S^{-1}). \qquad [11.15]$$

with: $C' = -P^{-1}C$. Hence, the ${}^{\gamma}T^{\alpha}$ components are the ones of a 1-contravariant vector valued linear tensor $\boldsymbol{T}$. Let T be the matrix of which ${}^{\gamma}T^{\alpha}$ is the element at the intersection of the γ-th row and α-th column. In matrix form, [11.14] reads:

$$T' = S^{-1} T \, P^{-T}. \qquad [11.16]$$

11.4. Force–mass tensor of a 1D material body

Taking into account the structure of the space–time and the submanifold $\mathcal{N}$ of dimension 2, the 2×4 matrix T is decomposed by blocks:

$$T = \begin{pmatrix} \rho_l & p^T \\ p_t & -F_\star^T \end{pmatrix}, \qquad [11.17]$$

where ρ_l, p_t are scalar, $p, F_\star \in \mathbb{R}^3$. From now on, we consider $\boldsymbol{T}$ as a Galilean tensor by restriction of its transformation law to Galilean transformations. Owing to their decompositions [1.15] and [11.12], the transformation law [11.16] itemizes in:

$$\boxed{\rho_l' = \rho_l,} \qquad [11.18]$$

$$\boxed{p' = R^T (p - \rho_l \, u),} \qquad [11.19]$$

$$\boxed{p_t' = p_t - \rho_l \, u_t,} \qquad [11.20]$$

$$\boxed{F_\star' = R^T \left(F_\star + p_t u + u_t p - \rho_l \, u_t u \right).} \qquad [11.21]$$

By projecting [11.19] onto n' and by taking into account [11.11], we recover [11.20], then p_t is nothing else but the tangential component of p. It is worth noting that Galilean transformations preserve the component ρ_l. To find the other invariants of $\boldsymbol{T}$, we follow the method of section 3.1.1. Starting in any Galilean coordinate system X, we choose the Galilean boost $u = p \, / \, \rho_l$ which annihilates p', p_t' and reduces [11.21] to:

$$F_\star' = R^T \left(F_\star + \frac{1}{\rho_l} \, p_t p \right). \qquad [11.22]$$

This suggests casting a glance over the column:

$$F = F_\star + \frac{1}{\rho_l}\, p_t p. \qquad [11.23]$$

Taking into account [11.18], [11.19], [11.20] and [11.21], we obtain its transformation law:

$$F' = R^T F. \qquad [11.24]$$

The missing invariant of $\boldsymbol{T}$ is $\| F \|$.

Conversely, let us consider a Galilean coordinate system in which the tensor field $\boldsymbol{T}$ at a given point of coordinates X' has a *reduced form*:

$$T' = \begin{pmatrix} \rho_l & 0 \\ 0 & -F'^T \end{pmatrix},$$

In the spirit of the boost method initiated in section 3.1.2, we claim now the line element around the point x' is at rest at time t'. Let X be another Galilean coordinate system obtained from X' through a Galilean boost v combined with a rotation R. Applying the inverse transformation law of [11.16]:

$$T \; = S\, T' P^T,$$

we obtain, taking into account [11.24]:

$$p = \rho_l v, \qquad p_t = \rho_l v_t, \qquad F_\star = F - \rho_l v_t v.$$

The boost reveals the physical meaning of the components:

– by analogy with section 3.1.2, the invariant quantity ρ_l is interpreted as the *mass per unit length* or density;

– the quantity p, product of the density and velocity, is the *linear momentum* (per unit length);

– we recognize in [11.24] the transformation law [2.10] of forces. Thus, F can be identified to the *statical forces*, while:

$$F_\star = F - \rho_l v_t v,$$

are the *dynamical forces*.

DEFINITION 11.1.– The *force–mass tensor* $\boldsymbol{T}$ is structured into four components:

– the *density* ρ_l;

– the *linear momentum* p;

– the *tangential linear momentum* p_t;

– the *dynamical forces* $F_\star$.

The invariants of the stress–mass tensor are:

– the *density* ρ_l;

– the *statical force norm* $\| F \|$.

In matrix form, the force–mass tensor reads:

$$\boxed{T = \begin{pmatrix} \rho_l & p^T \\ p_t & -F_\star^T \end{pmatrix} = \begin{pmatrix} \rho_l & \rho_l v^T \\ \rho_l v_t & (\rho_l v_t v - F)^T \end{pmatrix}.} \qquad [11.25]$$

11.5. Full torsor of a 1D material body

According to [8.19], we have seen in statics that the torsor of a 3D continuum is entirely determined by the stress tensor $\boldsymbol{\sigma}$, reason for which it was denoted by $\boldsymbol{\tau}_\sigma$ (Cauchy's continuum). We said that it was a stress torsor. This structure was extended to the dynamical torsor of a 3D continuum $\boldsymbol{\tau}_T$ completely depending on the stress–mass tensor $\boldsymbol{T}$ (generalized Cauchy continuum).

On the other hand, the torsor of an arch [4.1] is completely arbitrary with components F and M independent of each other (in other words, it is not a force torsor) and *a fortiori* the dynamical torsor $\boldsymbol{\tau}$ of a 1D material body is not entirely determined by the force–mass tensor $\boldsymbol{T}$. Hence, components ${}^\gamma J^{\alpha\beta}$ are independent of ${}^\gamma T^\beta$. Such media are called *Cosserat continua* [COS 07]. So, it is for 2D material bodies but we will not treat them (we leave it to the readers to do).

In order to detail the transformation law [11.16], we introduce:

– the matrix ${}^\gamma J$ of which ${}^\gamma J^{\alpha\beta}$ is the element at the intersection of the α-th row and β-th column;

– the row ${}^\gamma T$ of which ${}^\gamma T^\beta$ is the β-th element.

Hence, the transformation law [11.16] can be decomposed into two steps:

– Calculate the intermediate matrices:

$${}^\gamma J^* = P^{-1}\, {}^\gamma J\, P^{-T} + C'\, {}^\gamma T^* - ({}^\gamma T^*)^T\, C'^T,$$

where

$$ {}^{\gamma}T^{*} = {}^{\gamma}T\,P^{-T}, \qquad [11.26] $$

– Calculate the new component matrices ${}^{\tau}J' = {}^{\gamma}J^{*}\,{}^{\tau}_{\gamma}(S^{-1})$ that, taking into account [11.12], itemize in:

$$ {}^{0}J' = {}^{0}J^{*}, \qquad {}^{1}J' = {}^{1}J^{*} - u_t\,{}^{0}J^{*}. $$

As ${}^{\gamma}T$ are the rows of [11.17], [11.26] and [1.15] give:

$$ {}^{0}T^{*} = \left(\rho_l \; (p - \rho_l u)^T R \right), $$

$$ {}^{1}T^{*} = \left(p_t \; -(F_\star + p_t u)^T R \right). $$

Also, we put:

$$ {}^{0}J = \begin{pmatrix} 0 & -q^T \\ q & -j\,(l) \end{pmatrix}, \qquad {}^{1}J = \begin{pmatrix} 0 & -l_\star^T \\ l_\star & -j\,(M_\star) \end{pmatrix}, \qquad [11.27] $$

Hence, the transformation law [11.16] itemizes in:

$$ \boxed{q' = R^T q + \rho_l k' - \tau_0' p',} \qquad [11.28] $$

$$ \boxed{l' = R^T(l + u \times q) + k' \times p',} \qquad [11.29] $$

$$ \boxed{l_\star' = R^T(l_\star + \tau_0'(F_\star + p_t u)) + p_t k' - u_t q',} \qquad [11.30] $$

$$ \boxed{M_\star' = R^T(M_\star + u \times l_\star) - k' \times (R^T(F_\star + p_t u)) - u_t l',} \qquad [11.31] $$

where p' is given by [11.19].

11.6. Equations of motion of a continuum of arbitrary dimension

We now want to calculate the covariant differential of the continuum torsor:

$$\tilde{\nabla}_{\vec{d\xi}}\,\boldsymbol{\tau} = \tilde{\nabla}_{\vec{d\xi}}\,({}^{\gamma}\boldsymbol{\tau}\,{}_{\gamma}\vec{\boldsymbol{\eta}}) = (\tilde{\nabla}_{U\,\vec{d\xi}}\,{}^{\gamma}\boldsymbol{\tau})\,{}_{\gamma}\vec{\boldsymbol{\eta}} + {}^{\gamma}\boldsymbol{\tau}\,(\tilde{\nabla}_{\vec{d\xi}}\,{}_{\gamma}\vec{\boldsymbol{\eta}}).$$

Taking into account [14.36], we have:

$$\tilde{\nabla}_{\vec{d\xi}}\,\boldsymbol{\tau} = (\tilde{\nabla}_{U\,\vec{d\xi}}\,{}^{\gamma}\boldsymbol{\tau} + {}^{\gamma}_{\rho}\Gamma\,{}^{\rho}\boldsymbol{\tau})\,{}_{\gamma}\vec{\boldsymbol{\eta}}. \qquad [11.32]$$

Calculating the first term of the right-hand member is similar to the one [10.20] of the scalar valued dynamical torsor (add left-hand indices γ in [10.20], [10.21] and [10.22]):

$$\tilde{\nabla}_{U\vec{d\xi}}\,{}^{\gamma}\boldsymbol{\tau} = \tilde{\nabla}_{U\,d\xi}\,{}^{\gamma}T^{\beta}\,(\boldsymbol{X}_0 \otimes \vec{\boldsymbol{e}}_{\beta} - \vec{\boldsymbol{e}}_{\beta} \otimes \boldsymbol{X}_0) + \tilde{\nabla}_{U\,d\xi}\,{}^{\gamma}J^{\alpha\beta}\,\vec{\boldsymbol{e}}_{\alpha} \otimes \vec{\boldsymbol{e}}_{\beta}, \qquad [11.33]$$

with:

$$\tilde{\nabla}_{U\,d\xi}\,{}^{\gamma}T^{\beta} = d^{\gamma}T^{\beta} + \Gamma^{\beta}_{\rho}\,{}^{\gamma}T^{\rho}, \qquad [11.34]$$

$$\tilde{\nabla}_{U\,d\xi}\,{}^{\gamma}J^{\alpha\beta} = d\,{}^{\gamma}J^{\alpha\beta} + \Gamma^{\alpha}_{\rho}\,{}^{\gamma}J^{\rho\beta} + \Gamma^{\beta}_{\rho}\,{}^{\gamma}J^{\alpha\rho} + \Gamma^{\alpha}_{A}\,{}^{\gamma}T^{\beta} - {}^{\gamma}T^{\alpha}\Gamma^{\beta}_{A}. \qquad [11.35]$$

Hence, there exists a field $\tilde{\nabla}\,\boldsymbol{\tau}$ of 1-covariant and 3-contravariant affine tensors such that:

$$\tilde{\nabla}_{\vec{d\xi}}\,\boldsymbol{\tau} = (\tilde{\nabla}\,\boldsymbol{\tau}) \cdot \vec{d\boldsymbol{\xi}}. \qquad [11.36]$$

Expanding [11.34] and [11.35] gives:

$$\tilde{\nabla}_{U\,d\xi}\,{}^{\gamma}T^{\beta} = \left(\frac{\partial({}^{\gamma}T^{\beta})}{\partial({}^{\mu}\xi)} + {}_{\mu}U^{\sigma}\,\Gamma^{\beta}_{\sigma\rho}\,{}^{\gamma}T^{\rho}\right) d({}^{\mu}\xi),$$

$$\tilde{\nabla}_{U\,d\xi}\,{}^{\gamma}J^{\alpha\beta} = \left(\frac{\partial({}^{\gamma}J^{\alpha\beta})}{\partial({}^{\mu}\xi)} + {}_{\mu}U^{\sigma}\,\left(\Gamma^{\alpha}_{\sigma\rho}\,{}^{\gamma}J^{\rho\beta} + \Gamma^{\beta}_{\sigma\rho}\,{}^{\gamma}J^{\alpha\rho}\right.\right.$$
$$\left.\left. + \Gamma^{\alpha}_{A\sigma}\,{}^{\gamma}T^{\beta} - {}^{\gamma}T^{\alpha}\Gamma^{\beta}_{A\sigma}\right)\right) d({}^{\mu}\xi).$$

Due to the latter two expressions, we transform [11.33] and, taking into account:

$${}^{\gamma}_{\rho}\Gamma(d\xi) = {}^{\gamma}_{\mu\rho}\Gamma\,d({}^{\mu}\xi),$$

the expression [11.32] can be expressed as a linear function of:

$$d({}^{\mu}\xi) = {}^{\mu}\boldsymbol{\eta}\,(\vec{d\boldsymbol{\xi}}) = {}^{\mu}\boldsymbol{\eta} \cdot \vec{d\boldsymbol{\xi}}.$$

Comparing with [11.36] leads to:

$$\tilde{\nabla}\boldsymbol{\tau} = \left[{}_{\mu}\tilde{\nabla}\,{}^{\gamma}T^{\beta}\,(\boldsymbol{X}_0 \otimes \vec{\boldsymbol{e}}_\beta - \vec{\boldsymbol{e}}_\beta \otimes \boldsymbol{X}_0) + {}_{\mu}\tilde{\nabla}\,{}^{\gamma}J^{\alpha\beta}\,\vec{\boldsymbol{e}}_\alpha \otimes \vec{\boldsymbol{e}}_\beta\right]\,{}_{\gamma}\vec{\boldsymbol{\eta}} \otimes {}^{\mu}\boldsymbol{\eta}, \quad [11.37]$$

with:

$${}_{\mu}\tilde{\nabla}\,{}^{\gamma}T^{\beta} = \frac{\partial({}^{\gamma}T^{\beta})}{\partial({}^{\mu}\xi)} + {}^{\gamma}_{\mu\rho}\Gamma\,{}^{\rho}T^{\beta} + {}^{\gamma}T^{\rho}\,{}_{\mu}U^{\sigma}\,\Gamma^{\beta}_{\sigma\rho},$$

$${}_{\mu}\tilde{\nabla}\,{}^{\gamma}J^{\alpha\beta} = \frac{\partial({}^{\gamma}J^{\alpha\beta})}{\partial({}^{\mu}\xi)} + {}^{\gamma}J^{\rho\beta}\,{}_{\mu}U^{\sigma}\,\Gamma^{\alpha}_{\sigma\rho} + {}^{\gamma}J^{\alpha\rho}\,{}_{\mu}U^{\sigma}\,\Gamma^{\beta}_{\sigma\rho} + {}^{\gamma}_{\mu\rho}\Gamma\,{}^{\rho}J^{\alpha\beta}$$
$$+\,{}_{\mu}U^{\sigma}\,\Gamma^{\alpha}_{A\sigma}\,{}^{\gamma}T^{\beta} - {}^{\gamma}T^{\alpha}\,{}_{\mu}U^{\sigma}\,\Gamma^{\beta}_{A\sigma}. \quad [11.38]$$

By contraction of the two latter factors in [11.37], we obtain the *covariant divergence* of the continuum torsor:

$$\tilde{\boldsymbol{Div}}\,\boldsymbol{\tau} = {}_{\gamma}\tilde{\nabla}\,{}^{\gamma}T^{\beta}\,(\boldsymbol{X}_0 \otimes \vec{\boldsymbol{e}}_\beta - \vec{\boldsymbol{e}}_\beta \otimes \boldsymbol{X}_0) + {}_{\gamma}\tilde{\nabla}\,{}^{\gamma}J^{\alpha\beta}\,\vec{\boldsymbol{e}}_\alpha \otimes \vec{\boldsymbol{e}}_\beta, \quad [11.39]$$

with:

$${}_{\gamma}\tilde{\nabla}\,{}^{\gamma}T^{\beta} = \frac{\partial({}^{\gamma}T^{\beta})}{\partial({}^{\gamma}\xi)} + {}^{\gamma}_{\gamma\rho}\Gamma\,{}^{\rho}T^{\beta} + {}^{\gamma}T^{\rho}\,{}_{\gamma}U^{\sigma}\,\Gamma^{\beta}_{\sigma\rho}, \quad [11.40]$$

$${}_{\gamma}\tilde{\nabla}\,{}^{\gamma}J^{\alpha\beta} = \frac{\partial({}^{\gamma}J^{\alpha\beta})}{\partial({}^{\gamma}\xi)} + {}^{\gamma}J^{\rho\beta}\,{}_{\gamma}U^{\sigma}\,\Gamma^{\alpha}_{\sigma\rho} + {}^{\gamma}J^{\alpha\rho}\,{}_{\gamma}U^{\sigma}\,\Gamma^{\beta}_{\sigma\rho} + {}^{\gamma}_{\gamma\rho}\Gamma\,{}^{\rho}J^{\alpha\beta}$$
$$+\,{}_{\gamma}U^{\sigma}\,\Gamma^{\alpha}_{A\sigma}\,{}^{\gamma}T^{\beta} - {}^{\gamma}T^{\alpha}\,{}_{\gamma}U^{\sigma}\,\Gamma^{\beta}_{A\sigma}. \quad [11.41]$$

We are now able to generalize the free-coordinate equation of motion [10.32] to continua of arbitrary dimension:

$$\boxed{\tilde{\boldsymbol{Div}}\,\boldsymbol{\tau} + \boldsymbol{\tau}_{\vec{\boldsymbol{H}}} = 0} \quad [11.42]$$

where $\boldsymbol{\tau}_{\vec{\boldsymbol{H}}}$ is the torsor [10.29] of the resultant of other forces (i.e. different from the gravitation).

11.7. Equation of motion of 1D material bodies

We claimed that the motions of 1D continua are [11.42] which, taking into account [11.39], are structured into two groups:

$${}_{\gamma}\tilde{\nabla}\,{}^{\gamma}T^{\beta} + H^{\beta} = 0, \quad {}_{\gamma}\tilde{\nabla}\,{}^{\gamma}J^{\alpha\beta} + G^{\alpha\beta} = 0. \quad [11.43]$$

11.7.1. *First group of equations of motion*

Taking into account its structure [11.25] in Galilean coordinate systems, the force–mass tensor reads in tensor notations:

$$ {}^0T^0 = \rho_l, \qquad {}^0T^i = \rho_l v^i, \qquad {}^1T^0 = \rho_l v_t, \qquad {}^1T^i = \rho_l v_t v^i - F^i. \quad [11.44] $$

Excluding thrusts, we assume that the other forces are modeled likewise [10.48]:

$$ H^0 = 0, \qquad H^j = -f^j, $$

where f^j are the components of the external force of section 4.1.1. The non vanishing Christoffel's symbols representing the Galilean gravitation of the space–time are given by [10.49]:

$$ \Gamma^j_{00} = -g^j, \qquad \Gamma^j_{0k} = \Gamma^j_{k0} = \Omega^j_k. \quad [11.45] $$

According to [11.13], the unique non-vanishing Christoffel's symbols representing the 1D gravitation of $\mathcal{N}$ are ${}^1_{00}\Gamma = -g_t$ and we have:

$$ {}^{\gamma}_{\gamma\rho}\Gamma = 0 \quad [11.46] $$

The components of the tangent map [11.8] are given by [11.5]:

$$ {}_0U^0 = 1, \qquad {}_1U^0 = 0, \qquad {}_0U^i = v^i - v_t\, n^i, \qquad {}_1U^i = n^i. \quad [11.47] $$

Taking into account [11.40], the first group of [11.48] reads:

$$ {}_{\gamma}\tilde{\nabla}\, {}^{\gamma}T^{\beta} + H^{\beta} = \frac{\partial({}^{\gamma}T^{\beta})}{\partial({}^{\gamma}\xi)} + {}^{\gamma}_{\gamma\rho}\Gamma\, {}^{\rho}T^{\beta} + {}^{\gamma}T^{\rho}\, {}_{\gamma}U^{\sigma}\, \Gamma^{\beta}_{\sigma\rho} + H^{\beta} = 0. \quad [11.48] $$

By putting $\beta = 0$ which corresponds to the time coordinate, we obtain, owing to [11.46]:

$$ {}_{\gamma}\tilde{\nabla}\, {}^{\gamma}T^0 + H^0 = \frac{\partial}{\partial t}({}^0T^0) + \frac{\partial}{\partial s}({}^1T^0) = \frac{\partial \rho_l}{\partial t} + \frac{\partial}{\partial s}(\rho_l\, v_t) = 0. \quad [11.49] $$

For the spatial coordinates, we put $\beta = i$ that gives, taking into account the vanishing terms:

$$ {}_{\gamma}\tilde{\nabla}\, {}^{\gamma}T^i + H^i = \frac{\partial({}^{\gamma}T^i)}{\partial({}^{\gamma}\xi)} + {}^{\gamma}_{\gamma\rho}\Gamma\, {}^{\rho}T^i + {}^{\gamma}T^{\rho}\, {}_{\gamma}U^{\sigma}\, \Gamma^i_{\sigma\rho} - f^i = 0. $$

$$ {}_{\gamma}\tilde{\nabla}\, {}^{\gamma}T^i + H^i = \frac{\partial}{\partial t}({}^0T^i) + \frac{\partial}{\partial s}({}^1T^i) + {}^{\gamma}T^0\, {}_{\gamma}U^0\, \Gamma^i_{00} $$
$$ + {}^{\gamma}T^0\, {}_{\gamma}U^j\, \Gamma^i_{j0} + {}^{\gamma}T^j\, {}_{\gamma}U^0\, \Gamma^i_{0j} - f^i = 0. $$

Let us detail the calculation of the third to fifth terms of the latter expression, taking into account [11.44], [11.45] and [11.47]:

$$
\begin{aligned}
{}^{\gamma}T^{0}\,{}_{\gamma}U^{0}\,\Gamma^{i}_{00} &= -\rho_l g^{i}, \\
{}^{\gamma}T^{0}\,{}_{\gamma}U^{j}\,\Gamma^{i}_{j0} &= ({}^{0}T^{0}\,{}_{0}U^{j} + {}^{1}T^{0}\,{}_{1}U^{j})\,\Gamma^{i}_{j0} \\
&= (\rho_l(v^{j} - v_t n^{j}) + \rho_l v_t n^{j})\Omega^{i}_{j} = \rho_l v^{j}\Omega^{i}_{j}, \\
{}^{\gamma}T^{j}\,{}_{\gamma}U^{0}\,\Gamma^{i}_{0j} &= {}^{0}T^{j}\,{}_{0}U^{0}\,\Gamma^{i}_{0j} = \rho_l v^{j}\Omega^{i}_{j}.
\end{aligned}
$$

Finally, we obtain:

$${}_{\gamma}\tilde{\nabla}\,{}^{\gamma}T^{i} + H^{i} = \frac{\partial}{\partial t}(\rho_l v^{i}) + \frac{\partial}{\partial s}(\rho_l v_t v^{i} - F^{i}) - \rho_l(g^{i} - 2\,\Omega^{i}_{j}v^{j}) - f^{i} = 0.$$

After differentiation and taking into account [11.49], we obtain:

$$\rho_l\frac{\partial v^{i}}{\partial t} + \rho_l v_t\frac{\partial v^{i}}{\partial s} = \frac{\partial F^{i}}{\partial s} + f^{i} + \rho_l(g^{i} - 2\,\Omega^{i}_{j}v^{j}).$$

In short, the first group of [11.43] is recast as [11.49] and the latter relation that can be formulated in matrix notation and we obtain:

♢ *balance of mass*

$$\boxed{\frac{\partial \rho_l}{\partial t} + \frac{\partial}{\partial s}(\rho_l v_t) = 0} \qquad [11.50]$$

♡ *balance of linear momentum*

$$\boxed{\rho_l\left[\frac{\partial v}{\partial t} + \frac{\partial v}{\partial s}\,v_t\right] = \frac{\partial F}{\partial s} + f + \rho_l(g - 2\,\Omega\times v)} \qquad [11.51]$$

11.7.2. *Multiscale analysis*

As in section 4.1.1, we hope to study slender 3D bodies idealized by 1D material bodies but which are now solids or fluids in dynamical situations. At each instant, we model the geometry of the slender body by a mean line and, assigned to each point of the mean line, a cross-section. In order to generalize section 9.4.1 to the dynamics, we introduce a *projection map* from $T_{\boldsymbol{X}}\mathcal{M}$ into $T_{\boldsymbol{\xi}}\mathcal{N}$ where $\boldsymbol{X} = i(\boldsymbol{\xi})$:

$$\overrightarrow{d\boldsymbol{\xi}} = \boldsymbol{\Pi}\,\overrightarrow{d\boldsymbol{X}}, \qquad [11.52]$$

represented in the Galilean coordinate system X by:

$$d\xi = \Pi \, dX,$$

where the 2×4 matrix Π is defined by:

$$\Pi = \begin{pmatrix} 1 & 0 \\ 0 & n^T \end{pmatrix}. \qquad [11.53]$$

Reasoning as in section 11.2 and owing to [11.11], we can verify that:

$$S \, \Pi' = \Pi \, P,$$

where Π' represents the map $\mathbf{\Pi}$ in another Galilean coordinate system X'. It is also worth to observe that:

$$\Pi \, U = 1_{\mathbb{R}^2},$$

that represents in the coordinate system the intrinsic relation:

$$\mathbf{\Pi} \boldsymbol{U} = 1_{T_\xi \mathcal{N}}.$$

Now we state the rule to obtain the components ${}^\gamma T^\alpha$ of the 1D material body by projecting the components $T^{\beta\alpha}$ of the 3D medium and integrating on the cross-section $\mathcal{S}$ according to the rule:

$${}^\gamma T^\alpha = \int_{\mathcal{S}} {}^\gamma \Pi_\beta \, T^{\beta\alpha} \, d\mathcal{S}. \qquad [11.54]$$

To avoid mismatch of notation, we introduce the 4×4 matrix $\bar{T}$ of which $T^{\beta\alpha}$ is the element at the intersection of the β-th row and the α-th column while, as previously, the components ${}^\gamma T^\alpha$ are gathered into the matrix T. The previous rule reads:

$$T = \int_{\mathcal{S}} \Pi \, \bar{T} \, d\mathcal{S}. \qquad [11.55]$$

According to [10.42], the value of the stress–mass tensor at position $\bar{x}$ of the considered cross-section is:

$$\bar{T} = \begin{pmatrix} \rho & \rho \, \bar{v}^T \\ \rho \, \bar{v} & \rho \, \bar{v} \, \bar{v}^T - \sigma \end{pmatrix},$$

where $\bar{v}$ is the velocity at the current position $\bar{x}$ in $\mathcal{S}$. Taking into account the structure [11.25] of the force–mass tensor, [11.55] gives by identification:

$$\rho_l = \rho\,\mathcal{S}, \quad \rho_l v = \int_{\mathcal{S}} \rho\,\bar{v}\, d\mathcal{S}, \quad \rho_l v_t = \int_{\mathcal{S}} \rho\,\bar{v}_t\, d\mathcal{S}, \qquad [11.56]$$

$$\rho_l v_t v - F = \int_{\mathcal{S}} (\rho\,\bar{v}_t \bar{v} - \sigma\, n)\, d\mathcal{S}, \qquad [11.57]$$

The first relation of [11.56] gives the link between the mass by unit length ρ_l and the density ρ. The second one defines the velocity v of the force–mass tensor as the mean velocity on the cross-section weighted by the density. The third one is clearly the projection of the previous one on the unit tangent vector n to the curve. To interprete [11.57], we introduce the fluctuation of velocity on the cross-section with respect to the mean value:

$$\bar{w} = \bar{v} - v,$$

of null mean value on the cross-section:

$$\int_{\mathcal{S}} \rho\,\bar{w}\, d\mathcal{S} = 0,$$

because of [11.56]. Therefore, [11.57] gives the expression of the force:

$$F = \int_{\mathcal{S}} (\sigma\, n - \rho\,\bar{w}_t \bar{w})\, d\mathcal{S}.$$

In statics, the force is reduced to the first term and we recognize the former relation of [9.37] (simple change of notation $n = e_1$). In dynamics, the static force is completed by a second term due to velocity fluctuations accross the section. However, it is worth to note that for solids the fluctuations are negligeable, excepted in case of hight velocity (crash, explosion and so on).

In the same spirit as [11.54], we state the rule to obtain the components ${}^{\gamma}J^{\alpha\beta}$ of the 1D material body by projecting the components $J^{\alpha\beta\rho}$ of the 3D medium and integrating on the cross-section $\mathcal{S}$ according to the rule:

$${}^{\gamma}J^{\alpha\beta} = \int_{\mathcal{S}} {}^{\gamma}\Pi_{\rho}\, J^{\alpha\beta\rho}\, d\mathcal{S}, \qquad [11.58]$$

where the components of the projection map [11.52] are given by [11.53]:

$${}^{0}\Pi_0 = 1, \qquad {}^{0}\Pi_i = 0, \qquad {}^{1}\Pi_0 = 0, \qquad {}^{1}\Pi_i = n_i, \qquad [11.59]$$

with $n_i = \delta_{ij} n^j$. Next, we calculate the components $q, l, l_\star$ and $M_\star$ in two steps:

– Let $\boldsymbol{X}$ the event occuring at time t and at the current position $\boldsymbol{P}$ in the cross-section $\mathcal{S}$ corresponding to the point $\boldsymbol{Q}$ on the mean line (as in section 9.4.1). For a generalized Cauchy medium, the components $J^{\alpha\beta\gamma}$ in the affine frame $(\boldsymbol{X}, (\vec{e}_\alpha))$ vanish. Nevertheless, we must to take care that the component $J^{\beta\alpha\rho}$ must be calculated at the position $\boldsymbol{Q}$ before integrating, that requires a change of origin from $\boldsymbol{X}$ to the event $\bar{\boldsymbol{X}}$ occuring at time t and at position $\boldsymbol{Q}$ without basis change. Hence we apply the transformation law [10.25] with $J^{\alpha\beta\gamma} = 0$ in the old affine frame, $P = 1_{\mathbb{R}^4}$ and:

$$C'^0 = 0, \qquad C'^i = \bar{x}^i,$$

where $\bar{x}^i$ are the components of the vector $\overrightarrow{\boldsymbol{QP}}$, that gives in the new affine frame:

$$J'^{\alpha\beta\rho} = C'^\alpha\, T^{\beta\rho} - T^{\alpha\rho}\, C'^\beta,$$

next we remove the prime. In particular, we have:

$$J^{i0\rho} = \bar{x}^i T^{0\rho}, \qquad J^{ij\rho} = \bar{x}^i T^{j\rho} - \bar{x}^j T^{i\rho}.$$

– According to [11.27], $q^i = {}^0J^{i0}$ and combining with [11.58], [11.59], [10.47] and the previous condition, we have:

$$q^i = \int_{\mathcal{S}} J^{i00}\, dS = \int_{\mathcal{S}} \rho\, \bar{x}^i\, dS,$$

In a similar way, we obtain:

$$l^i_\star = {}^1J^{i0} = \int_{\mathcal{S}} J^{i0j} n_j\, dS = \int_{\mathcal{S}} \rho\, \bar{v}_t \bar{x}^i\, dS,$$

and, if (ikl) is a cyclic permutation of (123):

$$l^i = {}^0J^{kl} = \int_{\mathcal{S}} J^{kl0}\, dS = \int_{\mathcal{S}} \rho\,(\bar{x}^k \bar{v}^l - \bar{x}^l \bar{v}^k)\, dS,$$

$$M^i_\star = {}^1J^{kl} = \int_{\mathcal{S}} J^{klj} n_j\, dS$$

$$= \int_{\mathcal{S}} \left[\bar{x}^k(\rho \bar{v}_t \bar{v}^l - (\sigma\, n)^l) - \bar{x}^l(\rho \bar{v}_t \bar{v}^k - (\sigma\, n)^k)\right]\, dS.$$

In matrix form, these formula read:

$$q = \int_{\mathcal{S}} \rho\, \bar{x}\, d\mathcal{S}, \quad l = \int_{\mathcal{S}} \bar{x} \times \rho\, \bar{v}\, d\mathcal{S}, \quad l_{\star} = \int_{\mathcal{S}} \rho\, \bar{v}_t \bar{x}\, d\mathcal{S},$$

$$M_{\star} = \int_{\mathcal{S}} \bar{x} \times (\rho \bar{v}_t \bar{v} - (\sigma\, n))\, d\mathcal{S}.$$

In statics, the moment is reduced to the second term and, putting $M_{\star} = -M$, we recognize the latter relation of [9.37] (simple change of notation $n = e_1$).

11.7.3. *Secong group of equations of motion*

These components must satisfy the second group of equations of motion [11.43] that we hope to detail under slightly restrictive assumptions:

– It is convenient to choose the position $\boldsymbol{Q}$ on the mean line as the *mass-center of the cross-section* $\mathcal{S}$, because the q component vanishes:

$$q^i = 0. \qquad [11.60]$$

– Excluding thrusts, we assume that the resultant moment of other forces is modeled likewise [5.51]:

$$G = \begin{pmatrix} 0 & 0 \\ 0 & j(m) \end{pmatrix}, \qquad [11.61]$$

where m is the exterior moment of section 4.1.1.

– For a *proper coordinate system* (defined in section 3.1.2), [5.55] is [5.55] with $C^\alpha = 0$:

$$\Gamma^\alpha_{A\beta} = \delta^\alpha_\beta.$$

Taking into account the previous relation, [11.41] and [11.46], the second group of equations of motion reads:

$$\begin{aligned} {}_\gamma\tilde{\nabla}\, {}^\gamma J^{\alpha\beta} + G^{\alpha\beta} = \frac{\partial({}^\gamma J^{\alpha\beta})}{\partial({}^\gamma\xi)} + {}^\gamma J^{\rho\beta}\, {}_\gamma U^\sigma\, \Gamma^\alpha_{\sigma\rho} + {}^\gamma J^{\alpha\rho}\, {}_\gamma U^\sigma\, \Gamma^\beta_{\sigma\rho} \\ + {}_\gamma U^\alpha\, {}^\gamma T^\beta - {}^\gamma T^\alpha\, {}_\gamma U^\beta + G^{\alpha\beta} = 0. \qquad [11.62] \end{aligned}$$

By picking spatial coordinate indices $\alpha = i$ and $\beta = j$ and taking into account the skew-symmetry of J components, one has:

$$\frac{\partial({}^0J^{ij})}{\partial t} + \frac{\partial({}^1J^{ij})}{\partial s} + {}^\gamma J^{\rho j}\,{}_\gamma U^\sigma\,\Gamma^i_{\sigma\rho} - {}^\gamma J^{\rho i}\,{}_\gamma U^\sigma\,\Gamma^j_{\sigma\rho}$$
$$+ {}_\gamma U^i\,{}^\gamma T^j - {}^\gamma T^i\,{}_\gamma U^j + G^{ij} = 0. \quad [11.63]$$

Let us detail the calculation of the third term, taking into account ${}_1U^0 = 0$ and the unique non vanishing Christoffel's symbols are [11.45]:

$${}^\gamma J^{\rho j}\,{}_\gamma U^\sigma\,\Gamma^i_{\sigma\rho} = {}^0J^{0j}\,{}_0U^0\,\Gamma^i_{00} + {}^0J^{lj}\,{}_0U^0\,\Gamma^i_{0l} + {}^0J^{0j}\,{}_0U^l\,\Gamma^i_{l0} + {}^1J^{0j}\,{}_1U^l\,\Gamma^i_{l0},$$

which becames, owing to [11.44] and [11.47]:

$${}^\gamma J^{\rho j}\,{}_\gamma U^\sigma\,\Gamma^i_{\sigma\rho} = -\,{}^0J^{0j}\,g^i + {}^0J^{lj}\,\Omega^i_l + {}^0J^{0j}\,{}_0(v^l - v_t n^l)\,\Omega^i_l + {}^1J^{0j}\,n^l\,\Omega^i_l,$$

The fourth term of [11.63] can be deduced from the previous one by a simple swap of i for j. The fifth and sixth terms:

$${}_\gamma U^i\,{}^\gamma T^j - {}^\gamma T^i\,{}_\gamma U^j = {}_0U^i\,{}^0T^j - {}^0T^i\,{}_1U^j + {}_1U^i\,{}^1T^j - {}^1T^i\,{}_1U^j,$$

are, taking into account [11.44] and [11.47], reduced to:

$${}_\gamma U^i\,{}^\gamma T^j - {}^\gamma T^i\,{}_\gamma U^j = n^j F^i - n^i F^j.$$

Hence [11.63] becomes:

$$\frac{\partial({}^0J^{ij})}{\partial t} + \frac{\partial({}^1J^{ij})}{\partial s} + {}^0J^{0i}g^j - {}^0J^{0j}g^i + (v^l - v_t n^l)\,(\Omega^i_l\,{}^0J^{j0} - \Omega^j_l\,{}^0J^{i0})$$
$$+ n^l(\Omega^i_l\,{}^1J^{0j} - \Omega^j_l\,{}^1J^{0i}) + \Omega^i_l\,{}^0J^{lj} - \Omega^j_l\,{}^0J^{li} + n^j F^i - n^i F^j = 0.$$

Taking into account [11.27] and [11.60], we have, if (kij) is a cyclic permutation of (123):

$$\frac{\partial M^k_\star}{\partial s} + \frac{\partial l^k}{\partial t} + l^i_\star\Omega^j_l n^l - l^j_\star\Omega^i_l n^l + \Omega^i_l\,{}^0J^{lj} - \Omega^j_l\,{}^0J^{li} + n^j F^i - n^i F^j + G^{ij} = 0. \quad [11.64]$$

Likewise, by picking $\alpha = i,\ \beta = 0$ in [11.62] and taking into account [11.45] and [11.61], one has:

$$\frac{\partial({}^0J^{i0})}{\partial t} + \frac{\partial({}^1J^{i0})}{\partial s} + {}^\gamma J^{\rho 0}\,{}_\gamma U^\sigma\,\Gamma^i_{\sigma\rho} + {}_\gamma U^i\,{}^\gamma T^0 - {}^\gamma T^i\,{}_\gamma U^0 = 0.$$

Taking into account the skew-symmetry of J components and ${}_1U^0 = 0$, it remains:

$$\frac{\partial({}^0J^{i0})}{\partial t} + \frac{\partial({}^1J^{i0})}{\partial s} + {}^0J^{j0}\,{}_0U^0\,\Gamma^i_{0j} + {}_0U^i\,{}^0T^0 - {}^0T^i\,{}_0U^0 + {}_1U^i\,{}^1T^0 = 0.$$

Owing to [11.27], [11.44] and [11.47], we obtain:

$$\frac{\partial q^i}{\partial t} + \Omega^i_j q^j + \frac{\partial l^i_\star}{\partial s} = 0,$$

hence, taking into account the simplification [11.60]:

$$\frac{\partial l^i_\star}{\partial s} = 0.$$

Coming back to the vector analysis notation and taking into account [11.61], the previous equation and [11.64] read:

♠ *Balance of passage:*

$$\boxed{\frac{\partial l_\star}{\partial s} = 0} \qquad [11.65]$$

♣ *Balance of angular momentum:*

$$\boxed{\frac{\partial l}{\partial t} + \Omega \times l + l_\star \times (\Omega \times n) = -\frac{\partial M_\star}{\partial s} + n \times F + m.} \qquad [11.66]$$

In Statics, $l = l_\star = 0$, $M_\star = -M$ and the previous equation is reduced to the local moment equilibrium equations of arches [4.3] (simple change of notation $n = U$):

$$\frac{dM}{ds} + n \times F + m = 0.$$

12

More About Calculus of Variations

12.1. Calculus of variation and tensors

Although we have well progressed through the continuum mechanics, let us have a backward look at the variational formulation of the particle dynamics presented in Chapter 6 without apparent link with the tensors. Considering the 4-velocity $\vec{U}$ of which the components in Galilean coordinate systems are:

$$U = \begin{pmatrix} 1 \\ v \end{pmatrix},$$

and the Galilean symmetric 2-covariant linear tensor $\boldsymbol{G}$ represented by the matrix [10.17]:

$$G = \begin{pmatrix} -2\phi & A^T \\ A & 1_{\mathbb{R}^3} \end{pmatrix},$$

expressed in terms of potentials of the Galilean gravitation, it is easy to see that the Lagrangian [6.13] of a particle moving within the Galilean gravitation field can read using contracted products:

$$\mathcal{L} = \frac{m}{2} \vec{U} \cdot \boldsymbol{G} \cdot \vec{U} = \frac{m}{2} G_{\alpha\beta} U^\alpha U^\beta, \qquad [12.1]$$

where:

$$G_{ij} = \delta_{ij}, \qquad G_{0i} = G_{i0} = A_i, \qquad G_{00} = -2\,\phi. \qquad [12.2]$$

Euler–Lagrange equations [6.7] read in indicial notations:

$$\frac{d}{dt}\left(\frac{\partial \mathcal{L}}{\partial \dot{v}^i}\right) - \frac{\partial \mathcal{L}}{\partial x^i} = \frac{d}{dt}\left(\frac{\partial \mathcal{L}}{\partial \dot{U}^i}\right) - \frac{\partial \mathcal{L}}{\partial X^i} = 0,$$

where the Latin index i runs only from 1 to 3. Applying them to the Lagrangian [12.1], we have:

$$m\left(\frac{d}{dt}\left(G_{\alpha i}U^\alpha + G_{i\beta}U^\beta\right) - \frac{\partial G_{\alpha\beta}}{\partial X^i}U^\alpha U^\beta\right) = 0.$$

Renaming the dummy indices in the former term and taking into account the symmetry of $\boldsymbol{G}$, we have:

$$m\left(2\frac{d}{dt}\left(G_{i\beta}U^\beta\right) - \frac{\partial G_{\alpha\beta}}{\partial X^i}U^\alpha U^\beta\right) = 0.$$

Differentiating the first term gives:

$$m\left(G_{i\beta}\dot{U}^\beta + U^\alpha\frac{\partial G_{i\beta}}{\partial X^\alpha}U^\beta - \frac{1}{2}\frac{\partial G_{\alpha\beta}}{\partial X^i}U^\alpha U^\beta\right) = 0.$$

Splitting the second term into two balanced ones and renaming dummy indices, we obtain:

$$m\left(G_{i\beta}\dot{U}^\beta + [\alpha\beta, i]\ U^\alpha U^\beta\right) = 0. \qquad [12.3]$$

introducing the symbols:

$$[\alpha\beta,\rho] = \frac{1}{2}\left(\frac{\partial G_{\rho\beta}}{\partial X^\alpha} + \frac{\partial G_{\rho\alpha}}{\partial X^\beta} - \frac{\partial G_{\alpha\beta}}{\partial X^\rho}\right),$$

which are obviously symmetric:

$$[\rho\alpha,\beta] = [\alpha\rho,\beta].$$

Using $A_i = \delta_{ij}A^j$ and taking into account [12.2], the explicit calculation of these symbols gives:

$$[00,0] = -\frac{\partial\phi}{\partial t}, \qquad [00,i] = \frac{\partial\phi}{\partial x^i} + \frac{\partial A_i}{\partial t}, \qquad [0i,0] = -\frac{\partial\phi}{\partial x^i}, \qquad [12.4]$$

$$[0i,j] = \frac{1}{2}\left(\frac{\partial A_j}{\partial x^i} - \frac{\partial A_i}{\partial x^j}\right), \qquad [ij,0] = \frac{1}{2}\left(\frac{\partial A_j}{\partial x^i} + \frac{\partial A_i}{\partial x^j}\right). \qquad [12.5]$$

Remarking that $U^0 = 1$, Euler–Lagrange equations in the form [12.3] are reduced to:

$$m\left(G_{ij}\dot{U}^j + [00,i] + [0j,i]\ U^j + [j0,i]\ U^j\right)$$
$$= m\left(\delta_{ij}\dot{v}^j + \frac{\partial\phi}{\partial x^i} + \frac{\partial A_i}{\partial t} + \left(\frac{\partial A_j}{\partial x^i} - \frac{\partial A_i}{\partial x^j}\right)v^j\right) = 0,$$

which reads in matrix notations:

$$m\left(\dot{v} + grad\,\phi + \frac{\partial A}{\partial t} + (curl\,A) \times v\right) = 0$$

Taking into account [6.14], we recover once again the equation [3.46] of motion:

$$m\dot{v} = m\,(g - 2\,\Omega \times v).$$

12.2. Action principle for the dynamics of continua

Our goal now is to deduce some other conservation equations, namely the balance of energy and linear momentum, from a variational principle. In Chapter 11, the motion of the continuum is modeled through a field defined on the space–time:

$$(t, x) \mapsto s' = \kappa(t, x).$$

So, the equation of the trajectory of the particle identified by s' is:

$$s' = \kappa(t, x).$$

Its gradient $\partial s'/\partial X$ is given by [10.64], then the relevant variables for the constitutive law are:

– the velocity:

$$v = -F\,\frac{\partial s'}{\partial t} = -\frac{\partial x}{\partial s'}\,\frac{\partial s'}{\partial t}, \tag{12.6}$$

– the right Cauchy strains:

$$C = F^T F = \left(\frac{\partial x}{\partial s'}\right)^T \frac{\partial x}{\partial s'}. \tag{12.7}$$

This suggests to consider a Lagrangian of this form:

$$\mathcal{L} : \mathbb{R}^4 \times \mathbb{R}^3 \times \mathbb{R}^{3\times 4} \longrightarrow \mathbb{R} : (X, s', z) \longmapsto \lambda = \mathcal{L}(t, s', z).$$

and the corresponding action principle:

$$\alpha\,[s'] = \int_\Omega \mathcal{L}\left(X, s', \frac{\partial s'}{\partial X}\right) d^4X,$$

where the Lagrangian depends on the field s' and its first derivatives, defined on a bounded open subset Ω of the space–time $\mathbb{R}^4$. As we are only interested in what follows by the variational equations in the interior of Ω, we consider simple boundary conditions with the value of s' imposed on $\partial\Omega$.

In order to obtain the conservation identities, we use a special form of the calculus of variation (see Comment 1, section 12.5). The new viewpoint which consists of performing variations not only on the field s' and its derivatives but also on the variable X. To clairfy them, we consider a new parametrization given by a regular map $X = \psi\,(Y)$ of class C^1 and we perform the variation of the function ψ, the new variable being Y. After calculating the variation of the action, we will consider the particular case where the function ψ is the identity of Ω. Hence, we start with:

$$\alpha\,[X, s'] = \int_{\Omega'} \mathcal{L}\left(\psi(Y), s', \frac{\partial s'}{\partial Y}\frac{\partial Y}{\partial X}\right)\,\det\left(\frac{\partial X}{\partial Y}\right) d^4Y,$$

where $\Omega' = \psi^{-1}(\Omega)$ and the fields of the functional are now both X and s'. For the sake of easiness, we introduce the 3-row:

$$F = -\frac{\partial\mathcal{L}}{\partial s'},$$

the 4-row:

$$H = \frac{\partial\mathcal{L}}{\partial X},$$

and the 4×3 matrix:

$$P = \frac{\partial\mathcal{L}}{\partial\left(\frac{\partial s'}{\partial X}\right)}.$$

The variation of the action reads:

$$\delta\alpha = \int_{\Omega'} [\left(Tr\left(P\,\delta\left(\frac{\partial s'}{\partial Y}\frac{\partial Y}{\partial X}\right)\right) - F\,\delta s' + H\,\delta X\right)\,\det\left(\frac{\partial X}{\partial Y}\right)$$

$$+\,\mathcal{L}\,\delta\,(\det\left(\frac{\partial X}{\partial Y}\right)]\,d^4Y. \qquad [12.8]$$

First, we calculate the variation of the field derivative in terms of the derivative of its variation:

$$\delta\left(\frac{\partial s'}{\partial Y}\frac{\partial Y}{\partial X}\right) = \delta\left(\frac{\partial s'}{\partial Y}\right)\frac{\partial Y}{\partial X} + \frac{\partial s'}{\partial Y}\,\delta\,\left(\frac{\partial X}{\partial Y}\right)^{-1}$$

$$\delta\left(\frac{\partial s'}{\partial Y}\frac{\partial Y}{\partial X}\right) = \frac{\partial}{\partial Y}(\delta s')\,\frac{\partial Y}{\partial X} - \frac{\partial s'}{\partial Y}\frac{\partial Y}{\partial X}\frac{\partial}{\partial Y}(\delta X)\,\frac{\partial Y}{\partial X}. \qquad [12.9]$$

Incidentally, it is worth noting that when $X = Y$:

$$\delta\left(\frac{\partial s'}{\partial X}\right) = \frac{\partial}{\partial X}(\delta s') - \frac{\partial s'}{\partial X}\,\frac{\partial}{\partial X}(\delta X).$$

This formula shows that, unlike the usual rule used in the classical calculus of variation (see [6.4]), the derivative symbols $\partial/\partial X$ and δ may not be permuted in the present approach. Next, owing to [14.31], we have:

$$\delta\left(\det\left(\frac{\partial X}{\partial Y}\right)\right) = Tr\left(\frac{\partial}{\partial Y}(\delta X)\, adj\left(\frac{\partial X}{\partial Y}\right)\right). \qquad [12.10]$$

Introducing the expressions [12.9] and [12.10] into the variation of the action [12.8] gives:

$$\delta\alpha = \int_{\Omega'} [Tr\left(P\,\frac{\partial}{\partial Y}(\delta s')\,\det\left(\frac{\partial X}{\partial Y}\right)\frac{\partial Y}{\partial X}\right) - \det\left(\frac{\partial X}{\partial Y}\right)(F\,\delta s' + H\,\delta X)$$

$$+Tr\left(\mathcal{L}\,\frac{\partial}{\partial Y}(\delta X)\, adj\left(\frac{\partial X}{\partial Y}\right) - P\,\frac{\partial s'}{\partial X}\,\frac{\partial}{\partial Y}(\delta X)\,\det\left(\frac{\partial X}{\partial Y}\right)\frac{\partial Y}{\partial X}\right)]\, d^4Y. \qquad [12.11]$$

Taking into account [7.8] and introducing the 4×4 matrix:

$$T = P\,\frac{\partial s'}{\partial X} - \mathcal{L}\,1_{\mathbb{R}^4}, \qquad [12.12]$$

the variation of the action [12.11] becomes:

$$\delta\alpha = \int_{\Omega'} [\,Tr\left(adj\left(\frac{\partial X}{\partial Y}\right) P\,\frac{\partial}{\partial Y}(\delta s')\right) - \det\left(\frac{\partial X}{\partial Y}\right)(F\,\delta s' + H\,\delta X)$$

$$-Tr\left(adj\left(\frac{\partial X}{\partial Y}\right) T\,\frac{\partial}{\partial Y}(\delta X)\right)]\, d^4Y. \qquad [12.13]$$

Owing to [14.22], we integrate by part in [12.13]. Taking into account the fact that the values of s' and X are imposed on the boundary, the surface integrals vanish and we obtain:

$$\delta\alpha = \int_{\Omega'} (-\left[div_Y\left(adj\left(\frac{\partial X}{\partial Y}\right) P\right) + \det\left(\frac{\partial X}{\partial Y}\right) F\right]\delta s'$$

$$+\left[div_Y\left(adj\left(\frac{\partial X}{\partial Y}\right) T\right) + \det\left(\frac{\partial X}{\partial Y}\right) H\right]\delta X)\, d^4Y,$$

where the index of div indicates with respect to which variable we differentiate. Finally, considering the particular case where $X = Y$, the variational principle reads:

$$\delta\alpha = \int_{\Omega} (-[\,div_X P + F]\,\delta s' + [div_X T \,+ H]\,\delta X)\, d^4 X = 0.$$

The variation of s' and X being arbitrary, we obtain the equations of variation:

$$\begin{aligned} &div_X P + F = 0, \\ &div_X\, T + H = 0. \end{aligned} \qquad [12.14]$$

The first equation leads to a nonlinear partial derivative system which can be used to determine the unknown field s'. The last one gives extra conservation conditions (see Comment 2, section 12.5). In the next section, we physically interprete it.

12.3. Explicit form of the variational equations

Now, we are interested in particular continuous media for which ones the Lagrangian depends on the first partial derivatives of s' through the velocity and configuration:

$$\mathcal{L}\left(X, s', \frac{\partial s'}{\partial X}\right) = \mathcal{L}\,(s', v, C)\,.$$

Its differential is:

$$\delta\mathcal{L} = Tr\left(\frac{\partial \mathcal{L}}{\partial C}\,\delta C\right) + \frac{\partial \mathcal{L}}{\partial v}\,\delta v. \qquad [12.15]$$

Moreover, differentiating the right Cauchy strains [12.7] and taking into account [14.33]:

$$\delta C = \delta F^T F + F^T \delta F = -(F^T \delta(F^{-1})^T F^T F + F^T F\,\delta(F^{-1})\,F),$$

$$\delta C = -\left(F^T \delta\left(\frac{\partial s'}{\partial x}\right)^T C + C\,\delta\left(\frac{\partial s'}{\partial x}\right) F\right),$$

and differentiating the velocity [12.6]:

$$\delta v = -\delta\left[\left(\frac{\partial s'}{\partial x}\right)^{-1}\right]\frac{\partial s'}{\partial t} - \frac{\partial x}{\partial s'}\,\delta\left(\frac{\partial s'}{\partial t}\right) = \frac{\partial x}{\partial s'}\,\delta\left(\frac{\partial s'}{\partial x}\right)\frac{\partial x}{\partial s'}\frac{\partial s'}{\partial t}$$
$$- \frac{\partial x}{\partial s'}\,\delta\left(\frac{\partial s'}{\partial t}\right),$$

$$\delta v = F\left(\delta\left(\frac{\partial s'}{\partial x}\right) v + \delta\left(\frac{\partial s'}{\partial t}\right)\right),$$

next replacing both former expressions into [12.15], we have:

$$\delta\mathcal{L} = Tr\left(\frac{\partial\mathcal{L}}{\partial C}\,F^T\delta\left(\frac{\partial s'}{\partial x}\right)^T C\right) + Tr\left(\frac{\partial\mathcal{L}}{\partial C}\,C\,\delta\left(\frac{\partial s'}{\partial x}\right)^T F\right)$$

$$-\frac{\partial\mathcal{L}}{\partial v}\,F\,\delta\left(\frac{\partial s'}{\partial x}\right)\,v - \frac{\partial\mathcal{L}}{\partial v}\,F\,\delta\left(\frac{\partial s'}{\partial t}\right). \qquad [12.16]$$

By simple manipulations, taking into account [7.4] and the symmetry of C, then of $\partial\mathcal{L}/\partial C$, the first two terms of the right-hand side are equal to:

$$-Tr\left(F\,\frac{\partial\mathcal{L}}{\partial C}\,C\,\delta\left(\frac{\partial s'}{\partial x}\right)\right).$$

Moreover, the third term reads:

$$-Tr\left(v\,\frac{\partial\mathcal{L}}{\partial v}\,F\,\delta\left(\frac{\partial s'}{\partial x}\right)\right).$$

Thus, expression [12.3] reads:

$$\delta\mathcal{L} = -Tr\left(\left(2\,F\,\frac{\partial\mathcal{L}}{\partial C}\,C + v\,\frac{\partial\mathcal{L}}{\partial v}\,F\right)\,\delta\left(\frac{\partial s'}{\partial x}\right)\right).$$

$$-\frac{\partial\mathcal{L}}{\partial v}\,F\,\delta\left(\frac{\partial s'}{\partial t}\right). \qquad [12.17]$$

On the other hand, introducing the 4-row:

$$P_t = \frac{\partial\mathcal{L}}{\partial\left(\frac{\partial s'}{\partial t}\right)},$$

and the 3×3 matrix:

$$P_r = \frac{\partial\mathcal{L}}{\partial\left(\frac{\partial s'}{\partial x}\right)},$$

we can express the differential of the Lagrangian as:

$$\delta\mathcal{L} = P_t\,\delta\left(\frac{\partial s'}{\partial t}\right) + Tr\left(P_r\,\delta\left(\frac{\partial s'}{\partial x}\right)\right). \qquad [12.18]$$

Comparing [12.17] and [12.18] leads to:

$$P_t = -\frac{\partial \mathcal{L}}{\partial v} F,$$

$$P_r = -2\, F \frac{\partial \mathcal{L}}{\partial C} C - v \frac{\partial \mathcal{L}}{\partial v} F. \qquad [12.19]$$

Inspired by the Lagrangian [6.13] of a particle subjected to a Galilean gravitation and introducing the reversible energy potential, we define the Lagrangian as:

$$\mathcal{L} = \frac{\rho}{2} \parallel v \parallel^2 + \rho\, A \cdot v - \rho \phi - W\,(s', C),$$

$$\boxed{\mathcal{L} = \rho \left(\frac{1}{2} \parallel v \parallel^2 + A \cdot v - \phi - e_{int}(s', C) \right),} \qquad [12.20]$$

where ρ is given by [10.56] and e_{int} is the specific internal energy. The case of the barotropic fluid can be easily obtained by considering that the reversible energy potential W depends on the right Cauchy strains through $\det(C)$. The expression [12.20] can be factorized as:

$$\mathcal{L}\,(s', v, C) = \rho\,(s', C)\, L\,(s', v, C), \qquad [12.21]$$

where ρ is defined by [10.58]. Let us remark that applying [14.32] gives:

$$\frac{\partial \rho}{\partial C} = -\frac{\rho}{2}\, C^{-1}.$$

Thus, differentiating [12.21], we have:

$$\frac{\partial \mathcal{L}}{\partial C} = L \frac{\partial \rho}{\partial C} + \rho \frac{\partial L}{\partial C} = -\frac{\mathcal{L}}{2} C^{-1} + \rho \frac{\partial L}{\partial C}.$$

Substituting this expression into [12.19], it holds:

$$P_r = -2 \rho\, F \frac{\partial L}{\partial C} C + \left(\mathcal{L}\, 1_{\mathbb{R}^3} - v \frac{\partial \mathcal{L}}{\partial v} \right) F. \qquad [12.22]$$

Moreover, [12.18] can be written in a more compact form:

$$\delta \mathcal{L} = Tr \left(P\, \delta \left(\frac{\partial s'}{\partial X} \right) \right),$$

with:

$$P = \begin{pmatrix} P_t \\ P_r \end{pmatrix}.$$

Analogously to [6.17], we introduce the *generalized linear momentum* represented by the 3-column:

$$\pi = grad_v \mathcal{L} = \rho\,(v + A), \qquad [12.23]$$

and:

$$\mathcal{H} = \frac{\partial \mathcal{L}}{\partial v}\, v - \mathcal{L} = \rho\left(\frac{1}{2} \parallel v \parallel^2 + \phi + e_{int}\right), \qquad [12.24]$$

recovering the Hamiltonian density [10.81]. Finally, the 4×4 matrix [12.12] can be decomposed by block:

$$T = \begin{pmatrix} P_t \dfrac{\partial s'}{\partial t} - \mathcal{L} & P_t \dfrac{\partial s'}{\partial x} \\ & \\ P_r \dfrac{\partial s'}{\partial t} & P_r \dfrac{\partial s'}{\partial x} - \mathcal{L} 1_3 \end{pmatrix}.$$

Next, we calculate each block of T. For instance, owing to [12.22], we have:

$$P_r \frac{\partial s'}{\partial x} - \mathcal{L} 1_3 = \sigma - v\, \frac{\partial \mathcal{L}}{\partial v},$$

with the symmetric 3×3 matrix:

$$\sigma = -2\rho\, F\, \frac{\partial L}{\partial C}\, F^T = 2\rho\, F\, \frac{\partial e_{int}}{\partial C}\, F^T,$$

identified to the spatial stresses, according to [10.75]. Finally, the matrix is structured as (see Comment 3, section 12.5):

$$T = \begin{pmatrix} \mathcal{H} & -\pi^T \\ \mathcal{H} v - \sigma v & \sigma - v\, \pi^T \end{pmatrix} \qquad [12.25]$$

12.4. Balance equations of the continuum

Next, we show that the variational equation [12.14] leads to the balance of energy and momentum:

THEOREM 12.1.– If T satisfied the equation of variation [12.14]:

$$div_X\, T + H = 0,$$

then, we have:

1) ♢ the balance of linear momentum: $\rho \dfrac{dv}{dt} = (div\, \sigma)^T + \rho\, (g - 2\,\Omega \times v)$;

2) ♡ the balance of energy: $\dfrac{\partial \mathcal{H}}{\partial t} + div\, (\mathcal{H} v - \sigma v) = \rho \left(\dfrac{\partial \phi}{\partial t} - \dfrac{\partial A}{\partial t} \cdot v \right).$

PROOF.– ♢ Balance of linear momentum. Owing to [12.20], we have:

$$F = \left(\frac{\partial \mathcal{L}}{\partial t}, \frac{\partial \mathcal{L}}{\partial x} \right) = \left(\rho \left(v \cdot \frac{\partial A}{\partial t} - \frac{\partial \phi}{\partial t} \right), \rho \left(v^T \frac{\partial A}{\partial x} - \frac{\partial \phi}{\partial x} \right) \right). \qquad [12.26]$$

We calculate the divergence of the matrix T using [14.14]. Calculating the divergence of the last column of [12.25] and owing to [14.17] and [12.14] gives:

$$\begin{aligned} &- \frac{\partial}{\partial t} (\rho\, (v + A)^T) + div\, \sigma - div\, (\rho v)\, (v + A)^T \\ &- \rho v^T grad\, (v + A) + \frac{\partial \mathcal{L}}{\partial x} = 0, \end{aligned} \qquad [12.27]$$

or, expanding the first term:

$$-\rho\, \frac{\partial}{\partial t} (v + A)^T - \left(\frac{\partial \rho}{\partial t} + div\, (\rho v) \right)\, (v + A)^T - \rho v^T grad\, (v + A) + div\, \sigma + \frac{\partial \mathcal{L}}{\partial x} = 0.$$

Taking into account [12.26] and theorem 10.11, it holds that:

$$-\rho \left(\frac{\partial v^T}{\partial t} + v^T grad\, v \right) + div\, \sigma - \rho \left(\frac{\partial \phi}{\partial x} + \frac{\partial A}{\partial t} \right) + \rho v^T \left(\frac{\partial A}{\partial x} - grad\, A \right) = 0.$$

By transposition, we have, owing to [7.43] and [6.14]:

$$-\rho \left(\frac{\partial v}{\partial t} + \frac{\partial v}{\partial x}\, v \right) + div\, \sigma + \rho\, \left(g + 2 v^T j\, (\Omega) \right) = 0,$$

which leads to the balance of linear momentum ♢:

$$\rho \frac{dv}{dt} = (div\, \sigma)^T + \rho\, (g - 2\Omega \times v)\,, \qquad [12.28]$$

♡ *Balance of energy.* Calculating the divergence of the first column of [12.25], [12.14] gives:

$$\frac{\partial \mathcal{H}}{\partial t} + div\ (\mathcal{H}v - \sigma v) + \frac{\partial \mathcal{L}}{\partial x} = 0.$$

Taking into account [12.26], we obtain the balance of energy:

$$\frac{\partial \mathcal{H}}{\partial t} + div\ (\mathcal{H}v - \sigma v) = \rho \left(\frac{\partial \phi}{\partial t} - \frac{\partial A}{\partial t} \cdot v \right),$$

which achieves the proof. ■

The attentive readers will observe that we recover – by a method entirely different from that of the previous chapter – the balance of linear momentum [11.51] in the absence of volume forces f_v and the balance of energy [11.84] for hyperelastic materials.

12.5. Comments for experts

COMMENT 1.– We perform a special form of the calculus of variation on the jet space of order one.

COMMENT 2.– This conservation equation is obtained in the spirit of Noether's theorem.

COMMENT 3.– Matrix T is the analogous in Galilean mechanics of the energy-momentum tensor in relativistic one.

13

Thermodynamics of Continua

13.1. Introduction

Before addressing the thermodynamics of continua, background ideas of thermodynamics are briefly recalled. Inspired by Carnot's works, Clausius showed in 1865 that the ratio Q_R/θ, where Q_R is the amount of heat absorbed in an isothermal and reversible process by a thermodynamic system at the absolute temperature θ, is a state function:

$$\mathcal{S} = \frac{Q_R}{\theta}, \tag{13.1}$$

which he called the entropy. In 1877, Boltzmann threw a new light on this abstract physical quantity, by defining in statistical mechanics the entropy as proportional to the logarithm of the number of microscopic configurations that result in the macroscopic description of the system. To overcome the limitations of the previous approaches originally based on the study of thermal engines, in 1908 Caratheodory proposed an axiomatic approach. He considered reversible processes with varying temperature. While the elementary heat supply δQ_R is not integrable, the elementary variation of entropy:

$$d\mathcal{S} = \frac{1}{\theta}\, \delta Q_R,$$

so is. Then, the difference of entropy between two states of the system A and B does not depend on the path to go from A to B:

$$\mathcal{S}(B) - \mathcal{S}(A) = \int_A^B \frac{1}{\theta}\, \delta Q_R.$$

The first law of the thermodynamics, formalized through the heat-friction experiments of Joule in 1843, claims that a thermodynamical system can store and supply energy but its total energy is conserved:

$$dE_{int} + dK = \delta W + \delta Q,$$

where E_{int} is the internal energy of the system, K is the kinetic one, δW is the work done by surroundings and δQ is the element of heat absorbed by the system.

Reversible processes are ideal concepts but, in realistic situations, a part of the energy is lost or dissipated due to internal frictions producing heat. The second law of thermodynamics, originally stated by Clausius, claims that, for the reversible and irreversible processes, the total production of entropy is positive:

$$\frac{\delta Q}{\theta} = \frac{\delta Q_R}{\theta} - \frac{\delta Q_I}{\theta} = dS - \frac{\delta Q_I}{\theta} \geq 0,$$

the equality being reached only for the reversible processes. In this formula, the sign before δQ_I is conventionally chosen because this irreversible heat is produced by the system itself.

The previous concepts were initially introduced to describe the behavior of systems, independently of the mechanics of continua, but these two topics can be discussed in spirit of Truesdell's ideas [TRU 60] and those of his school. The basic idea is to apply the concepts of the thermodynamics to any volume element of a continuum to obtain local versions of the two principles consistent with Galileo's principle of relativity 1.1.

13.2. An extra dimension

For not introducing early too much complexity, we begin with modeling the thermodynamics in situations where the gravitation can be neglected. For a particle in uniform straight motion, classical integrals of the motion – mass, linear and angular momenta – were revealed as components of the dynamical torsor (law 3.1) but there is a noticeable absent one, the kinetic energy:

$$e = \frac{1}{2} m \parallel v \parallel^2 . \qquad [13.2]$$

The difficulty to recover it is deep and overcoming it needs a strong change of viewpoint. The cornerstone idea is to add to the space-time an extra dimension roughly speaking linked to the energy by a three step method that we will be going to present in a heuristic way:

– As we are concerned with the uniform straight motion, we do not consider provisionally the gravitation. We start with a fictitious 5-dimensional affine space $\hat{\mathcal{U}}$ containing the space-time $\mathcal{U}$. We claim that any point $\hat{\boldsymbol{X}}$ of $\hat{\mathcal{U}}$ can be represented in some suitable coordinate systems by a column:

$$\hat{X} = \begin{pmatrix} X \\ z \end{pmatrix} \in \mathbb{R}^5,$$

in such a way that the space-time is identified to the subspace $\hat{X}^4 = z = 0$ of $\hat{\mathcal{U}}$ and X gathers Galilean coordinates.

– We wish to build a group of affine transformations $d\hat{X}' \mapsto d\hat{X} = \hat{P}\, d\hat{X}' + \hat{C}$ of $\mathbb{R}^5$ which are Galilean when acting on the space-time only. Clearly, the 5×5 matrix $\hat{P}$ is structured as:

$$\hat{P} = \begin{pmatrix} P & 0 \\ \Phi & \alpha \end{pmatrix},$$

where the 4-row Φ and the scalar α have to take an appropriate physical meaning.

– It is worth to notice that under a Galilean coordinate change $X' \longmapsto X$ characterized by a boost u and a rotation R, using the velocity addition formula [1.13], its transformation law is:

$$e = \frac{1}{2} m \parallel u + R v' \parallel^2 = \frac{1}{2} m \parallel u \parallel^2 + m\, u \cdot (R v') + \frac{1}{2} m \parallel v' \parallel^2,$$

that is:

$$e = \frac{1}{2} m \parallel u \parallel^2 + m\, u \cdot (R v') + e'. \qquad [13.3]$$

Next, we claim that the extra coordinate is linked to the energy as follows:

$$dz = \frac{e}{m}\, dt, \qquad dz' = \frac{e'}{m}\, dt'. \qquad [13.4]$$

The division by m is guided by the fact that we wish the extra coordinate being universal, independent of the mass of particle moving in the space-time. According to $dt = dt'$ and $dx' = v' dt'$, we obtain:

$$dz = \frac{1}{2} \parallel u \parallel^2 dt' + u^T R\, dx' + dz'.$$

On this basis, we state:

DEFINITION 13.1.– The *Bargmannian transformations* are affine transformations $d\hat{X}' \mapsto d\hat{X} = \hat{P}\, d\hat{X}' + \hat{C}$ of $\mathbb{R}^5$ such that:

$$\hat{P} = \begin{pmatrix} 1 & 0 & 0 \\ u & R & 0 \\ \frac{1}{2} \parallel u \parallel^2 & u^T R & 1 \end{pmatrix}. \quad [13.5]$$

It is straightforward to verify that the set of the Bargmannian transformations is a subgroup of $\mathbb{A}ff(5)$ called *Bargmann's group* and is denoted by $\mathbb{B}$ in the sequel (see Comment 1, section 13.9). The subgroup of Bargmannian linear transformations is denoted by $\mathbb{B}_0$. In particular, the inverse of [13.5] is:

$$\hat{P}^{-1} = \begin{pmatrix} 1 & 0 & 0 \\ -R^T u & R^T & 0 \\ \frac{1}{2} \parallel u \parallel^2 & -u^T & 1 \end{pmatrix}. \quad [13.6]$$

Of course, the calculus may be organized as in section 1.3.4 by working in $\mathbb{R}^6$. The Bargmannian transformation looks like a linear transformation $d\tilde{X} = \tilde{P}\, d\tilde{X}'$ if the column $d\hat{X}$ and $\hat{a} = (\hat{C}, \hat{P})$ are represented, respectively, by:

$$\tilde{X} = \begin{pmatrix} 1 \\ d\hat{X} \end{pmatrix} \in \mathbb{R}^6 \qquad \tilde{P} = \begin{pmatrix} 1 & 0 \\ \hat{C} & \hat{P} \end{pmatrix},$$

with:

$$\tilde{P} = \begin{pmatrix} 1 & 0 & 0 & 0 \\ \tau & 1 & 0 & 0 \\ k & u & R & 0 \\ \eta & \frac{1}{2} \parallel u \parallel^2 & u^T R & 0 \end{pmatrix}, \quad \tilde{P}^{-1} = \begin{pmatrix} 1 & 0 & 0 & 0 \\ \tau' & 1 & 0 & 0 \\ k' & -R^T u & R^T & 0 \\ \eta' & \frac{1}{2} \parallel u \parallel^2 & -u^T & 0 \end{pmatrix}. \quad [13.7]$$

DEFINITION 13.2.– In the absence of gravitation, the coordinate systems of $\hat{\mathcal{U}}$ which are deduced one from the other by Bargmanian transformations are called *Bargmannian coordinate systems* (see Comment 2, section 13.9).

In theorem 1.1, we define Galilean transformations as preserving some objects – uniform straight motions, durations, distances and angles, oriented volumes – but what Bargmannian transformations preserve? Combining [13.2] and [13.4] leads to:

$$\parallel dx \parallel^2 - 2\, dz\, dt = 0,$$

in every Bargmannian coordinate system. The left-hand member is a quadratic form in $d\hat{X}$ that suggests to introduce a symmetric 2-covariant tensor $\hat{\mathbf{G}}$ represented in Bargmannian coordinate systems by:

$$\hat{G} = \begin{pmatrix} 0 & 0 & -1 \\ 0 & 1_{\mathbb{R}^3} & 0 \\ -1 & 0 & 0 \end{pmatrix}. \qquad [13.8]$$

As it is regular, $\hat{\mathbf{G}}$ is a covariant metric tensor. It is preserved by Bargmannian linear transformations:

$$\forall \hat{P} \in \mathbb{B}_0, \qquad \hat{P}^T \hat{G}\, \hat{P} = \hat{G}. \qquad [13.9]$$

It is easy to verify that:

$$\hat{G}^2 = 1_{\mathbb{R}^5},$$

which proves the contravariant metric tensor $\hat{\mathbf{G}}^{-1}$ is represented by the same matrix as the covariant one:

$$\hat{G}^{-1} = \hat{G}.$$

13.3. Temperature vector and friction tensor

As subgroup of the affine group $\mathbb{A}ff(5)$, Bargmann's group naturally acts onto the tensors by restriction of their transformation laws. The $\mathbb{B}$-tensors are called *Bargmannian tensors*. To begin with, let us study the Bargmannian vectors $\hat{\vec{W}}$ represented by a 5-column:

$$\hat{W} = \begin{pmatrix} W \\ \zeta \end{pmatrix} = \begin{pmatrix} \beta \\ w \\ \zeta \end{pmatrix}, \qquad [13.10]$$

where $W \in \mathbb{R}^4$, $w \in \mathbb{R}^3$ and $\beta, \zeta \in \mathbb{R}$. According to [13.6], their transformation law [7.24] gives:

$$\beta' = \beta, \qquad w' = R^T(w - \beta u), \qquad \zeta' = \zeta - w \cdot u + \frac{\beta}{2} \parallel u \parallel^2 . \qquad [13.11]$$

Bargmannian transformations leave β invariant and there is no trouble to put β instead of β' in the following. Through a method similar to that in section 3.1.1, we search the other invariants of $\hat{\vec{W}}$ in the case that β does not vanish. Starting in any Bargmannian coordinate system $\hat{X}$ and we choose the Galilean boost:

$$u = \frac{w}{\beta},$$

which annihilates w' and reduces the last component to the generally non-vanishing expression:

$$\zeta' = \zeta - \frac{1}{2\,\beta} \parallel w \parallel^2 .$$

This suggests that we consider the quantity:

$$\zeta_{int} = \zeta - \frac{1}{2\,\beta} \parallel w \parallel^2 .$$

Taking into account [13.11], we verify that it is invariant by Bargmannian transformation. Conversely, let us consider a Bargmannian coordinate system $\hat{X}'$ in which the vector $\hat{\vec{W}}$ has a *reduced form*:

$$\hat{W}' = \begin{pmatrix} \beta \\ 0 \\ \zeta_{int} \end{pmatrix} .$$

In the spirit of the boost method initiated in section 3.1.2, we now claim the considered elementary volume is at rest. Let X be another Galilean coordinate system obtained from X' through a Galilean boost v. Applying the inverse transformation law of [13.6] with a Galilean boost v, we obtain:

$$\hat{W} = \begin{pmatrix} \beta \\ \beta\, v \\ \zeta_{int} + \frac{\beta}{2} \parallel v \parallel^2 \end{pmatrix} .$$

As $w = \beta\, v$, the last components become:

$$\zeta = \zeta_{int} + \frac{1}{2\beta} \parallel w \parallel^2 \qquad [13.12]$$

It is worth noting that under Bargmannian transformations, β is invariant. It is independent of the coordinate system and, for reasons that will appear later, we claim the following.

DEFINITION 13.3.– When $\beta = 1 / \theta = 1 / k_B T$ is the *reciprocal temperature*, where k_B is Boltzmann's constant and T is the absolute temperature, $\hat{\vec{W}}$ is called the *temperature vector*.

Let us also observe that:

$$\vec{W} = \beta \vec{U}.$$

The temperature 4-vector $\vec{W}$ is represented by a column decomposed by block as:

$$W = \begin{pmatrix} \beta \\ w \end{pmatrix}. \qquad [13.13]$$

DEFINITION 13.4.– The *friction tensor* is the 1-covariant and 1-contravariant mixed tensor:

$$\mathrm{f} = \nabla \vec{W}.$$

As the gravitation is provisionally neglected, we assume for the moment these latter ones vanish. Taking into account [14.37], it is represented in a Galilean coordinate system by the 4×4 matrix:

$$f = \nabla W = \frac{\partial W}{\partial X} = \begin{pmatrix} \dfrac{\partial \beta}{\partial t} & \dfrac{\partial \beta}{\partial x} \\ \dfrac{\partial\, w}{\partial t} & \dfrac{\partial\, w}{\partial x} \end{pmatrix}. \qquad [13.14]$$

13.4. Momentum tensors and first principle

DEFINITION 13.5.– A *momentum tensor* is a 1-covariant tensor $\hat{T}$ on the 5-dimensional space $\hat{\mathcal{U}}$ with vector values in the space-time $\mathcal{U}$, represented in a Bargmannian coordinate system by a 4×5 matrix structured as follows:

$$\boxed{\hat{T} = \begin{pmatrix} \mathcal{H} & -p^T & \rho \\ k & \sigma_* & p \end{pmatrix},} \qquad [13.15]$$

where $\mathcal{H} \in \mathbb{R}$, $p, k \in \mathbb{R}^3$ and $\sigma_* \in \mathbb{M}_{33}^{symm}$.

In indicial notation, the components of $\hat{T}$ are:

$$\hat{T}_0^0 = \mathcal{H}, \qquad \hat{T}_i^0 = -\delta_{ik} p^k, \qquad \hat{T}_4^0 = \rho,$$
$$\hat{T}_0^j = k^j, \qquad \hat{T}_i^j = \sigma_{*i}^{\ j}, \qquad \hat{T}_4^j = p^j.$$

According to [14.4], the transformation law of $\hat{T}$ is:

$$\boxed{\hat{T}' = P^{-1}\, \hat{T}\, \hat{P},} \qquad [13.16]$$

itemizes in the already known relations [10.35], [10.36] and [10.37]:

$$\rho' = \rho, \qquad [13.17]$$

$$p' = R^T\,(p - \rho u), \qquad [13.18]$$

$$\sigma'_\star = R^T\,(\sigma_\star + u\,p^T + p\,u^T - \rho\,u\,u^T)\,R, \qquad [13.19]$$

completed by two extra rules:

$$\boxed{\mathcal{H}' = \mathcal{H} - u \cdot p + \frac{\rho}{2} \parallel u \parallel^2,} \qquad [13.20]$$

$$\boxed{k' = R^T\,(k - \mathcal{H}' u + \sigma_* u + \frac{1}{2} \parallel u \parallel^2 p).} \qquad [13.21]$$

It is worth noting that the hypothesis of symmetry of $\sigma_\star$ is consistent with the rule [13.19]. The components ρ, p and $\sigma_\star$ can be physically identified with the mass density, the linear momentum and the dynamical stresses. To interpret the other components, we intend to annihilate components of $\hat{T}$. As usual, we discuss only the case of non-zero mass density. Starting in any Bargmannian coordinate system $\hat{X}$, we choose the Galilean boost:

$$u = \frac{p}{\rho},$$

which annihilates p', reduces [13.19] to [10.37] and transforms [13.20] and [13.21] as follows:

$$\mathcal{H}' = \mathcal{H} - \frac{1}{2\rho} \parallel p \parallel^2,$$

$$k' = R^T \left(k - \mathcal{H}\frac{p}{\rho} + \sigma_\star \frac{p}{\rho} \right).$$

The components $\mathcal{H}'$ and k' obviously cannot be annihilated by a convenient choice of a rotation R. At the most we could diagonalize the symmetric matrix $\sigma_\star$ but it is not useful now. Next, we use the boost method of section 3.1.2. Let us consider a Bargmannian coordinate system in which the tensor field $\hat{\boldsymbol{T}}$ at a given point of coordinates $\hat{X}'$ has a *reduced form*:

$$\hat{T}' = \begin{pmatrix} \rho e_{int} & 0 & \rho \\ h' & \sigma' & 0 \end{pmatrix},$$

for an elementary volume around the point x' at rest at time t'. Let $\hat{X}$ be another Bargmannian coordinate system obtained from $\hat{X}'$ through a Galilean boost v combined with a rotation R. Applying the inverse transformation law of [13.16]:

$$\hat{T} = P\hat{T}'\hat{P}^{-1}, \qquad [13.22]$$

we obtain:

$$p = \rho v, \qquad \sigma_\star = \sigma - \rho v v^T,$$

$$\mathcal{H} = \rho \left(\frac{1}{2} \parallel v \parallel^2 + e_{int} \right), \qquad [13.23]$$

$$k = h + \mathcal{H}v - \sigma v, \qquad [13.24]$$

according to the transformation law [10.40] of the spatial stresses σ and provided that:

$$h = Rh'. \qquad [13.25]$$

The boost method turns out the physical meaning of the components:

– the quantities already identified in section 10.4.2, the mass density ρ and the dynamic stresses $\sigma_\star$;

– the Hamiltonian density $\mathcal{H}$ defined by [10.81], apart from the potential ϕ (the gravitation being considered only later);

– and the *energy flux* k composed of h – further identified to the *heat flux* –, the *Hamiltonian flux* $\mathcal{H}v$ and the *stress flux* σv.

We could name $\hat{\boldsymbol{T}}$ the stress-mass-energy-momentum tensor but for briefness we call it momentum tensor. Finally, it has the form:

$$\hat{T} = \begin{pmatrix} \mathcal{H} & -p^T & \rho \\ h + \mathcal{H}v - \sigma v & \sigma - vp^T & \rho v \end{pmatrix}. \qquad [13.26]$$

In the last column of [13.26], we spot the 4-*flux of mass*:

$$N = \rho\, U. \tag{13.27}$$

Therefore, we can write:

$$\hat{T} = \begin{pmatrix} T & N \end{pmatrix}, \tag{13.28}$$

with:

$$T = \begin{pmatrix} \mathcal{H} & -p^T \\ h + \mathcal{H} v - \sigma v & \sigma - v p^T \end{pmatrix}.$$

In fact, it is more convenient to express the momentum tensor as [13.26], accounting for the following proposition.

THEOREM 13.1.– The expression [13.26] of the momentum tensor is standard provided that σ and h are changing according, respectively, to the rules [10.40] and [13.25].

PROOF.– Matrix [13.26] can be recast as:

$$\hat{T} = \begin{pmatrix} \mathcal{H} & -p^T & \rho \\ h + \mathcal{H}\,\dfrac{p}{\rho} - \sigma\,\dfrac{p}{\rho} & \sigma - \dfrac{1}{\rho}\,p\,p^T & p \end{pmatrix}. \tag{13.29}$$

Owing to [13.17], [10.40] and [13.18], we have:

$$\sigma'_* = \sigma' - \frac{1}{\rho'}\,p'\,p'^T = R^T \sigma\, R^T - \frac{1}{\rho}\,R^T (p - \rho u)(p^T - \rho u^T)\,R,$$

and developing:

$$\sigma'_\star = R^T\,(\sigma - \frac{1}{\rho}\,p\,p^T + u\,p^T + p\,u^T - \rho\,u\,u^T)\,R,$$

which, owing to [10.39], is nothing else but the transformation law [13.19].

In a similar way, taking into account [13.17], [13.25], [10.40] and [13.18], it holds:

$$k' = h' + \mathcal{H}'\,\frac{p'}{\rho'} - \sigma'\,\frac{p'}{\rho'} = R^T h + \frac{\mathcal{H}'}{\rho}\,R^T (p - \rho u) - \frac{1}{\rho}\,R^T \sigma\,(p - \rho u).$$

Taking into account [13.20] gives:

$$k' = R^T h + \left(\mathcal{H} - u \cdot p + \frac{\rho}{2} \parallel u \parallel^2\right) R^T \frac{p}{\rho} + \mathcal{H}' R^T u - R^T \sigma \frac{p}{\rho} + R^T \sigma\, u,$$

and with some arrangements:

$$k' = R^T \left[h + \mathcal{H} \frac{p}{\rho} - \sigma \frac{p}{\rho} - \mathcal{H}' u + \left(\sigma - \frac{1}{\rho} p\, p^T\right) u + \frac{1}{2} \parallel u \parallel^2 p\right],$$

which, owing to [13.24] and [10.39], is nothing else but the transformation law [13.21]. ■

The advantage of the standard form [13.26] is that the transformation laws [10.40] and [13.25] for σ and h are easier to manipulate than the corresponding transformation laws [13.19] and [13.21] for $\sigma_\star$ and k.

Also, introducing:

$$\Pi = \left(\mathcal{H} \ -p^T\right), \qquad [13.30]$$

it is worth noting that the momentum [13.26] can be recast as:

$$T = U\, \Pi + \begin{pmatrix} 0 & 0 \\ h - \sigma v & \sigma \end{pmatrix}. \qquad [13.31]$$

Owing to [13.14], [13.31] and the symmetry of σ leads to:

$$Tr\,(T\, f) = \Pi\, \frac{\partial W}{\partial X}\, U + Tr\, \left(\sigma\, \left(grad_s\, w - \frac{1}{2}\left(v\, \frac{\partial \beta}{\partial x} + grad\, \beta\, v^T\right)\right)\right)$$
$$+ h \cdot grad\, \beta. \qquad [13.32]$$

The first principle of thermodynamics claims that the total energy of a system is conserved. We are now able to propose an enhanced local version including the balance of mass and the equation of the motion (balance of linear momentum). It is based on the following result.

THEOREM 13.2.– If $\hat{T}$ is divergence free:

$$div_X\, \hat{T} = 0,$$

then, we have:

– ◇ *balance of mass:* $\dfrac{\partial \rho}{\partial t} + div\, (\rho\, v) = 0\,;$

– ♡ *balance of linear momentum*: $\rho\left[\frac{\partial v}{\partial t}+\frac{\partial v}{\partial x}v\right]=(div\,\sigma)^T$;

– ♠ *balance of energy*: $\frac{\partial \mathcal{H}}{\partial t}+div\,(h+\mathcal{H}v-\sigma v)=0.$

PROOF.– To calculate the divergence of the 4×5 matrix $\hat{T}$, we use [14.21] and some simple transformations already done to establish Euler's equations of motion (law 10.9) and that will not be repeated here. ■

Leaving provisionally the volume force f_v and the gravitation aside to compare, it can be observed that these three balance equations were previously found but by distinct ways:

– in section 10.5, we directly obtained the former two conditions starting from the study of the dynamical torsor of a three-dimensional (3D) medium. The balance of energy was further deduced from them;

– in section 12.4, the balance of mass was *a priori* assumed and the latter two equations were recovered by a variational principle.

In the thermodynamic framework, the three balance equations are obtained together due to the extra fifth dimension. On this basis and involving the gravitation, we state the *first principle of the thermodynamics*:

PRINCIPLE 13.1.– The momentum tensor of a continuum is covariant divergence free:

$$\boxed{\boldsymbol{Div\,\hat{T}=0.}} \qquad [13.33]$$

The covariant form of equation makes it consistent with Galileo's principle of relativity 1.1. The principle is general in the sense that it is valid for both reversible and dissipative continua. We are now going to successively describe these two kinds of media.

13.5. Reversible processes and thermodynamical potentials

To model the reversible processes, we need a new hypothesis inspired from the concept of potential as developed in section 10.7, claiming that:

– these phenomena can be represented due to a scalar function ζ of the particle s', its first partial derivative $\partial s'/\partial X$ and the temperature vector W, called *Planck's potential* (or *Massieu's potential*);

– according to the principle of material indifference 10.6, ζ depends on $\partial s'/\partial X$ through right Cauchy strains $C = F^T F$.

On this basis, we prove the following proposition.

THEOREM 13.3.– If ζ is a smooth function of s', C and W, then:

$$T_R = U\,\Pi_R + \begin{pmatrix} 0 & 0 \\ -\sigma_R v & \sigma_R \end{pmatrix}, \qquad [13.34]$$

with:

$$\Pi_R = -\rho\,\frac{\partial \zeta}{\partial W}, \qquad [13.35]$$

$$\sigma_R = -\frac{2\rho}{\beta}\,F\,\frac{\partial \zeta}{\partial C}\,F^T, \qquad [13.36]$$

is such that:

1) $\diamondsuit$ $Tr\left(\hat{T}_R \nabla \hat{W}\right) = 0$;

2) $\heartsuit$ $T_R\,U = -\rho\left(\frac{\partial \zeta}{\partial W}\,U\right)\,U$;

3) $\spadesuit$ $\hat{T}_R = \left(\,T_R\;N\,\right)$ represents a momentum tensor $\hat{\boldsymbol{T}}_R$;

4) $\clubsuit$ $\hat{T}_R\,\hat{W} = \left(\zeta - \frac{\partial \zeta}{\partial W}\,W\right)\,N.$

PROOF.– Taking into account [13.10], [13.14] and [13.28], condition $\diamondsuit$ reads:

$$(\nabla \zeta)\,N = -Tr\,(T_R f), \qquad [13.37]$$

or, in the absence of gravitation:

$$\frac{\partial \zeta}{\partial X}\,N = -Tr\,(T_R f) \qquad [13.38]$$

On the other hand, owing to [13.27], we have:

$$\frac{\partial \zeta}{\partial X}\, N = \rho\, (\frac{\partial \zeta}{\partial t} + \frac{\partial \zeta}{\partial x}\, v) = \rho \frac{d\zeta}{dt}.$$

As ζ depends on X through s', C and W, we have:

$$\frac{\partial \zeta}{\partial X}\, N = \rho\, \left(\frac{\partial \zeta}{\partial s'}\frac{ds'}{dt} + \frac{\partial \zeta}{\partial W}\frac{dW}{dt} + Tr\left(\frac{\partial \zeta}{\partial C}\frac{dC}{dt} \right) \right). \qquad [13.39]$$

Owing to [10.11], the first term of the right-hand side vanishes. Taking into account [13.35], the second term is:

$$\rho\, \frac{\partial \zeta}{\partial W}\frac{dW}{dt} = -\Pi_R\, \frac{\partial W}{\partial X}\, U. \qquad [13.40]$$

Next, we have to transform the last term of [13.39]. Because of [10.78] and [13.36], we have:

$$\begin{aligned} \rho\, Tr\left(\frac{\partial \zeta}{\partial C}\frac{dC}{dt} \right) &= Tr\left(2\rho F \frac{\partial \zeta}{\partial C} F^T D \right) = -Tr\,(\sigma_R \beta\, D) \\ &= -Tr\,(\sigma_R \beta\, grad_s\, v), \end{aligned}$$

that, taking into account [7.38], leads to:

$$\rho\, Tr\left(\frac{\partial \zeta}{\partial C}\frac{dC}{dt} \right) = -Tr\left(\sigma_R \left(grad_s\, w - \frac{1}{2}\left(v\, \frac{\partial \beta}{\partial x} + grad\, \beta\, v^T \right) \right) \right). \qquad [13.41]$$

Introducing the expressions [13.40] and [13.41] into [13.39] gives [13.38] and proves $\diamondsuit$, owing to [13.32]. Moreover, owing to [13.34] and [13.35], it holds:

$$T_R\, U = U\, \Pi_R\, U + \begin{pmatrix} 0 & 0 \\ -\sigma_R v & \sigma_R \end{pmatrix} \begin{pmatrix} 1 \\ v \end{pmatrix} = (\Pi_R\, U)\; U = -\rho\, \left(\frac{\partial \zeta}{\partial W}\, U \right)\, U,$$

that proves $\heartsuit$. Statement $\spadesuit$ results of the fact that [13.34] has the standard form [13.31]. Consequently, taking into account [13.10] and [13.28], we have:

$$S = T_R\, W + \zeta\, N = \rho\, \left(\zeta - \beta \frac{\partial \zeta}{\partial W}\, U \right)\, U = \left(\zeta - \frac{\partial \zeta}{\partial W}\, W \right)\, N,$$

and $\clubsuit$ is satisfied. ■

Planck's potential ζ is a prototype of scalar functions called *thermodynamical potentials* and is derived as follows:

– Comparing [13.31], [13.34] and [13.35] allows writing $\Pi_R = \left(\mathcal{H}_R \; -p^T \right)$ with:

$$\mathcal{H}_R = -\rho \, \frac{\partial \zeta}{\partial \beta}, \qquad p = \rho \; grad_w \, \zeta. \qquad [13.42]$$

Taking into account [13.12], it holds that:

$$\mathcal{H}_R = -\rho \, \frac{\partial \zeta_{int}}{\partial \beta} + \frac{\rho}{2\beta^2} \parallel w \parallel^2,$$

which allows recovering [13.23] because $w = \beta v$ and provided that:

$$\boxed{e_{int} = -\frac{\partial \zeta_{int}}{\partial \beta}.} \qquad [13.43]$$

This potential, called *internal energy* (by unit volume), is a function of s', C and W as a derivative of ζ_{int}.

– According to theorem 10.3, the 4-vector $\vec{\boldsymbol{S}} = \hat{\mathbf{T}}_R \, \hat{\vec{\boldsymbol{W}}}$ is a 4-flux that reads for convenience:

$$\vec{\boldsymbol{S}} = \rho \, s \, \vec{\boldsymbol{U}} = s \, \vec{\boldsymbol{N}}$$

represented by a 4-column:

$$S = \hat{T}_R \, \hat{W}. \qquad [13.44]$$

Then, setting $S = s\,N$ and taking into account ♣ of theorem 13.3, we have:

$$s = \zeta - \frac{\partial \zeta}{\partial \beta} \, \beta - \frac{\partial \zeta}{\partial w} \, w = \zeta_{int} + \frac{1}{2\beta} \parallel w \parallel^2 - \left(\frac{\partial \zeta_{int}}{\partial \beta} - \frac{1}{2\beta^2} \parallel w \parallel^2 \right) \beta - \frac{1}{\beta} w^T w,$$

that leads to:

$$\boxed{s = \zeta_{int} - \beta \, \frac{\partial \zeta_{int}}{\partial \beta}}$$

This quantity is called *specific entropy* and $\vec{\boldsymbol{S}}$ is its 4-flux. Hence, $-s$ appears as Legendre's transform [7.40] of ζ_{int} with respect to β. The latter equation and [13.43] are called *state equations* of the continuum.

– Moreover, we introduce a new potential called *Helmholtz free energy* (by unit volume):

$$\boxed{\psi = -\frac{1}{\beta}\,\zeta_{int} = -\theta\,\zeta_{int}.} \qquad [13.45]$$

By simple calculations, we obtain:

$$-e_{int} = \theta\,\frac{\partial \psi}{\partial \theta} - \psi, \qquad [13.46]$$

$$-s = \frac{\partial \psi}{\partial \theta}.$$

Hence, $-e_{int}$ appears as Legendre's transform of the free energy $\psi\,(s', C, \theta)$ with respect to θ. It is a function of s', C and $-s$ such that:

$$\theta = \frac{\partial e_{int}}{\partial s}.$$

Finally, we can find a nice integral of the motion:

THEOREM 13.4.– For reversible processes, the 4-flux S is divergence free and the specific entropy s is an integral of the motion.

PROOF.– Taking into account [13.44] and [14.22], it holds that:

$$div\, S = (div\, \hat{T}_R)\, \hat{W} + Tr\left(\hat{T}_R \frac{\partial\, \hat{W}}{\partial X}\right).$$

The momentum tensor $\hat{T}_R$ satisfies condition $\diamondsuit$ of theorem 13.3 and, according to the first principle (law 13.33), it is divergence free. Then, the divergence of the 4-flux of specific entropy S vanishes and we have, owing to [7.41]:

$$div\, S = div\,(s\, N) = \frac{\partial s}{\partial X}\, N + s\, div\, N = 0.$$

However, as seen in theorem 13.2, the freeness of the divergence implies the balance of mass. As discussed in section 10.5, this condition means the flux of mass N is divergence free and the last term of the previous equation vanishes. Then, because of the definition 10.2 of the material derivative, we have:

$$div\, S = \frac{\partial s}{\partial X}\, N = \rho\,\frac{\partial s}{\partial X}\, U = \rho\,\frac{ds}{dt} = 0,$$

that achieves the proof. ■

13.6. Dissipative continuum and heat transfer equation

In section 13.4, we showed that the thermodynamical behavior of a continuum is modeled by the momentum tensor $\hat{\mathbf{T}}$. From theorem 13.1, we prove that $\hat{\mathbf{T}}_R$ is a momentum tensor. We define $\hat{\mathbf{T}}_I = \hat{\mathbf{T}} - \hat{\mathbf{T}}_R$, that amounts to introduce an additive decomposition of the momentum tensor:

$$\hat{\mathbf{T}} = \hat{\mathbf{T}}_R + \hat{\mathbf{T}}_I, \qquad [13.47]$$

into a reversible part $\hat{\mathbf{T}}_R$ defined by theorem 13.3 and an irreversible part $\hat{\mathbf{T}}_I$ of which we will now examine the detailed representations. Owing to [13.42], the momentum tensor $\hat{\mathbf{T}}_R$ given by [13.34] is represented by:

$$\hat{T}_R = \begin{pmatrix} \mathcal{H}_R & -p^T & \rho \\ \mathcal{H}_R v - \sigma_R v & \sigma_R - v p^T & \rho v \end{pmatrix}.$$

Subtracting the previous matrix to [12.25] leads to:

$$\hat{T}_I = \begin{pmatrix} \mathcal{H}_I & 0 & 0 \\ h + \mathcal{H}_I v - \sigma_I v & \sigma_I & 0 \end{pmatrix}, \qquad [13.48]$$

where:

– the Hamiltonian density $\mathcal{H}_I = \mathcal{H} - \mathcal{H}_R$ is the opposite of the *irreversible heat source* (by unit volume);

– the column h is the *heat flux*;

– the symmetric 3×3 matrix $\sigma_I = \sigma - \sigma_R$ represents dissipative stresses (for instance, due to viscous effects).

The transformation law [13.20] and [10.40] gives:

$$\mathcal{H}'_I = \mathcal{H}_I, \qquad [13.49]$$

$$\sigma'_I = R^T \sigma_I R. \qquad [13.50]$$

Introducing also for convenience the *specific irreversible heat source* q_I such that:

$$\mathcal{H}_I = -\rho q_I, \qquad [13.51]$$

we have $\mathcal{H} = \rho\eta$ where the definition [10.82] of the specific Hamiltonian must be modified for the additive decomposition [13.47] and no gravitation ($\phi = 0$):

$$\eta = \frac{1}{2} \parallel v \parallel^2 + e_{int} - q_I.$$

Taking into account [10.83] and [14.15], the balance energy (♠ of theorem 13.2) reads:

$$\rho\,\frac{d\eta}{dt} + div\,h - (div\,\sigma)\,v - Tr\left(\frac{\partial v}{\partial x}\right) = 0.$$

Taking into account the symmetry of σ, the definition [9.14] of the strain velocity D and the balance of linear momentum (♡ of theorem 13.2), the balance of energy becomes:

$$\rho\,\frac{de_{int}}{dt} = Tr\ (\sigma\,D) - div\,h + \rho\,\frac{dq_I}{dt}. \qquad [13.52]$$

With respect to the corresponding formula [10.79] for the reversible processes, the dissipative case exhibits two extra terms due to the heat flux and the irreversible energy source.

A cornerstone consequence is the heat transfer equation. By differentiation of [13.46], we have:

$$\rho\,\frac{de_{int}}{dt} = \rho\left(\frac{d\psi}{dt} - \frac{d\theta}{dt}\frac{\partial\psi}{\partial\theta} - \theta\,\frac{d}{dt}\left(\frac{\partial\psi}{\partial\theta}\right)\right).$$

Taking into account $\psi = \psi\,(s', C, \theta)$ and $ds'/dt = 0$, we have:

$$\rho\,\frac{de_{int}}{dt} = -\rho\,\theta\,\frac{\partial^2\psi}{\partial\theta^2}\frac{d\theta}{dt} + \rho\,Tr\left(B\,\frac{dC}{dt}\right), \qquad [13.53]$$

with:

$$B = \frac{\partial\psi}{\partial C} - \theta\,\frac{\partial}{\partial\theta}\left(\frac{\partial\psi}{\partial C}\right). \qquad [13.54]$$

On the other hand, owing to [13.12] and [13.45], we have:

$$\frac{\partial\zeta}{\partial C} = -\beta\,\frac{\partial\psi}{\partial C}.$$

Hence, [13.36] becomes:

$$\sigma_R = 2\rho\,F\,\frac{\partial\psi}{\partial C}\,F^T,$$

and consequently:

$$\sigma_R - \theta\,\frac{\partial\sigma_R}{\partial\theta} = 2\rho\,F\,B\,F^T.$$

Taking into account [10.78], the last term in [13.53] becomes:

$$\rho\, Tr\, \left(B\,\frac{dC}{dt}\right) = 2\,\rho\, Tr\, \left(F\,B\,F^T D\right) = Tr\left(\left(\sigma_R - \theta\,\frac{\partial \sigma_R}{\partial \theta}\right)\, D\right).$$

Introducing it into [13.53] gives:

$$\rho\,\frac{de_{int}}{dt} = \rho\, c_v\,\frac{d\theta}{dt} + Tr\,\left(\left(\sigma_R - \theta\,\frac{\partial \sigma_R}{\partial \theta}\right)\, D\right),$$

where:

$$c_v = \theta\,\frac{\partial s}{\partial \theta} = -\theta\,\frac{\partial^2 \psi}{\partial \theta^2},$$

is the *heat capacity at constant volume*. Combining with the form [13.52] of the balance of energy leads to the *equation of heat transfer*:

$$\rho\, c_v\,\frac{d\theta}{dt} = \theta\, Tr\,\left(\frac{\partial \sigma_R}{\partial \theta}\, D\right) + Tr\,\left(\sigma_I\, D\right) - div\, h + \rho\,\frac{dq_I}{dt}. \qquad [13.55]$$

The physical interpretation of this equation is that the variation of reversible thermal energy, at the left-hand member, is equal, at the right-hand member, to the contributions of each term to the dissipation due to:

– the reversible stress variation resulting from the temperature one;

– the dissipative stresses;

– the heat flux;

– the irreversible heat sources.

We are now able to state the *second principle of thermodynamics*.

PRINCIPLE 13.2.– The *local production of entropy* of a continuous medium characterized by fields of velocity vector $\vec{U}$, temperature vector $\hat{\vec{W}}$ and momentum tensor $\hat{\mathbf{T}}$ is non-negative:

$$\Phi = \boldsymbol{Div}\,\left(\hat{\boldsymbol{T}}\,\hat{\vec{\boldsymbol{W}}}\right) - \left(\mathrm{e}^0(\boldsymbol{f}(\vec{\boldsymbol{U}}))\right)\,\left(\mathrm{e}^0(\boldsymbol{T}_I(\vec{\boldsymbol{U}}))\right) \geq 0, \qquad [13.56]$$

and vanishes if and only if the process is reversible.

In this expression, $\mathbf{e}^0$ is the time arrow and $\vec{U}$ is the 4-velocity (see section 10.2). In terms of tensor fields, expression [13.56] is covariant, and then consistent with Galileo's principle of relativity 1.1. As a scalar field, the value of Φ is invariant. Without gravitation and in any Galilean coordinate system, the expression of the local production of entropy is:

$$\Phi = div\left(\hat{T}\,\hat{W}\right) - \left(\mathrm{e}^0\, f\, U\right)\,\left(\mathrm{e}^0\, T_I U\right) \geq 0, \qquad [13.57]$$

If the process is reversible, $T_I = 0$ and, because of theorem 13.4:

$$\Phi = div\left(\hat{T}_R\,\hat{W}\right) = div\, S = \rho\,\frac{ds}{dt} = 0,$$

thus the entropy is constant, which explains the name of Φ. Conversely, if the local production of entropy is positive, it cannot be proved that the process is dissipative, reason for which it is a principle – i.e. an axiom – and not a theorem.

Next, let us calculate explicitly the expression of the local production of entropy. Owing to [10.15], [13.14] and [1.12], the former factor of the second term of [13.57]:

$$\mathrm{e}^0\, f\, U = (1\,0)\begin{pmatrix} \dfrac{\partial\beta}{\partial t} & \dfrac{\partial\beta}{\partial x} \\ \dfrac{\partial\, w}{\partial t} & \dfrac{\partial\, w}{\partial x} \end{pmatrix}\begin{pmatrix} 1 \\ v \end{pmatrix} = \frac{\partial\beta}{\partial t} + \frac{\partial\beta}{\partial x}\, v = \frac{d\beta}{dt},$$

is invariant under any Galilean transformation. Besides, [13.48] gives:

$$\mathrm{e}^0\, T_I U = (1\,0)\begin{pmatrix} \mathcal{H}_I & 0 \\ h + \mathcal{H}_I v - \sigma_I v & \sigma_I \end{pmatrix}\begin{pmatrix} 1 \\ v \end{pmatrix} = \mathcal{H}_I,$$

which is a Galilean invariant too. Thus, the local production of entropy reads also:

$$\Phi = div\left(\hat{T}\,\hat{W}\right) - \mathcal{H}_I\,\frac{d\beta}{dt} \geq 0. \qquad [13.58]$$

Now, we establish a new expression of the production of entropy.

THEOREM 13.5.– If the momentum tensor $\hat{\mathbf{T}}$ is divergence free, the local production of entropy [13.57] is given by:

$$\Phi = h \cdot grad\,\beta + \beta\, Tr\,\left(\sigma_I\, D\right) \geq 0. \qquad [13.59]$$

PROOF.– Starting from [13.58], owing to [14.22] and the first principle 13.33, it holds:

$$\Phi = Tr\left(\hat{T}\,\frac{\partial \hat{W}}{\partial X}\right) - \mathcal{H}_I \frac{d\beta}{dt} = Tr\ (T\ f) + \frac{\partial \zeta}{\partial X} N + q_I\ \frac{d\beta}{dt}.$$

Because of theorem 13.3 ◇ or equivalently [13.38], we have:

$$\Phi = Tr\ (T\ f) - Tr\ (T_R\ f) + q_I\ \frac{d\beta}{dt} = Tr\ (T_I\ f) + q_I\ \frac{d\beta}{dt}.$$

Using expression [13.14] of the friction and [13.48] of the irreversible momentum tensor:

$$Tr\ (T_I\ f) = -q_I\left(\frac{\partial \beta}{\partial t} + \frac{\partial \beta}{\partial x}\,v\right) + \frac{\partial \beta}{\partial x}\,h + Tr\left(\sigma_I\left(\frac{\partial w}{\partial x}\right) - v\,\frac{\partial \beta}{\partial x}\right).$$

Owing to $w = \beta\, v$, it holds:

$$Tr\ (T_I\ f) = -q_I\ \frac{d\beta}{dt} + h.\ grad\ \beta + \beta\, Tr\left(\sigma_I\,\frac{\partial v}{\partial x}\right),$$

and because σ_I is symmetric:

$$Tr\ (T_I\ f) + q_I\ \frac{d\beta}{dt} = h.\ grad\ \beta + \beta\, Tr\ (\sigma_I\ D)\,,$$

that achieves the proof. ■

Through the relation:

$$\Phi = h \cdot a + Tr\ (\sigma_I A),$$

the interest of theorem [13.5] is turning out a correspondence between:

– *thermodynamic forces* (or *affinities*)

$$a = grad\ \beta$$

$$A = \beta\, grad_s v = \beta\, D,; \qquad [13.60]$$

– and the corresponding *thermodynamic fluxes* h, σ_I.

$\hat{\mathbf{T}}_I$ being represented by $\tau_I = (h,\ \sigma_I)$ and $\mathbf{f}$ by $\alpha = (a,\ A)$, this dual pairing reads:

$$\Phi = \langle \tau_I,\ \alpha \rangle\,.$$

13.7. Constitutive laws in thermodynamics

To define completely the dissipative processes of the material, we need an additional relation called the constitutive law. In the most simple situations, it is given by a map $g : \alpha \mapsto \tau_I$, or more explicitly in terms of fluxes and affinities:

$$(a, A) \mapsto (h, \sigma_I) = g\,(a, A).$$

Before discussing some aspects of the constitutive laws, we want to characterize the non-dissipative processes due to the following proposition:

THEOREM 13.6.– For a continuum occupying a connected domain, let $g : \alpha \mapsto \tau_I$ be a continuous map defining a constitutive law and verifying the second principle

$$\forall\, \alpha, \qquad \Phi = \langle g\,(\alpha),\, \alpha\rangle \geq 0, \qquad [13.61]$$

Then, if the friction tensor field is identically null:

– $\diamondsuit$ the temperature field is uniform and the motion of the continuum is rigid;

– $\heartsuit$ the heat conduction flux and the viscous stresses vanish.

PROOF.– As the friction is null, $a = grad\ \beta = 0$ then β and $\theta = 1/\beta$ are uniform on a connected domain. Besides, $A = \beta\, D = 0$ and $\beta > 0$, then $D = grad_s v = 0$. According to theorem 9.1, in a connected domain, there exist maps $t \mapsto v_0(t) \in \mathbb{R}^3$ and $t \mapsto \omega(t) \in \mathbb{R}^3$ such that:

$$v\,(t, r) = v_0(t) + \omega\,(t) \times r,$$

that defines a rigid motion of the continuum and proves $\diamondsuit$.

If λ is a real number, the condition [13.61] gives:

$$\langle g\,(\lambda\alpha),\, \lambda\alpha\rangle = \lambda\,\langle g\,(\lambda\alpha),\, \alpha\rangle \geq 0,$$

which means that λ and $\langle g\,(\lambda\alpha),\, \alpha\rangle$ have the same sign. As λ approaches 0, by continuity:

$$\langle g\,(0),\, \alpha\rangle = 0.$$

As this occurs for any α, it is possible only if $\tau_I = g\,(0) = 0$. Then h and σ_I vanish, which proves $\heartsuit$. ■

Our aim is now to find the explicit form of the constitutive law in relatively simple situations, for instance when the behavior of the continuum is isotropic and the law is linear. First, we have to discuss how the components of $\mathbf{f}$ and $\mathbf{T_I}$ change under Galilean and Bargmannian transformations. Let us consider a Galilean transformation with boost u and rotation R:

$$P = \begin{pmatrix} 1 & 0 \\ u & R \end{pmatrix}.$$

The transformation law of 1-covariant and 1-contravariant tensors gives for f:

$$f' = P^{-1} f\, P,$$

then:

$$\frac{\partial \beta'}{\partial t'} = \frac{\partial \beta}{\partial t} + \frac{\partial \beta}{\partial x}\, u,$$

$$\frac{\partial \beta'}{\partial x'} = \frac{\partial \beta}{\partial x}\, R, \qquad [13.62]$$

$$\frac{\partial w'}{\partial t'} = R^T \left(\frac{\partial w}{\partial t} + \frac{\partial w}{\partial x}\, u \right) - \left(\frac{\partial \beta}{\partial t} + \frac{\partial \beta}{\partial x}\, u \right) R^T u,$$

$$\frac{\partial w'}{\partial x'} = R^T \left(\frac{\partial w}{\partial x} - u\, \frac{\partial \beta}{\partial x} \right) R. \qquad [13.63]$$

By transposing relation [13.62], we have:

$$a' = R^T a. \qquad [13.64]$$

Also, taking into account [13.62], [13.63] and the velocity addition formula [1.13], we get:

$$\frac{\partial w'}{\partial x'} - v'\, \frac{\partial \beta'}{\partial x'} = R^T \left(\frac{\partial w}{\partial x} - v\, \frac{\partial \beta}{\partial x} \right) R.$$

Hence, the transformation law of [13.60] is:

$$A' = R^T A\, R. \qquad [13.65]$$

Now, we can determine the invariants of α. It is easy to verify that there are three eigenvalues of A, $\| a \|$, $\| A\, a \|$ and $a^T A\, a$.

The transformation laws [13.25] and [13.50] of h and σ_I are formally the same as the ones [13.64] and [13.65] of a and A. Analogously to α, the six independent invariants of τ_I are the three eigenvalues of σ_I, $\parallel h \parallel$, $\parallel \sigma_I h \parallel$ and $h^T \sigma_I \, h$. Once again, we can verify that the production of entropy is invariant:

$$h' \cdot a' + Tr\,(\sigma'_I A') = h \cdot a + Tr\,(\sigma_I A).$$

On this basis, we can construct constitutive laws. For instance, an isotropic linear law has the form:

$$h = k_1 a, \qquad [13.66]$$

$$\sigma_I = k_2\, Tr\,(A)\, I_{\mathbb{R}^3} + k_3\, A, \qquad [13.67]$$

where $k_1, k_2, k_3 \in \mathbb{R}$. Introducing [13.66] and [13.67] into the production of entropy [13.59] gives:

$$\Phi = k_1 \parallel a \parallel^2 + k_2\, (Tr\, A)^2 + k_3\, Tr\, \left(A^2\right),$$

which is satisfied if the following restrictions are imposed to the material parameters:

$$k_1 \geq 0, \qquad k_3 \geq 0, \qquad k_2 + \frac{k_3}{3} \geq 0.$$

LAW 13.1.– In terms of temperature, the constitutive laws are:

– *Fourier's law* or *law of heat conduction*:

$$\boxed{h = -k\, grad\, \theta,} \qquad [13.68]$$

where $k = k_1 \,/\, \theta^2 \geq 0$ is the *thermal conductivity*.

– *Newton's viscous flow law*:

$$\boxed{\sigma_I = \eta\, (div\; v)\, I_{\mathbb{R}^3} + 2\mu\, grad_s\, v,} \qquad [13.69]$$

where $\eta = k_2 \,/\, \theta$ and $\mu = k_3 \,/\, 2\,\theta \geq 0$ is the *dynamic viscosity*.

For simple fluids (in particular, water, air and gases as methane), the law can be simplified by assuming that the viscous stresses σ_I are traceless (*Stokes hypothesis*), leading to:

$$\sigma_I = 2\mu \left(grad_s\, v - \frac{1}{3}\, (div\; v)\, I_{\mathbb{R}^3} \right).$$

In many situations, the material parameters k and η are considered to be constant. For barotropic fluids, owing to [10.67] and [14.16], we have:

$$(div\,\sigma_R)^T = -grad\,q.$$

On the other hand, Newton's viscous flow law [13.69] combined with [14.23], [14.18] and [14.16] gives:

$$(div\,\sigma_I)^T = \mu\,\Delta v + \frac{\mu}{3}\,grad\,(div\,v).$$

Introducing the previous two expressions into the balance of linear momentum (theorem 13.2, ♡) with $\sigma = \sigma_R + \sigma_I$, we obtain *Navier–Stokes equations*:

$$\boxed{\rho\left[\frac{\partial v}{\partial t} + \frac{\partial v}{\partial x}\,v\right] = -grad\,q + \mu\,\Delta v + \frac{\mu}{3}\,grad\,(div\,v).} \qquad [13.70]$$

DEFINITION 13.6.– A fluid is *incompressible* if the mass of each volume element remains constant :

$$\frac{d\rho}{dt} = 0,$$

then the density ρ is an integral of the motion and, owing to [10.55]:

$$div\,v = 0.$$

Also, taking into account [14.32] and [10.77], J is an integral of the motion:

$$\frac{dJ}{dt} = J\,Tr\left(\frac{dF}{dt}\,F^{-1}\right) = J\,div\,v = 0.$$

Hence, in [10.76], the derivative has no sense and the pressure q is indeterminate. The *Navier–Stokes equations for incompressible flows* are:

$$\boxed{div\,v = 0, \qquad \rho\left[\frac{\partial v}{\partial t} + \frac{\partial v}{\partial x}\,v\right] = -grad\,q + \mu\,\Delta v.} \qquad [13.71]$$

13.8. Thermodynamics and Galilean gravitation

Until now, we have only been concerned with the uniform straight motion which can be described by the calculus of variation with a Lagrangian equal to the kinetic energy as mentioned in section 6.2. It was also seen that this expression of the Lagrangian is not general and, for a particle subjected to a Galilean gravitation, it must be replaced with [6.13]:

$$\mathcal{L}(t,x,v) = \frac{1}{2} m \parallel v \parallel^2 - m\,\phi + m\,A\cdot v,$$

containing the gravitation potentials ϕ and A. Let us introduce a coordinate system $\hat{X}'$ for which we have:

$$dz' = \frac{\mathcal{L}}{m}\,dt.$$

This extra coordinate z' has the physical dimension and the meaning of an *action by unit mass*. Taking into account [13.4], we obtain:

$$dz' = dz - \phi\,dt + A\cdot dx,$$

which can be completed by:

$$dt' = dt, \qquad dx' = dx,$$

to define a linear transformation:

$$d\hat{X}' = \hat{Q}^{-1}\,d\hat{X}.$$

where:

$$\hat{Q}^{-1} = \begin{pmatrix} 1 & 0 & 0 \\ 0 & 1_{\mathbb{R}^3} & 0 \\ -\phi & A^T & 1 \end{pmatrix}, \qquad \hat{Q} = \begin{pmatrix} 1 & 0 & 0 \\ 0 & 1_{\mathbb{R}^3} & 0 \\ \phi & -A^T & 1 \end{pmatrix}. \qquad [13.72]$$

What is the physical interpretation of these transformations? Before applying it, the particle is, in absence of gravitation, in uniform straight motion (USM). The effect of applying such a transformation is to embed the particle into the gravitation field. It is straightforward to verify that the set of such transformation matrix is an abelian subgroup of the affine group $\mathbb{A}ff(5)$. It is also worth to notice that, according to the transformation law [14.2] of 2-covariant tensors, Gram's matrix of the covariant metric tensor $\hat{\mathbf{G}}$ in the new coordinate system $\hat{X}'$ is given by:

$$\hat{G}' = \hat{Q}^T \hat{G}\,\hat{Q},$$

which gives for the metric embedded in the gravitation field:

$$\hat{G}' = \begin{pmatrix} -2\phi & A^T & -1 \\ A & 1_{\mathbb{R}^3} & 0 \\ -1 & 0 & 0 \end{pmatrix}. \qquad [13.73]$$

The attentive readers will observe that reducing this matrix to the space-time provides the one given by [10.17]. Our starting point is now to work in these coordinate systems $\hat{X}'$ that we call *Bargmannian coordinate systems* (see Comment 3, section 13.9). When equipped with the previous covariant metric, the 5-dimensional space $\hat{\mathcal{U}}$ is now a Riemannian manifold and $\mathcal{U}$ is a 4-dimensional submanifold thereof.

There exists one and only one symmetric covariant differential such that the covariant differential of the metric vanishes. The only non-vanishing quantities [14.44] are [12.4] and [12.5]. Using [14.45] and taking into account the definition [6.14] of Galilean gravitation potentials, the only non-vanishing Christoffel's symbols are:

$$\Gamma^j_{00} = -g^j, \qquad \Gamma^j_{0k} = \Gamma^j_{k0} = \Omega^j_k, \qquad [13.74]$$

$$\Gamma^4_{00} = \frac{\partial\phi}{\partial t} - A\cdot g, \qquad \Gamma^4_{ij} = \frac{1}{2}\left(\frac{\partial A_i}{\partial x^j} + \frac{\partial A_j}{\partial x^i}\right) = (grad_s A)^i_j, \qquad [13.75]$$

$$\Gamma^4_{0i} = \Gamma^4_{i0} = \frac{\partial\phi}{\partial x^i} - \frac{1}{2}\left(\frac{\partial A_i}{\partial x^j} - \frac{\partial A_j}{\partial x^i}\right)A^j = (grad\,\phi - \Omega\times A)^i. \qquad [13.76]$$

It is worth to observe that we recover [10.49]. In matrix form, the gravitation of the space $\hat{\mathcal{U}}$ reads:

$$\hat{\Gamma}(d\hat{X}) =$$

$$\begin{pmatrix} 0 & 0 & 0 \\ j(\Omega)\,dx - g\,dt & j(\Omega)\,dt & 0 \\ \left(\frac{\partial\phi}{\partial t} - A\cdot g\right)dt + (grad\,\phi - \Omega\times A)\cdot dx & [(grad\,\phi - \Omega\times A)\,dt - grad_s A\,dx]^T & 0 \end{pmatrix},$$

It is the expansion of the space-time gravitation [3.38] to the fifth dimension.

In a similar way, the thermodynamical tensors can be embedded into the gravitation field. Applying the matrix $\hat{Q}^{-1}$ given by [13.72] preserves β, w and embeds the ζ component in the gravitation, which reads omitting the bar:

$$\zeta = \zeta_{int} + \frac{1}{2\beta}\parallel w\parallel^2 - \beta\,\phi + A\cdot w.$$

Taking into account [14.37], the friction tensor is represented in a Galilean coordinate system by the 4×4 matrix:

$$f = \nabla W = \begin{pmatrix} \dfrac{\partial \beta}{\partial t} & \dfrac{\partial \beta}{\partial x} \\ \dfrac{\partial w}{\partial t} - \beta g + \Omega \times w & \dfrac{\partial w}{\partial x} + \beta j(\Omega) \end{pmatrix} . \qquad [13.77]$$

Let us calculate the expression of the momentum tensor embedded in the gravitation field. The transformation law [13.16] reads:

$$\hat{T}' = Q^{-1} \hat{T} \hat{Q},$$

where $\hat{T}$ is given by [13.26], $\hat{Q}$ is given by [13.72] and the corresponding Q is $1_{\mathbb{R}^4}$, which leads to:

$$\hat{T}' = \begin{pmatrix} \mathcal{H}' & -\pi^T & \rho \\ h + \mathcal{H}' v - \sigma v & \sigma - v \pi^T & \rho v \end{pmatrix}, \qquad [13.78]$$

where the following occur:

– the Hamiltonian density: $\mathcal{H}' = \rho \left(\frac{1}{2} \parallel v \parallel^2 + \phi + e_{int} - q_I \right)$;

– the generalized linear momentum: $\pi = \rho \, (v + A)$.

The attentive readers will observe that we recover as a particular case the expression [10.81] or [12.24] of the Hamiltonian density and the expression [12.23] of π. In the following, we are implicitly supposed to work in Bargmannian coordinate systems, but the prime will be omitted for the sake of easiness. Let us now examine the expression of the first principle of thermodynamics [13.33] in the presence of gravitation.

THEOREM 13.7.– If $\hat{T}$ is covariant divergence free:

$$Div_X \, \hat{T} = 0,$$

then, we have:

– ♢ *balance of mass*: $\dfrac{\partial \rho}{\partial t} + div \, (\rho v) = 0$;

– ♡ *balance of linear momentum*: $\rho \left[\dfrac{\partial v}{\partial t} + \dfrac{\partial v}{\partial x} v \right] = (div \, \sigma)^T + \rho \, (g - 2 \Omega \times v)$;

– ♠ *balance of energy*: $\dfrac{\partial \mathcal{H}}{\partial t} + div \, (h + \mathcal{H} v - \sigma v) = \rho \left(\dfrac{\partial \phi}{\partial t} - \dfrac{\partial A}{\partial t} \cdot v \right)$.

PROOF.– As the space-time $\mathcal{U}$ is a submanifold of the 5-dimensional space $\hat{\mathcal{U}}$, an event $\boldsymbol{X} \in \mathcal{U}$ also belongs to $\hat{\mathcal{U}}$. We consider a momentum field $\boldsymbol{X} \mapsto \hat{\boldsymbol{T}}(\boldsymbol{X})$ where $\hat{\boldsymbol{T}}(\boldsymbol{X})$ is a 1-covariant tensor on the tangent space $T_{\boldsymbol{X}}\hat{\mathcal{U}}$ with vector values in $T_{\boldsymbol{X}}\mathcal{U}$ (which can be identified to a linear map from $T_{\boldsymbol{X}}\hat{\mathcal{U}}$ to $T_{\boldsymbol{X}}\mathcal{U}$):

$$\hat{\boldsymbol{T}} : T_{\boldsymbol{X}}\hat{\mathcal{U}} \to T_{\boldsymbol{X}}\mathcal{U} : \hat{\vec{\boldsymbol{V}}} \mapsto \hat{\boldsymbol{T}}(\hat{\vec{\boldsymbol{V}}}). \qquad [13.79]$$

We wish to calculate its covariant divergence. In addition to convention 1.1, we adopt the extra one: Greek indices $\hat{\alpha}, \hat{\beta}, \hat{\gamma}$ and so on run over the five coordinate labels $0, 1, 2, 3, 4$. The basis $(\vec{e}_\alpha)$ of $T_{\boldsymbol{X}}\mathcal{U}$ is completed by $\vec{e}_4$ to build a basis $(\vec{e}_{\hat{\alpha}})$ of $T_{\boldsymbol{X}}\hat{\mathcal{U}}$ in which the momentum tensor is decomposed as:

$$\hat{\boldsymbol{T}} = \hat{T}^\gamma \vec{e}_\gamma, \qquad \hat{T}^\gamma = \hat{T}^\gamma_{\hat{\alpha}} \boldsymbol{e}^{\hat{\alpha}}.$$

Taking into account [14.37], its covariant differential is:

$$\nabla_{\overrightarrow{d\boldsymbol{X}}} \hat{\boldsymbol{T}} = \nabla_{\overrightarrow{d\boldsymbol{X}}} (\hat{T}^\gamma \vec{e}_\gamma) = (\nabla_{\overrightarrow{d\boldsymbol{X}}} \hat{T}^\gamma)\, \vec{e}_\gamma + \hat{T}^\gamma\, (\nabla_{\overrightarrow{d\boldsymbol{X}}}\, \vec{e}_\gamma) = (\nabla_{\overrightarrow{d\boldsymbol{X}}} \hat{T}^\gamma + \Gamma^\gamma_\rho \hat{T}^\rho)\, \vec{e}_\gamma,$$

where, owing to [14.37]:

$$\nabla_{\overrightarrow{d\boldsymbol{X}}} \hat{T}^\gamma = \nabla_{dX}\, (\hat{T}^\gamma_{\hat{\alpha}} \boldsymbol{e}^{\hat{\alpha}}) = d\hat{T}^\gamma_{\hat{\alpha}}\, \boldsymbol{e}^{\hat{\alpha}} + \hat{T}^\gamma_{\hat{\alpha}}\, \nabla_{dX}\, \boldsymbol{e}^{\hat{\alpha}} = (d\hat{T}^\gamma_{\hat{\alpha}} - \hat{T}^\gamma_{\hat{\beta}} \Gamma^{\hat{\beta}}_{\hat{\alpha}})\, \boldsymbol{e}^{\hat{\alpha}}.$$

Hence, we obtain:

$$\nabla_{\overrightarrow{d\boldsymbol{X}}} \hat{\boldsymbol{T}} = \left[\nabla_{dX} \hat{T}^\gamma_{\hat{\alpha}}\, \boldsymbol{e}^{\hat{\alpha}}\right] \vec{e}_\gamma,$$

with:

$$\nabla_{dX} \hat{T}^\gamma_{\hat{\alpha}} = d\hat{T}^\gamma_{\hat{\alpha}} + \Gamma^\gamma_\rho \hat{T}^\rho_{\hat{\alpha}} - \hat{T}^\gamma_{\hat{\beta}} \Gamma^{\hat{\beta}}_{\hat{\alpha}}\,;$$

Hence, there exists a field $\nabla \hat{\boldsymbol{T}}$ of 2-covariant and 1-contravariant tensors such that:

$$\nabla_{\overrightarrow{d\boldsymbol{X}}} \hat{\boldsymbol{T}} = (\nabla \hat{\boldsymbol{T}}) \cdot \overrightarrow{d\boldsymbol{X}}.$$

Using Christoffel's symbols [14.38], we have:

$$\nabla \hat{\boldsymbol{T}} = \left[\nabla_\sigma \hat{T}^\gamma_{\hat{\alpha}}\, \boldsymbol{e}^{\hat{\alpha}}\right] \vec{e}_\gamma \otimes \boldsymbol{e}^\sigma,$$

with:

$$\nabla_\sigma \hat{T}^\gamma_{\hat{\alpha}} = \frac{\partial \hat{T}^\gamma_{\hat{\alpha}}}{\partial X^\sigma} + \Gamma^\gamma_{\rho\sigma} T^\rho_{\hat{\alpha}} - \hat{T}^\gamma_{\hat{\beta}} \Gamma^{\hat{\beta}}_{\hat{\alpha}\sigma}.$$

By contraction, we define the *covariant divergence* of the momentum tensor:

$$\boldsymbol{Div}\, \hat{\boldsymbol{T}} = \nabla_\gamma \hat{T}^\gamma_{\hat{\alpha}}\, \boldsymbol{e}^{\hat{\alpha}},$$

with:

$$\nabla_\gamma \hat{T}^\gamma_{\hat{\alpha}} = \frac{\partial \hat{T}^\gamma_{\hat{\alpha}}}{\partial X^\gamma} + \Gamma^\gamma_{\rho\gamma} T^\rho_{\hat{\alpha}} - \hat{T}^\gamma_{\hat{\beta}} \Gamma^{\hat{\beta}}_{\hat{\alpha}\gamma}. \qquad [13.80]$$

In indicial notation, the components of $\hat{\boldsymbol{T}}$ are:

$$\hat{T}^0_0 = \mathcal{H}, \qquad \hat{T}^0_i = -\delta_{ik}\pi^k, \qquad \hat{T}^0_4 = \rho, \qquad [13.81]$$

$$\hat{T}^j_0 = h^j + \mathcal{H} v^j - \sigma^j_k v^k, \qquad \hat{T}^j_i = \sigma^j_i - v^j \delta_{ik}\pi^k, \qquad \hat{T}^j_4 = p^j. \qquad [13.82]$$

The first principle of the thermodynamics [13.33] reads:

$$\nabla_\gamma \hat{T}^\gamma_{\hat{\alpha}} = 0,$$

where Christoffel's symbols are given by [13.74], [13.75] and [13.76]. Putting $\hat{\alpha} = 4$ in the previous equation and taking into account the vanishing terms, we have:

$$\nabla_\gamma \hat{T}^\gamma_4 = \frac{\partial \hat{T}^\gamma_4}{\partial X^\gamma} = 0,$$

which allows us to recover the balance of mass $\Diamond$. Similarly, putting $\hat{\alpha} = i$ and taking into account the non-vanishing terms, it holds that:

$$\nabla_\gamma \hat{T}^\gamma_i = \frac{\partial \hat{T}^\gamma_i}{\partial X^\gamma} - \hat{T}^0_j \Gamma^j_{i0} - \hat{T}^0_4 \Gamma^4_{i0} - \hat{T}^j_4 \Gamma^4_{ij} = 0,$$

or, owing to the momentum components [13.81] and [13.82]:

$$-\frac{\partial}{\partial t}\left(\delta_{ik}\pi^k\right) + \frac{\partial}{\partial x^j}\left(\sigma^j_i - v^j \delta_{ik}\pi^k\right) + \delta_{jk}\pi^k \Omega^j_i - \rho\, \Gamma^4_{i0} - p^j \Gamma^4_{ij} = 0.$$

Owing to [14.17], it reads in matrix form:

$$-\frac{\partial}{\partial t}(\rho\,(v+A)^T) + div\,\sigma - div\,(\rho v)\,(v+A)^T - \rho v^T grad\,(v+A)$$
$$+\,\rho\,(v+A)^T j(\Omega) - \rho\,(grad\,\phi - \Omega \times A)^T + \rho\, v^T grad_s\, A = 0.$$

But, owing to [6.14], it holds that:

$$grad_s\, A = \frac{\partial A}{\partial x} - j(\Omega),$$

then, taking into account [12.26], we recover [12.27] and finishing the calculation as in the proof of theorem 12.1, we demonstrate the balance of linear momentum ♡.

Finally, putting $\hat{\alpha} = 0$ and taking into account the non-vanishing terms, it holds that:

$$\nabla_\gamma \hat{T}_0^\gamma = \frac{\partial \hat{T}_0^\gamma}{\partial X^\gamma} - \hat{T}_j^0 \Gamma_{00}^j - \hat{T}_j^i \Gamma_{i0}^4 - \hat{T}_4^0 \Gamma_{00}^4 - \hat{T}_4^j \Gamma_{j0}^4 = 0,$$

which reads in matrix form, after some simplifications:

$$\frac{\partial \mathcal{H}}{\partial t} + div\ (h + \mathcal{H} v - \sigma v) - \rho \left(v \cdot (g + grad\, \phi) + \frac{\partial \phi}{\partial t} \right) = 0.$$

But, owing to [6.14], it holds that:

$$v \cdot g + \frac{d\phi}{dt} = v \cdot (g + grad\, \phi) + \frac{\partial \phi}{\partial t} = \frac{\partial \phi}{\partial t} - v \cdot \frac{\partial A}{\partial t} \qquad [13.83]$$

which leads to the balance of energy ♠ and achieves the proof. ■

Theorem 13.3 concerning the reversible processes remains true. Only the demonstration of ◇ is different. Before calculating, it is worth noting that the covariant derivative of ζ is meaningful because ζ is not a scalar but a component of the temperature vector. We can verify that:

$$(\nabla \zeta)\, N = \rho \left(\frac{d\zeta}{dt} - \beta g \cdot v \right).$$

the remaining part of the calculation is straightforward taking into account the expression [13.77] of the friction embedded into the gravitation field.

For the dissipative continua, the equation of heat transfer [13.55] is slightly modified. We let the readers to show that [13.52] is replaced by with:

$$\rho \frac{d}{dt} (e_{int} + \phi) = Tr\ (\sigma\, D) - div\, h + \rho \frac{dq_I}{dt} + \rho \left(\frac{\partial \phi}{\partial t} - \frac{\partial A}{\partial t} \cdot v \right),$$

or, taking into account [13.83]:

$$\rho \frac{de_{int}}{dt} = Tr\ (\sigma\, D) + g \cdot v - div\ h + \rho \frac{dq_I}{dt}.$$

Next, the *equation of heat transfer with gravitation* is:

$$\boxed{\rho\, c_v \frac{d\theta}{dt} = \theta\, Tr \left(\frac{\partial \sigma_R}{\partial \theta} D \right) + Tr\ (\sigma_I\, D) + g \cdot v - div\ h + \rho \frac{dq_I}{dt}.} \quad [13.84]$$

Taking into account the expression [13.77] of the friction, readers can also easily verify that the expression [13.58] of the production of entropy is replaced with:

$$\Phi = Div \left(\hat{T}\, \hat{W} \right) - \mathcal{H}_I \frac{d\beta}{dt} \geq 0, \quad [13.85]$$

where now the covariant divergence in the first term occurs. To be consistent with Galileo's principle of relativity 1.1, theorem 13.5 is replaced with:

THEOREM 13.8.– If the momentum tensor $\hat{\mathbf{T}}$ is *covariant* divergence free, the local production of entropy [13.57] is given by [13.59]:

$$\Phi = h.\ grad\ \beta + \beta\, Tr\ (\sigma_I\ D) \geq 0.$$

PROOF.– According to the rule [14.41], we have:

$$\boldsymbol{Div}\, (\hat{\boldsymbol{T}} \cdot \hat{\boldsymbol{W}}) = (\boldsymbol{Div}\, \hat{\boldsymbol{T}}) \cdot \hat{\boldsymbol{W}} + \hat{\boldsymbol{T}} : \nabla \hat{\boldsymbol{W}},$$

or in local coordinates:

$$Div(\hat{T}\, \hat{W}) = (Div\, \hat{T})\, \hat{W} + Tr\, (\hat{T}\, \nabla \hat{W}),$$

hence, starting from [13.85] and owing to the first principle 13.33, it holds that:

$$\Phi = Tr \left(\hat{T} \nabla \hat{W} \right) - \mathcal{H}_I \frac{d\beta}{dt} = Tr\ (T\ f) + (\nabla \zeta)\, N + q_I \frac{d\beta}{dt}.$$

Because of theorem 13.3 ◊ or equivalently [13.37], we have:

$$\Phi = Tr\ (T_I\ f) + q_I \frac{d\beta}{dt},$$

and the proof follows the one of theorem 13.5, which achieves the proof. ■

As an exercise, readers can verify that, starting from [13.58] and using the balance of mass, the local production of entropy reads (see Comment 4, section 13.9 below):

$$\Phi = \rho \frac{ds}{dt} - \frac{\rho}{\theta} \frac{dq_I}{dt} + div \left(\frac{h}{\theta} \right) \geq 0. \quad [13.86]$$

13.9. Comments for experts

COMMENT 1.– Bargmann's group was introduced to solve problems of group quantization. In fact, Galileo's group is not quantizable [SOU 70, SOU 97] and the reason of this failure is cohomologic as discussed in section 17.4. For more explanation about the construction of Bargmann's group, the readers are referred to section 17.6. Although Bargmann's group was introduced for applications to quantum mechanics, it also turns out to be very useful in thermodynamics.

COMMENT 2.– This definition is meaningful only in the absence of gravitation as discussed in section 16.7.

COMMENT 3.– To know more about Bargmannian coordinate systems, the readers are referred to section 16.7.

COMMENT 4.– This relation is known in the literature as Clausius–Duhem inequality, but it seems to have first appeared in Truesdell's works [TRU 52, TRU 60].

14

Mathematical Tools

14.1. Group

A *group* is a set G together with an *operation* called the *group law*, which we will denote mutiplicatively by:

$$G \times G \to G : (a, b) \mapsto ab,$$

with the following properties:

– *associativity:* $(ab)c = a(bc)$;

– *existence of an identity element* e such that: $\forall a \in G, \qquad ae = ea = a$;

– *existence of an inverse element* a^{-1} for any $a \in G : a\,a^{-1} = a^{-1}a = e$.

If the operation is commutative, the group is called *abelian*. For instance, a linear space is an abelian group for the addition with zero as an identity element and the opposite vector as an inverse element. The set of regular $n \times n$ matrices is a group for the matrix product called the *linear group* and is denoted by $\mathbb{GL}(n)$. The set of the regular affine transformations of $\mathbb{R}^n$ is called the *affine group* and is denoted by $\mathbb{A}ff(n)$.

A subset H is a *subgroup* of G if it is also a group for the operation of G. For instance, $\mathbb{GL}(n)$ is a subgroup of $\mathbb{A}ff(n)$. The set of the 3×3 orthogonal matrices is a subgroup of $\mathbb{GL}(3)$ called the *orthogonal group* and is denoted by $\mathbb{O}(3)$. The set of the rotations of $\mathbb{R}^3$ is a subgroup of $\mathbb{O}(3)$ called the *special orthogonal group* and is denoted by $\mathbb{SO}(3)$. The Euclidean transformations (respectively, special Euclidean transformations) are affine transformation of $\mathbb{R}^3$ of which the linear part is an orthogonal transformation (respectively, a rotation). The set of Euclidean transformations (respectively, special Euclidean transformations) is a subgroup of $\mathbb{A}ff(3)$ and is denoted by $\mathbb{E}(3)$ (respectively, $\mathbb{SE}(3)$).

A *left* (respectively, *right*) *action* of G onto a set M is a map:

$$\Phi : G \times M \to M : (a, x) \mapsto x' = \Phi(a, x) = a \cdot x,$$

such that $a \cdot (b \cdot x) = (ab) \cdot x$ (respectively, $a \cdot (b \cdot x) = (ba) \cdot x$). A left action in which we substitute a for a^{-1} is a right action. For instance, the map $(P, A) \mapsto A' = P^{-1} A P$ is a right action of $\mathbb{GL}(n)$ onto $\mathbb{M}_{nn}$.

Hence, the group defines a family of *transformations* (or *symmetries*) of M. We said that G is a *transformation group* (or a *symmetry group*) of M. The *orbital map*:

$$\Phi_x : G \to M : a \mapsto x' = a \cdot x,$$

defines the *variance law*. The *orbit* of x is the value set of the orbital map, i.e. the set of all the elements of M which can be reached from x through a symmetry:

$$orb(x) = \{x' \quad s.t. \quad \exists a \in G \quad and \quad x' = a \cdot x\}.$$

The *isotropy group* of $x \in M$ is the subgroup of all transformations which fix x:

$$iso(x) = \{a \in G \quad s.t. \quad a \cdot x = x\}.$$

It can be viewed as the symmetry subgroup for x. A subgroup H of G naturally acts onto a set M by restriction to H of the action of G onto M.

A *linear* (respectively, *affine*) *representation* (of finite dimension n) of a group G is a map ρ from G into $\mathbb{GL}(n)$ (respectively, $\mathbb{A}ff(n)$) such that:

$$\forall a_1, a_2 \in G, \qquad \rho(a_1 a_2) = \rho(a_1)\, \rho(a_2).$$

14.2. Tensor algebra

14.2.1. *Linear tensors*

A *tensor* is an object:

– that assigns a set of scalars, called its *components*, to each linear frame S of a linear space $\mathcal{T}$ of finite dimension n;

– with a *transformation law* of these components, when changing of frames, which is a linear representation of $\mathbb{GL}(n)$.

A linear tensor can be constructed as a *multilinear map*, that is a map which is linear with respect to each of its arguments. Let $\mathcal{T}$ and $\mathcal{R}$ be two linear spaces of finite dimensions n and m. We will define the tensors on $\mathcal{T}$ with values in $\mathcal{R}$. $\mathcal{T}$ is called the *source space* and $\mathcal{R}$ is called the *target space*. If $\mathcal{R}$ is different from $\mathbb{R}$, the tensor is *vector valued*. If $\mathcal{R} = \mathbb{R}$, we often said more simply that they are tensors on $\mathcal{T}$. The linear tensors can be classified into three families.

The *p-covariant tensors* (or *covariant tensors of rank p*) are the multilinear maps:

$$\mathbf{T} : \overbrace{\mathcal{T} \times \mathcal{T} \times \cdots \mathcal{T}}^{p\ times} \to \mathcal{R} : (\vec{\mathbf{V}}_1, \vec{\mathbf{V}}_2, \cdots, \vec{\mathbf{V}}_p) \mapsto \mathbf{T}(\vec{\mathbf{V}}_1, \vec{\mathbf{V}}_2, \cdots, \vec{\mathbf{V}}_p).$$

As a set of maps from the n-ary cartesian power of $\mathcal{T}$ into the linear space $\mathcal{R}$, the set of the p-covariant tensors is a linear space for the operations defined by [7.25] and [7.26]. It is denoted by $(\otimes^p \mathcal{T}^*) \otimes \mathcal{R}$, or more simply by $\otimes^p \mathcal{T}^*$ if $\mathcal{R} = \mathbb{R}$. Hence, we generalize the concept of linear forms which are 1-covariants tensors and the one of linear maps from $\mathcal{T}$ into $\mathcal{R}$ which are 1-covariants tensors on $\mathcal{T}$ with values in $\mathcal{R}$.

The *q-contravariant tensors* (or *contravariant tensors of rank q*) are the multilinear maps:

$$\mathbf{T} : \overbrace{\mathcal{T}^* \times \mathcal{T}^* \times \cdots \mathcal{T}^*}^{q\ times} \to \mathcal{R} : (\mathbf{\Phi}_1, \mathbf{\Phi}_2, \cdots, \mathbf{\Phi}_q) \mapsto \mathbf{T}(\mathbf{\Phi}_1, \mathbf{\Phi}_2, \cdots, \mathbf{\Phi}_q).$$

The set of the q-contravariant tensors is a linear space and is denoted by $(\otimes^q \mathcal{T}) \otimes \mathcal{R}$ or more simply by $\otimes^q \mathcal{T}$ if $\mathcal{R} = \mathbb{R}$. The 1-contravariants tensors are linear forms from $\mathcal{T}^*$ into $\mathbb{R}$, then the elements $\hat{\mathbf{V}}$ of the dual space $\mathcal{T}^{**}$ of $\mathcal{T}^*$. It is called the *bidual* space and has the same dimension as $\mathcal{T}^*$ then as $\mathcal{T}$. For any $\vec{\mathbf{V}} \in \mathcal{T}$, the map $\hat{\mathbf{V}}$ defined by $\hat{\mathbf{V}}(\mathbf{\Phi}) = \mathbf{\Phi}(\vec{\mathbf{V}})$ is an element of the bidual verifying:

$$\widehat{\mathbf{V} + \mathbf{U}} = \hat{\mathbf{V}} + \hat{\mathbf{U}}, \qquad \widehat{\lambda \mathbf{V}} = \lambda \hat{\mathbf{V}}, \qquad [14.1]$$

and the map $\vec{\mathbf{V}} \mapsto \hat{\mathbf{V}}$ is one-to-one from $\mathcal{T}$ into $\mathcal{T}^{**}$. Hence, the 1-contravariants tensors can be identified to the vectors. Owing to [7.27], we have:

$$\hat{\mathbf{e}}_i(\mathbf{\Phi}) = \mathbf{\Phi}(\vec{\mathbf{e}}_i) = \Phi_i.$$

The *mixed p-covariant* and *q-contravariant tensors* are the multilinear maps:

$$\mathbf{T} : \overbrace{\mathcal{T} \times \cdots \mathcal{T}}^{p\ times} \times \overbrace{\mathcal{T}^* \times \cdots \mathcal{T}^*}^{q\ times} \to \mathcal{R} :$$

$$(\vec{\mathbf{V}}_1, \cdots, \vec{\mathbf{V}}_p, \mathbf{\Phi}_1, \cdots, \mathbf{\Phi}_q) \mapsto \mathbf{T}(\vec{\mathbf{V}}_1, \cdots, \vec{\mathbf{V}}_p, \mathbf{\Phi}_1, \cdots, \mathbf{\Phi}_q).$$

The order in which the arguments appear in $\mathbf{T}$ must be specified. To simplify, we choose here to order the arguments starting with all the elements of $\mathcal{T}$ and following with the ones of $\mathcal{T}^*$. The set of the p-covariant and q-contravariant tensors is a linear space and is denoted by $(\otimes^p\mathcal{T}^*)\otimes(\otimes^q\mathcal{T})\otimes\mathcal{R}$ or more simply by $(\otimes^p\mathcal{T}^*)\otimes(\otimes^q\mathcal{T})$ if $\mathcal{R}=\mathbb{R}$. Hence, we generalize the concept of linear maps from $\mathcal{T}$ into itself which are mixed 1-covariant and 1-contravariants tensors through the identification of the linear map $\mathbf{A}$ with the tensor $\hat{\mathbf{A}}$ defined by $\hat{\mathbf{A}}(\mathbf{\Phi},\vec{\mathbf{U}})=\mathbf{\Phi}(\mathbf{A}(\vec{\mathbf{U}}))$.

Let $\mathbf{T}$ and $\mathbf{T}'$ be two arbitrary tensors. For instance, we will suppose that $\mathbf{T}$ is 1-covariant and 1-contravariant and $\mathbf{T}'$ is 1-covariant. It is clear that the scalar $\mathbf{T}(\mathbf{\Phi},\vec{\mathbf{U}})\,\mathbf{T}'(\vec{\mathbf{V}})$ linearly depends on $\mathbf{\Phi},\vec{\mathbf{U}}$ and $\vec{\mathbf{V}}$. Thus, the map $\mathbf{T}''$ defined by $\mathbf{T}''(\mathbf{\Phi},\vec{\mathbf{U}},\vec{\mathbf{V}})=\mathbf{T}(\mathbf{\Phi},\vec{\mathbf{U}})\,\mathbf{T}'(\vec{\mathbf{V}})$ is a 2-covariant and 1-contravariant tensor. $\mathbf{T}''$ is called the *tensor product* of $\mathbf{T}$ and $\mathbf{T}'$ and is denoted by $\mathbf{T}\otimes\mathbf{T}'$. The generalization of the definition to arbitrary tensors is straightforward. The tensor product is associative but is not in general commutative. It is distributive over the addition.

Let us consider an arbitrary tensor that we will suppose, for instance 3-covariant and 1-contravariant:

$$\mathbf{T}(\vec{\mathbf{U}},\mathbf{\Phi},\vec{\mathbf{V}},\vec{\mathbf{W}})$$

Let us choose a basis S of $\mathcal{T}$. According to the linearity of $\mathbf{T}$, we have:

$$\mathbf{T}(\vec{\mathbf{U}},\mathbf{\Phi},\vec{\mathbf{V}},\vec{\mathbf{W}})=\mathbf{T}(\sum_i U^i\vec{\mathbf{e}}_i,\sum_j \Phi_j\mathbf{e}^j,\sum_k V^k\vec{\mathbf{e}}_k,\sum_l W^l\vec{\mathbf{e}}_l),$$

$$\mathbf{T}(\vec{\mathbf{U}},\mathbf{\Phi},\vec{\mathbf{V}},\vec{\mathbf{W}})=\sum_{ijkl} U^i\Phi_jV^kW^l\mathbf{T}(\vec{\mathbf{e}}_i,\mathbf{e}^j,\vec{\mathbf{e}}_k,\vec{\mathbf{e}}_l).$$

With the convention of summation on the repeated indices which will be used in the following except explicit mention of the contrary, it holds:

$$\mathbf{T}(\vec{\mathbf{U}},\mathbf{\Phi},\vec{\mathbf{V}},\vec{\mathbf{W}})=U^i\Phi_jV^kW^lT^j_{ikl}\ ,$$

where the n^4 scalars $T^j_{ikl}=\mathbf{T}(\vec{\mathbf{e}}_i,\mathbf{e}^j,\vec{\mathbf{e}}_k,\vec{\mathbf{e}}_l)$ are called the *components* of the tensor $\mathbf{T}$ in the basis S. Observing that:

$$U^i\Phi_jV^kW^l=\mathbf{e}^i(\vec{\mathbf{U}})\,\hat{\mathbf{e}}_j(\mathbf{\Phi})\,\mathbf{e}^k(\vec{\mathbf{V}})\,\mathbf{e}^l(\vec{\mathbf{W}})=(\mathbf{e}^i\otimes\vec{\mathbf{e}}_j\otimes\mathbf{e}^k\otimes\mathbf{e}^l)\,(\vec{\mathbf{U}},\mathbf{\Phi},\vec{\mathbf{V}},\vec{\mathbf{W}}),$$

we obtain:

$$\mathbf{T}=T^j_{ikl}\ \mathbf{e}^i\otimes\vec{\mathbf{e}}_j\otimes\mathbf{e}^k\otimes\mathbf{e}^l.$$

The n^4 tensor products $\mathbf{e}^i \otimes \vec{\mathbf{e}}_j \otimes \mathbf{e}^k \otimes \mathbf{e}^l$ form a basis of the linear space of the 3-covariant and 1-contravariant tensors. Let T'^{r}_{mst} be the components of the tensor in another basis $S' = S\,P$. The components of the tensor are modified according to the *transformation law:*

$$T'^{r}_{mst} = P^i_m (P^{-1})^r_j P^k_s P^l_t \, T^j_{ikl}.$$

It defines a right action of $\mathbb{GL}(n)$ on the set of the *component system* $T = (T^j_{ikl})_{1 \le i,j,k,l \le n}$ identified to $\mathbb{R}^{n^4}$. It is a linear representation of $\mathbb{GL}(n)$ of dimension n^4:

$$T' = \rho(P)\,T.$$

Fixing the value of the arguments $\vec{\mathbf{U}}$ and $\vec{\mathbf{W}}$ in the previous mixed tensor, the scalar:

$$\lambda = \mathbf{T}(\vec{\mathbf{U}}, \mathbf{\Phi}, \vec{\mathbf{V}}, \vec{\mathbf{W}}),$$

linearly depends on $\mathbf{\Phi}$ and $\vec{\mathbf{V}}$. Hence, there exists a 1-covariant and 1-contravariant tensor $\check{\mathbf{T}}$ such that $\lambda = \check{\mathbf{T}}(\mathbf{\Phi}, \vec{\mathbf{V}})$, then a linear map $\mathbf{A}$ such that $\check{\mathbf{T}} = \hat{\mathbf{A}}$:

$$\forall \mathbf{\Phi} \in \mathcal{T}^*, \vec{\mathbf{V}} \in \mathcal{T}, \qquad \mathbf{T}(\vec{\mathbf{U}}, \mathbf{\Phi}, \vec{\mathbf{V}}, \vec{\mathbf{W}}) = \mathbf{\Phi}(\mathbf{A}(\vec{\mathbf{V}})).$$

Owing to [7.30] and using the convention of summation, the trace of $\mathbf{A}$:

$$Tr(\mathbf{A}) = \mathbf{e}^r(\mathbf{A}(\vec{\mathbf{e}}_r)) = \mathbf{T}(\vec{\mathbf{U}}, \mathbf{e}^r, \vec{\mathbf{e}}_r, \vec{\mathbf{W}}),$$

does not depend on the choice of the basis and linearly depend on $\vec{\mathbf{U}}$ and $\vec{\mathbf{W}}$, then it is a 2-covariant tensor $\bar{\mathbf{T}}$ defined by:

$$\bar{\mathbf{T}}(\vec{\mathbf{U}}, \vec{\mathbf{W}}) = \mathbf{T}(\vec{\mathbf{U}}, \mathbf{e}^r, \vec{\mathbf{e}}_r, \vec{\mathbf{W}}),$$

independent of the choice of the basis and is called a *contracted tensor* of $\mathbf{T}$. Its components are:

$$\bar{T}_{mt} = T^r_{mrt},$$

hence, the rule: we give the same value r to a superior index and to an inferior index, next we sum for r varying from 1 to n.

Let us consider a linear map $\mathbf{R}$ mapping a 2-covariant tensor:

$$\mathbf{T} = T_{kl}\mathbf{e}^k \otimes \mathbf{e}^l,$$

onto a 2-contravariant tensor:

$$\mathbf{R}(\mathbf{T}) = T_{kl}\mathbf{R}(\mathbf{e}^k \otimes \mathbf{e}^l).$$

Thus, putting $\hat{R}^{ijkl} = \mathbf{R}(\mathbf{e}^k \otimes \mathbf{e}^l)(\mathbf{e}^i \otimes \mathbf{e}^j)$, its components are:

$$[\mathbf{R}(\mathbf{T})]^{ij} = \hat{R}^{ijkl} T_{kl}.$$

Denoting by $\hat{\mathbf{R}}$ the tensor admitting the $\hat{R}^{ijkl}$ as components, we have:

$$[\mathbf{R}(\mathbf{T})]^{ij} = \left[\hat{\mathbf{R}} \otimes \mathbf{T}\right]^{ijkl}_{kl},$$

hence, the tensor $\mathbf{R}(\mathbf{T})$ is obtained by contracting the tensor $\hat{\mathbf{R}} \otimes \mathbf{T}$ twice. The operation $(\hat{\mathbf{R}}, \mathbf{T}) \mapsto \mathbf{R}(\mathbf{T})$ is called a *contracted product*. The twice contracted product is often simply denoted by $\hat{\mathbf{R}} : \mathbf{T}$ and, similarly, the once contracted product is simply denoted by $\hat{\mathbf{R}} \cdot \mathbf{T}$. In particular, the contracted product of a 1-covariant tensor $\boldsymbol{\Phi}$ and a 1-contravariant tensor $\vec{\mathbf{V}}$ is the value of the linear form $\boldsymbol{\Phi}$ for the vector $\vec{\mathbf{V}}$:

$$\boldsymbol{\Phi} \cdot \vec{\mathbf{V}} = \boldsymbol{\Phi}(\vec{\mathbf{V}}).$$

Conversely to *free indices* i, j, the summation indices k, l can be renamed and for this reason they are called *dummy indices*.

The representation of the vector valued tensors is similar. For instance, let us consider a 2-covariant tensor $\mathbf{T}$. Let $(\vec{\eta}_\alpha)$ be a basis of the target space $\mathcal{R}$. Hence, it reads:

$$\mathbf{T} = \mathbf{T}^\alpha \vec{\eta}_\alpha,$$

with:

$$\mathbf{T}^\alpha = T^\alpha{}_{ij}\mathbf{e}^i \otimes \mathbf{e}^j.$$

Finally, let us say some words about two important types of tensors. Let T be the $n \times n$ matrix of which the element at the intersection of the *i*-th line and *j*-th column is the component T_{ij} of a 2-covariant tensor $\mathbf{T}$. Then, the transformation law reads in matrix form:

$$T' = P^T T\, P. \qquad [14.2]$$

A 2-covariant tensor $\mathbf{T}$ is *symmetric* (respectively, *skew-symmetric*) if:

$$\forall \vec{\mathbf{V}}_1, \vec{\mathbf{V}}_2 \in \mathcal{T}, \quad \mathbf{T}(\vec{\mathbf{V}}_1, \vec{\mathbf{V}}_2) = \mathbf{T}(\vec{\mathbf{V}}_2, \vec{\mathbf{V}}_1) \quad \text{(respectively,} \quad \mathbf{T}(\vec{\mathbf{V}}_1, \vec{\mathbf{V}}_2) = -\mathbf{T}(\vec{\mathbf{V}}_2, \vec{\mathbf{V}}_1)),$$

hence:

$$T_{ij} = T_{ji} \qquad \text{(respectively,} \quad T_{ij} = -T_{ji}).$$

Let T be the $n \times n$ matrix of which the element at the intersection of the i-th line and j-th column is the component T^{ij} of a 2-contravariant tensor $\mathbf{T}$. Then, the transformation law reads in matrix form:

$$T' = P^{-1} T \, P^{-T}. \qquad [14.3]$$

A 2-contravariant tensor $\mathbf{T}$ is *symmetric* (respectively, *skew-symmetric*) if:

$$\forall \mathbf{\Phi}_1, \mathbf{\Phi}_2 \in \mathcal{T}^*, \quad \mathbf{T}(\mathbf{\Phi}_1, \mathbf{\Phi}_2) = \mathbf{T}(\mathbf{\Phi}_2, \mathbf{\Phi}_1) \quad \text{(resp. } \mathbf{T}(\mathbf{\Phi}_1, \mathbf{\Phi}_2) = -\mathbf{T}(\mathbf{\Phi}_2, \mathbf{\Phi}_1)),$$

hence:

$$T^{ij} = T^{ji} \qquad \text{(resp.} \quad T^{ij} = -T^{ji}).$$

With the previous conventions, the contracted product of a 2-contravariant tensor $\mathbf{R}$ and a 2-covariant tensor $\mathbf{T}$ is the dot product [7.5] of the matrices R and T gathering their respective components:

$$\mathbf{R} : \mathbf{T} = R^{ij} T_{ij} = R : T.$$

The extension of tensor Algebra to vector valued tensors is straightforward. For instance, a 1-covariant tensor $\boldsymbol{T}$ defined on the linear space $\mathcal{T}$ of dimension n with vector values in the linear space $\mathcal{R}$ of dimension p is a linear map from $\mathcal{T}$ into $\mathcal{R}$. Let $(\vec{\mathbf{e}}_j)$ be a basis of $\mathcal{T}$ and $(\boldsymbol{\eta}^i)$ a cobasis of $\mathcal{R}$. If $\vec{\mathbf{V}} \doteq \mathbf{T}(\vec{\mathbf{U}})$, then because of the linearity of $\boldsymbol{T}$ and $\boldsymbol{\eta}^i$:

$$V^i = \boldsymbol{\eta}^i(\boldsymbol{T}(U^j \vec{\mathbf{e}}_j)) = T^i_j U^j,$$

where the np components $T^i_j = \boldsymbol{\eta}^i(\boldsymbol{T}(\vec{\mathbf{e}}_j))$ of the tensor are the elements of the matrix T representing the linear map in the considered basis. The contracted product of the vector valued tensor:

$$\boldsymbol{T} = T^i_j \vec{\boldsymbol{\eta}}_i \otimes \boldsymbol{e}^j,$$

with the vector $\vec{\mathbf{U}}$ is the value of the linear map $\boldsymbol{T}$ for the vector $\vec{\mathbf{U}}$:

$$\boldsymbol{T} \cdot \vec{\mathbf{U}} = \boldsymbol{T}(\vec{\mathbf{U}}).$$

Let P (respectively, Q) be the transformation matrix of the change between $(\vec{e}_j)$ and a new basis $(\vec{e}'_j)$ (respectively, $(\vec{\eta}_i)$ and $(\vec{\eta}'_i)$). As the linear map is represented in the new basis by the *equivalent matrix* [7.29]:

$$T' = Q^{-1} T P, \qquad [14.4]$$

the transformation law of the tensor reads in indicial notation:

$$T'^r_{\ s} = (Q^{-1})^r_i P^j_s T^i_j \qquad [14.5]$$

14.2.2. *Affine tensors*

An *affine tensor* is an object:

– that assigns a set of *components* to each affine frame f of an affine space $\mathcal{AT}$ of finite dimension n;

– with a *transformation law*, when changing of frames, which is an affine or a linear representation of $\mathbb{A}ff(n)$.

With this definition, the affine tensors are a natural generalization of the classical tensors defined at the previous section and that we will call *linear tensors*, these last ones being trivial affine tensors for which the affine transformation $a = (C, P)$ acts through its linear part $P = lin(a)$.

An affine tensor can be constructed as a map which is affine or linear with respect to each of its arguments. As the linear tensors, the affine ones can be classified into three families.

The basic *p-covariant affine tensors* (or *covariant affine tensors of rank* p) are the multiaffine maps:

$$\mathbf{T} : \overbrace{\mathcal{AT} \times \mathcal{AT} \times \cdots \mathcal{AT}}^{p\ times} \rightarrow \mathcal{R} : (\mathbf{a}_1, \mathbf{a}_2, \cdots, \mathbf{a}_p) \mapsto \mathbf{T}(\mathbf{a}_1, \mathbf{a}_2, \cdots, \mathbf{a}_p).$$

The set of the p-covariant affine tensors is a linear space and is denoted by $(\otimes^p A^*\mathcal{T}) \otimes \mathcal{R}$ or more simply by $\otimes^p A^*\mathcal{T}$ if $\mathcal{R} = \mathbb{R}$. They generalize the affine forms which are 1-covariant affine tensors. Other kinds of affine tensors can be generated by taking linear parts. For instance, from a 2-covariant affine tensor $\mathbf{T}$, we can derive other 2-covariant affine tensors:

– the two linear parts, $lin_1(\mathbf{T}) : \mathcal{T} \times A\mathcal{T} \to \mathbb{R}$ and $lin_2(\mathbf{T}) : A\mathcal{T} \times \mathcal{T} \to \mathbb{R}$ such that:

$$(lin_1(\mathbf{T}))(\overrightarrow{\mathbf{a}_1\mathbf{b}_1}, \mathbf{a}_2) = \mathbf{T}(\mathbf{b}_1, \mathbf{a}_2) - \mathbf{T}(\mathbf{a}_1, \mathbf{a}_2),$$

$$(lin_2(\mathbf{T}))(\mathbf{a}_1, \overrightarrow{\mathbf{a}_2\mathbf{b}_2}) = \mathbf{T}(\mathbf{a}_1, \mathbf{b}_2) - \mathbf{T}(\mathbf{a}_1, \mathbf{a}_2).$$

– by taking the linear part of $lin_1(\mathbf{T})$ as a function of its second argument (or the linear part of $lin_2(\mathbf{T})$ as a function of its first argument), we obtain the *bilinear part* of $\mathbf{T}$, a 2-covariant linear tensor such that:

$$(lin_{12}\mathbf{T})(\overrightarrow{\mathbf{a}_1\mathbf{b}_1}, \overrightarrow{\mathbf{a}_2\mathbf{b}_2}) = \mathbf{T}(\mathbf{b}_1, \mathbf{b}_2) - \mathbf{T}(\mathbf{b}_1, \mathbf{a}_2) - \mathbf{T}(\mathbf{a}_1, \mathbf{b}_2) + \mathbf{T}(\mathbf{a}_1, \mathbf{a}_2).$$

The basic *q-contravariant affine tensors* (or *contravariant affine tensors of rank* q) are the multilinear maps:

$$\mathbf{T} : \overbrace{A^*\mathcal{T} \times A^*\mathcal{T} \times \cdots A^*\mathcal{T}}^{q\ times} \to \mathcal{R} : (\mathbf{\Psi}_1, \mathbf{\Psi}_2, \cdots, \mathbf{\Psi}_q) \mapsto \mathbf{T}(\mathbf{\Psi}_1, \mathbf{\Psi}_2, \cdots, \mathbf{\Psi}_q).$$

The set of the q-contravariant tensors is a linear space and is denoted by $(\otimes^q A^{**}\mathcal{T}) \otimes \mathcal{R}$ or more simply by $\otimes^q A^{**}\mathcal{T}$ if $\mathcal{R} = \mathbb{R}$. Particular attention is paid to the most simple ones, the 1-contravariant affine tensors $\boldsymbol{T}$ with scalar values of which the set is the dual $A^{**}\mathcal{T} = (A^*\mathcal{T})^*$ of the affine dual space, then a linear space of dimension $(n+1)$. As above, we claim that $\boldsymbol{T}(\mathbf{\Psi}) = \mathbf{\Psi}(\boldsymbol{T})$. We denote by $\mathbf{1}$ the constant function defined by $\mathbf{1}(\boldsymbol{a}) = 1$. Among the elements of $A^{**}\mathcal{T}$, we focus our attention on the two following kinds:

– For any $\vec{\boldsymbol{U}} \in \mathcal{T}$, the map $\hat{\boldsymbol{U}}$ defined by $\hat{\boldsymbol{U}}(\mathbf{\Psi}) = (lin(\mathbf{\Psi}))(\vec{\boldsymbol{U}})$ is such that $\mathbf{1}(\hat{\boldsymbol{U}}) = \hat{\boldsymbol{U}}(\mathbf{1}) = 0$ and verifies [14.1]. Hence, the vectors of $\mathcal{T}$ can be identified to the elements of the linear subspace of equation $\mathbf{1}(\boldsymbol{T}) = 0$.

– For any $\boldsymbol{a} \in A\mathcal{T}$, the map $\hat{\boldsymbol{a}}$ defined by $\hat{\boldsymbol{a}}(\mathbf{\Psi}) = \mathbf{\Psi}(\boldsymbol{a})$ is such that $\mathbf{1}(\boldsymbol{a}) = 1$. If $\boldsymbol{b} = \boldsymbol{a} + \vec{\boldsymbol{U}}$, we have

$$\hat{\boldsymbol{b}}(\mathbf{\Psi}) = \mathbf{\Psi}(\boldsymbol{a} + \vec{\boldsymbol{U}}) = \mathbf{\Psi}(\boldsymbol{a}) + (lin(\mathbf{\Psi}))(\vec{\boldsymbol{U}}) = \hat{\boldsymbol{a}}(\mathbf{\Psi}) + \hat{\boldsymbol{U}}(\mathbf{\Psi}) = (\hat{\boldsymbol{a}} + \hat{\boldsymbol{U}})(\mathbf{\Psi})$$

As the affine form $\mathbf{\Psi}$ is arbitrary, this defines an action $(\hat{\boldsymbol{a}}, \hat{\boldsymbol{U}}) \mapsto \hat{\boldsymbol{a}} + \hat{\boldsymbol{U}}$. Hence, the points of $A\mathcal{T}$ can be identified to the elements of the affine subspace of equation $\mathbf{1}(\boldsymbol{T}) = 1$.

Unlike the bidual $\mathcal{T}^{**}$ which can be identified to $\mathcal{T}$, the vector space $A^{**}\mathcal{T}$ is distinct from the affine space $A\mathcal{T}$ but contains it, reason for which it is called the *vector hull* of $A\mathcal{T}$.

The *mixed p-covariant* and *q-contravariant affine tensors* are the p-affine and q-linear maps:

$$\mathbf{T} : \overbrace{\mathcal{T} \times \cdots \mathcal{T}}^{p\ times} \times \overbrace{\mathcal{T}^* \times \cdots \mathcal{T}^*}^{q\ times} \rightarrow \mathcal{R} :$$

$$(\vec{\mathbf{V}}_1, \cdots, \vec{\mathbf{V}}_p, \mathbf{\Phi}_1, \cdots, \mathbf{\Phi}_q) \mapsto \mathbf{T}(\vec{\mathbf{V}}_1, \cdots, \vec{\mathbf{V}}_p, \mathbf{\Phi}_1, \cdots, \mathbf{\Phi}_q).$$

The order in which the arguments appear in $\mathbf{T}$ must be specified. To simplify, we choose here to order the arguments starting with all the points of $A\mathcal{T}$ and following with all the forms of $A^*\mathcal{T}$. The set of the p-covariant and q-contravariant affine tensors is a linear space and is denoted by $(\otimes^p A^*\mathcal{T}) \otimes (\otimes^q A^{**}\mathcal{T}) \otimes \mathcal{R}$ or more simply by $(\otimes^p A^*\mathcal{T}) \otimes (\otimes^q A^{**}\mathcal{T})$ if $\mathcal{R} = \mathbb{R}$. For instance, the linear maps from $A^*\mathcal{T}$ into $\mathcal{T}^*$ are mixed 1-covariant and 1-contravariants affine tensors through the identification of the linear map $\boldsymbol{\mu}$ with the tensor $\hat{\boldsymbol{\mu}}$ defined by:

$$\hat{\boldsymbol{\mu}}(\vec{\mathbf{U}}, \mathbf{\Psi}) = (\boldsymbol{\mu}(\mathbf{\Psi}))\vec{\mathbf{U}}. \qquad [14.6]$$

The generalization to the affine tensors of the concepts of tensor product, contracted tensor and product is straightforward. For instance, the tensor product of a point $\boldsymbol{a}$ and a vector $\vec{\boldsymbol{U}}$ is the 2-contravariant affine tensor $\boldsymbol{a} \otimes \vec{\boldsymbol{U}}$ such that $(\boldsymbol{a} \otimes \vec{\boldsymbol{U}})\,(\mathbf{\Psi}_1, \mathbf{\Psi}_2) = \hat{\mathbf{a}}(\mathbf{\Psi}_1)\,\hat{\boldsymbol{U}}(\mathbf{\Psi}_2)$.

It is worth noting that putting for a 1-covariant affine tensor:

$$\mathbf{\Psi} = \chi \mathbf{1} + \Phi_i \boldsymbol{e}^i, \qquad [14.7]$$

we recover [7.36] with the convention:

$$\mathbf{e}^i(\boldsymbol{a}) = \mathbf{e}^i(\overrightarrow{\boldsymbol{a}_0\boldsymbol{a}}). \qquad [14.8]$$

Let us now consider a 2-covariant affine tensor $\mathbf{T}$ and an affine frame f of origin $\mathbf{a}_0$ and basis S. Hence:

$$\boldsymbol{T}(\boldsymbol{a}_1, \boldsymbol{a}_2) = \boldsymbol{T}(\boldsymbol{a}_0 + \overrightarrow{\boldsymbol{a}_0\boldsymbol{a}_1}, \boldsymbol{a}_0 + \overrightarrow{\boldsymbol{a}_0\boldsymbol{a}_2}),$$

$$\boldsymbol{T}(\boldsymbol{a}_1, \boldsymbol{a}_2) = \boldsymbol{T}(\boldsymbol{a}_0, \boldsymbol{a}_0) + (lin_1(\boldsymbol{T}))(\overrightarrow{\boldsymbol{a}_0\boldsymbol{a}_1}, \boldsymbol{a}_0) + (lin_2(\boldsymbol{T}))(\boldsymbol{a}_0, \overrightarrow{\boldsymbol{a}_0\boldsymbol{a}_2}) + (lin_{12}(\boldsymbol{T}))(\overrightarrow{\boldsymbol{a}_0\boldsymbol{a}_1}, \overrightarrow{\boldsymbol{a}_0\boldsymbol{a}_2}).$$

Taking into account $\overrightarrow{\boldsymbol{a}_0\boldsymbol{a}_1} = U_1^i \vec{e}_i$, $\overrightarrow{\boldsymbol{a}_0\boldsymbol{a}_2} = U_2^i \vec{e}_i$ and the linearity, we have:

$$\boldsymbol{T}(\boldsymbol{a}_1, \boldsymbol{a}_2) = T_{00} + U_1^i T_{i0} + U_2^j T_{0j} + U_1^i U_2^j T_{ij},$$

where:

$$T_{00} = \boldsymbol{T}(\boldsymbol{a}_0, \boldsymbol{a}_0), \qquad T_{i0} = (lin_1(\boldsymbol{T}))(\vec{\boldsymbol{e}}_i, \boldsymbol{a}_0),$$
$$T_{0j} = (lin_2(\boldsymbol{T}))(\boldsymbol{a}_0, \vec{\boldsymbol{e}}_j), \qquad T_{ij} = (lin_{12}(\boldsymbol{T}))(\vec{\boldsymbol{e}}_i, \vec{\boldsymbol{e}}_j).$$

are called the *components* of the affine tensor. Observing that $\mathbf{1}(\boldsymbol{a}_1) = \mathbf{1}(\boldsymbol{a}_2) = 1$, $U_1^i = \mathbf{e}^i(\boldsymbol{a}_1)$ and $U_2^j = \mathbf{e}^j(\boldsymbol{a}_2)$ with the convention [14.8], we obtain:

$$\boldsymbol{T} = T_{00}\, \mathbf{1} \otimes \mathbf{1} + T_{i0}\, \mathbf{e}^i \otimes \mathbf{1} + T_{0j}\, \mathbf{1} \otimes \mathbf{e}^j + T_{ij}\, \mathbf{e}^i \otimes \mathbf{e}^j.$$

If the points $\mathbf{a}_1$ and $\mathbf{a}_2$ are, respectively, represented in a given affine frame f by the $(n+1)$-columns:

$$\tilde{U}_1 = \begin{pmatrix} 1 \\ U_1 \end{pmatrix}, \qquad \tilde{U}_2 = \begin{pmatrix} 1 \\ U_2 \end{pmatrix},$$

gathering the components of $\mathbf{T}$ into a $(n+1) \times (n+1)$ matrix $\tilde{T}$, its value for $\mathbf{a}_1 = f(U_1)$ and $\mathbf{a}_2 = f(U_2)$ is $\tilde{U}_1^T \tilde{T}\, \tilde{U}_2$. Then, the transformation law reads in matrix form:

$$\tilde{T}' = \tilde{P}^T \tilde{T}\, \tilde{P}.$$

A 2-covariant affine tensor $\mathbf{T}$ is *skew-symmetric* if:

$$\forall \mathbf{a}_1, \mathbf{a}_2 \in \mathcal{AT}, \qquad \mathbf{T}(\mathbf{a}_1, \mathbf{a}_2) = -\mathbf{T}(\mathbf{a}_2, \mathbf{a}_1).$$

In a similar way, we can study a 2-contravariant affine tensor $\mathbf{T}$. Taking into account the decomposition [14.7] and the linearity, its value reads:

$$\boldsymbol{T}(\boldsymbol{\Psi}, \bar{\boldsymbol{\Psi}}) = \boldsymbol{T}(\chi \mathbf{1} + \Phi_i \mathbf{e}^i, \bar{\chi} \mathbf{1} + \bar{\Phi}_j \mathbf{e}^j)$$
$$\boldsymbol{T}(\boldsymbol{\Psi}, \bar{\boldsymbol{\Psi}}) = \chi \bar{\chi}\, T^{00} + \Phi_i \bar{\chi}\, T^{i0} + \chi \bar{\Phi}_j\, T^{0j} + \Phi_i \bar{\Phi}_j\, T^{ij}.$$

where we defined the components of $\mathbf{T}$:

$$T^{00} = \boldsymbol{T}(\mathbf{1}, \mathbf{1}), \quad T^{i0} = \boldsymbol{T}(\mathbf{e}^i, \mathbf{1}), \quad T^{0j} = \boldsymbol{T}(\mathbf{1}, \mathbf{e}^j), \quad T^{ij} = \boldsymbol{T}(\mathbf{e}^i, \mathbf{e}^j).$$

Observing that $\chi = \boldsymbol{\Psi}(\boldsymbol{a}_0) = \hat{\mathbf{a}}_0(\boldsymbol{\Psi})$ and $\Phi_i = \boldsymbol{\Phi}(\vec{\boldsymbol{e}}_i) = (lin(\boldsymbol{\Psi}))(\vec{\boldsymbol{e}}_i) = \hat{\mathbf{e}}_i(\boldsymbol{\Psi})$, it holds:

$$\boldsymbol{T} = T^{00}\, \boldsymbol{a}_0 \otimes \boldsymbol{a}_0 + T^{i0}\, \vec{\boldsymbol{e}}_i \otimes \boldsymbol{a}_0 + T^{0j}\, \boldsymbol{a}_0 \otimes \vec{\boldsymbol{e}}_j + T^{ij}\, \vec{\boldsymbol{e}}_i \otimes \vec{\boldsymbol{e}}_j.$$

If the affine forms $\mathbf{\Psi}$ and $\bar{\mathbf{\Psi}}$ are, respectively, represented in a given affine frame f by the $(n+1)$-rows:

$$\tilde{\Psi} = \begin{pmatrix} \chi & \Phi \end{pmatrix}, \qquad \tilde{\bar{\Psi}} = \begin{pmatrix} \bar{\chi} & \bar{\Phi} \end{pmatrix},$$

gathering the components of $\mathbf{T}$ into a $(n+1)\times(n+1)$ matrix $\tilde{T}$, its value is $\tilde{\Psi}\,\tilde{T}\,\tilde{\bar{\Psi}}^T$. Then, the transformation law reads in matrix form:

$$\tilde{T}' = \tilde{P}^{-1}\tilde{T}\,\tilde{P}^{-T}. \qquad [14.9]$$

A 2-contravariant affine tensor $\mathbf{T}$ is *skew-symmetric* if:

$$\forall \mathbf{\Psi}_1, \mathbf{\Psi}_2 \in A^*\mathcal{T}, \qquad \boldsymbol{T}(\mathbf{\Psi}_1, \mathbf{\Psi}_2) = -\boldsymbol{T}(\mathbf{\Psi}_2, \mathbf{\Psi}_1).$$

The contracted product of a 2-contravariant affine tensor $\mathbf{R}$ and a 2-covariant affine tensor $\mathbf{T}$ is the dot product [7.5] of the matrices $\tilde{R}$ and $\tilde{T}$ gathering their respective components:

$$\mathbf{R} : \mathbf{T} = R^{ij} T_{ij} = \tilde{R} : \tilde{T}.$$

The readers can easily verify that mixed 1-covariant and 1-contravariants affine tensors $\hat{\boldsymbol{\mu}}$ defined by [14.6] can be decomposed with respect to the affine frame f according to:

$$\hat{\boldsymbol{\mu}} = \mu_{0i}\, \boldsymbol{e}^i \otimes \boldsymbol{a}_0 + \mu_i^j\, \boldsymbol{e}^i \otimes \vec{\boldsymbol{e}}_j,$$

where the affine components of $\hat{\boldsymbol{\mu}}$ are defined by:

$$\mu_{0i} = \hat{\boldsymbol{\mu}}(\vec{\mathbf{e}}_i, \mathbf{1}), \qquad \mu_i^j = \hat{\boldsymbol{\mu}}(\vec{\mathbf{e}}_i, \boldsymbol{e}^j).$$

14.2.3. *G-tensors and Euclidean tensors*

A subgroup G of $\mathbb{A}ff(n)$ naturally acts onto the affine tensors by restriction to G of their transformation laws. Let F_G be a set of affine frames of which G is a transformation group. The elements of F_G are called *G-frames*. A *G-tensor* is an object:

– that assigns a set of *components* to G-frame f;

– with a *transformation law*, when changing of frames, which is an affine or a linear representation of G.

The corresponding basis of a G-frame is called a *G-basis*.

For instance, the linear tensors are $\mathbb{GL}(n)$-tensors and the *Euclidean tensors* of $\mathbb{R}^3$ are $\mathbb{E}(3)$-tensors. As they are often used, let us give some more details in a slightly more general framework.

The *(covariant) metric tensor* $\boldsymbol{G}$ on a linear space $\mathcal{T}$ is a symmetric 2-covariant tensor which is non-degenerate:

$$\forall \vec{\boldsymbol{U}} \in \mathcal{T}, \qquad \boldsymbol{G}(\vec{\boldsymbol{U}}, \vec{\boldsymbol{V}}) = 0 \quad \Leftrightarrow \quad \vec{\boldsymbol{V}} = \vec{\boldsymbol{0}}.$$

The value of the metric tensor for $\vec{\boldsymbol{U}}$ and $\vec{\boldsymbol{V}}$ is called their *scalar product* and is denoted by $\vec{\boldsymbol{U}} \cdot \vec{\boldsymbol{V}}$. The symmetric regular matrix G gathering the components $G_{ij} = \vec{e}_i \cdot \vec{e}_j$ is called *Gram's matrix*. Thus, it holds:

$$\boldsymbol{G}(\vec{\boldsymbol{U}}, \vec{\boldsymbol{V}}) = \vec{\boldsymbol{U}} \cdot \vec{\boldsymbol{V}} = U^T G\, V,$$

where U and V are the columns gathering, respectively, the components of $\vec{\boldsymbol{U}}$ and $\vec{\boldsymbol{V}}$. The transformation law reads in matrix form:

$$G' = P^T G\, P.$$

A linear space equipped with a metric tensor is called an *Euclidean space*. For instance, $\mathbb{R}^n$ and $\mathbb{M}_{nn}$ equipped with the dot product are Euclidean spaces.

To every vector $\vec{\boldsymbol{U}}$ is associated one and only one linear form $\vec{\boldsymbol{V}} \mapsto \boldsymbol{G}(\vec{\boldsymbol{U}}, \vec{\boldsymbol{V}})$ denoted by $\boldsymbol{U}^*$. The covariant components of $\boldsymbol{U}^*$ depend on the contravariant components of $\vec{\boldsymbol{U}}$ through the operation of *lowering the index*:

$$U_i = G_{ij} U^j. \qquad [14.10]$$

The elements G^{ij} of the inverse G^{-1} of Gram's matrix are the components of a 2-contravariant tensor $\boldsymbol{G}^{-1}$ called *contravariant metric tensor*, hence the reverse operation of *raising the index*:

$$U^i = G^{ij} U_j.$$

Let $\boldsymbol{A}$ be a linear map from an Euclidean space $\mathcal{T}_0$ into another one $\mathcal{T}$. Its *adjoint* (with respect to the scalar product) is the linear map $\boldsymbol{A}^*$ from $\mathcal{T}$ into $\mathcal{T}_0$ such that:

$$\forall \vec{\boldsymbol{U}} \in \mathcal{T}, \quad \forall \vec{\boldsymbol{V}} \in \mathcal{T}_0, \qquad \vec{\boldsymbol{U}} \cdot (\boldsymbol{A}\, \vec{\boldsymbol{V}}) = (\boldsymbol{A}^* \vec{\boldsymbol{U}}) \cdot \vec{\boldsymbol{V}}.$$

If $\boldsymbol{A}$ is represented by the matrix A in basis of $\mathcal{T}_0$ and $\mathcal{T}$, $\boldsymbol{A}^*$ is represented by:

$$A^* = G_0^{-1} A^T G. \qquad [14.11]$$

We verify that:

$$(\boldsymbol{A} + \boldsymbol{A}')^* = \boldsymbol{A}^* + \boldsymbol{A}'^*, \qquad (\boldsymbol{A}\,\boldsymbol{B})^* = \boldsymbol{B}^* \boldsymbol{A}^*, \qquad (\boldsymbol{A}^*)^* = \boldsymbol{A}.$$

In particular, if $\mathcal{T}_0 = \mathbb{R}$, the linear map $\boldsymbol{U} : \mathbb{R} \to \mathcal{T} : \lambda \mapsto \lambda \vec{\boldsymbol{U}}$ can be identified to the vector $\vec{\boldsymbol{U}}$ and $\boldsymbol{U}^*$ is the unique linear form associated with $\vec{\boldsymbol{U}}$ with respect to the metric since [14.11] degenerates into:

$$U^* = U^T G,$$

which is the matrix form of [14.10]. Another particular case of interest is when $\mathcal{T}_0 = \mathcal{T}$ then [14.11] reads:

$$A^* = G^{-1} A^T G. \qquad [14.12]$$

We verify that:

$$Tr(\boldsymbol{A}^*) = Tr(\boldsymbol{A}).$$

The linear map is *self-adjoint* (respectively, *anti-self-adjoint*) if:

$$\boldsymbol{A} = \boldsymbol{A}^* \qquad (\text{resp.} \quad \boldsymbol{A} = -\boldsymbol{A}^*).$$

An *orthogonal basis* is a basis for which Gram's matrix is diagonal. The number p (respectively, $q = p - n$) of positive (respectively, negative) elements of the diagonal does not depend on the choice of the orthogonal basis. The couple (p, q) is called the *signature* of the metric. An *orthonormal basis* is a basis for which the diagonal form Δ of Gram's matrix is compound of p elements equal to $+1$ and q elements equal to -1 , the other ones vanishing. The set $\mathbb{O}(p, q)$ of the transformation matrices P such that:

$$P^T \Delta P = \Delta,$$

is a subgroup of $\mathbb{GL}(n)$. The set of $\mathbb{E}(p, q)$ of the affine transformations $a = (C, P)$ where $P \in \mathbb{O}(p, q)$ is a subgroup of $\mathbb{A}ff(n)$. Considering the $\mathbb{E}(p, q)$-frames f as being the frames of which the linear part $S = lin(f)$ is orthonormal, we define the $\mathbb{E}(p, q)$-tensors.

If the metric is *positive* ($p = n$), $\mathbb{O}(p, q)$ (respectively, $\mathbb{E}(p, q)$) is simply denoted by $\mathbb{O}(n)$ (respectively, $\mathbb{E}(n)$), which allows us to recover as particular case the definition of Euclidean tensors. Therefore, in an orthonormal basis, Gram's matrix is the identity of $\mathbb{R}^n$, which can read $G_{ij} = \delta_{ij}$ where the components of the identity matrix are denoted by Kronecker's covariant symbols δ_{ij}, hence U^i and U_i have the same value.

14.3. Vector analysis

14.3.1. *Divergence*

Let A be a square matrix field such that $def(A) \subset \mathbb{R}^n, val(A) \subset \mathbb{M}_{nn}$. The *divergence* of A is the field $div\, A \in (\mathbb{R}^n)^*$ of n-rows such that for every uniform vector field $k(x) = C^{te} \in \mathbb{R}^n$:

$$(div\, A)\, k = div\, (A\, k). \tag{14.13}$$

Choosing k as the key-columns, we deduce:

$$div\, (A_1, \cdots, A_n) = (div\, A_1, \cdots, div\, A_n). \tag{14.14}$$

For any scalar field λ, any vector fields $u, v \in \mathbb{R}^n$ and any square matrix field $A \in \mathbb{M}_{nn}$, it holds:

$$div\, (A\, v) = (div\, A)\, v + Tr\left(A\, \frac{\partial v}{\partial x}\right), \tag{14.15}$$

$$div\, (\lambda\, A) = \lambda\, div\, A + \frac{\partial \lambda}{\partial x}\, A, \tag{14.16}$$

$$div\, (uv^T) = (div\, u)\, v^T + u^T grad\, v \tag{14.17}$$

$$div\left(\frac{\partial v}{\partial x}\right) = \frac{\partial}{\partial x}(div\, v). \tag{14.18}$$

For any open domain $\mathcal{V}$ of $\mathbb{R}^n$ with suitable regularity assumptions and any C^1 vector field v and square matrix field A, we have *Green formulae* (or *divergence formulae*):

$$\int_{\mathcal{V}} div\, v\, d\mathcal{V} = \int_{\partial\mathcal{V}} n^T v\, dS, \tag{14.19}$$

$$\int_{\mathcal{V}} div\, A\, d\mathcal{V} = \int_{\partial\mathcal{V}} n^T A\, dS, \tag{14.20}$$

the column n representing the unit normal vector to $\partial\mathcal{V}$, pointing away from $\mathcal{V}$.

The definition [14.13] of the divergence of a square matrix can be easily generalized to matrices of arbitrary dimensions. Let A be a matrix field such that $def(A) \subset \mathbb{R}^n, val(A) \subset \mathbb{M}_{np}$. The *divergence* of A is the field $div\, A \in (\mathbb{R}^p)^*$ of p-rows such that for every uniform vector field $k(x) = C^{te} \in \mathbb{R}^p$:

$$(div\, A)\, k = div\, (A\, k).$$

Choosing k as the key-columns, we deduce:

$$div\, (A_1, \cdots, A_p) = (div\, A_1, \cdots, div\, A_p). \qquad [14.21]$$

For any vector field $v \in \mathbb{R}^p$, and any matrix field $A \in \mathbb{M}_{np}$, it holds:

$$div\, (A\, v) = (div\, A)\, v + Tr\left(A\, \frac{\partial v}{\partial x}\right), \qquad [14.22]$$

14.3.2. *Laplacian*

The *laplacian* of a scalar field is the scalar field:

$$\Delta\lambda = div\, (grad\, \lambda).$$

Let v be a *vector field*, such that $val(v), def(v) \subset \mathbb{R}^n$. Its *laplacian* is the vector field:

$$\Delta v = (div\, (grad\, v))^T. \qquad [14.23]$$

14.3.3. *Vector analysis in* $\mathbb{R}^3$ *and curl*

For any matrix field $x \longmapsto A(x) \in \mathbb{M}_{33}$ of class C^1, we call *curl* of A the matrix field $x \longmapsto curl\, A(x) \in \mathbb{M}_{nn}$ such that for any $dx, \delta x$:

$$(\delta x)^T dA - (dx)^T \delta A = (dx \times \delta x)^T curl\, A, \qquad [14.24]$$

where dA (respectively, δA) is the infinitesimal variation of A resulting from dx (respectively, δx). Alternatively, $curl\, A$ is the matrix field such that for any uniform column field $k(x) = C^{te}$, it holds:

$$curl\, (A\, k) = (curl\, A)\, k.$$

Choosing k as the key-columns, we deduce:

$$curl\,(A_1, A_2, A_3) = (curl\,A_1, curl\,A_2, curl\,A_3). \qquad [14.25]$$

For any vector field $x \longmapsto v(x) \in \mathbb{R}^3$ and any matrix field $x \longmapsto A(x) \in \mathbb{M}_{33}$, we have:

$$curl\,(j(v)) = \frac{\partial v}{\partial x} - (div\,v)\,1_{\mathbb{R}^3}, \qquad [14.26]$$

$$curl\left(\frac{\partial v}{\partial x}\right) = \frac{\partial}{\partial x}\,(curl\,v), \qquad [14.27]$$

$$curl\,(grad\,v) = 0, \qquad [14.28]$$

$$Tr\,(curl\,A) = div\,(j^{-1}(A^T - A)). \qquad [14.29]$$

14.4. Derivative with respect to a matrix

Let $f : \mathbb{M}_{np} \to \mathbb{R} : M \mapsto f(M)$ be a scalar valued matrix function of class C^1. Its derivative is a $p \times n$ matrix $\partial f / \partial M$ defined by:

$$df = Tr\left(\frac{\partial f}{\partial M}\,dM\right) = Tr\left(dM\,\frac{\partial f}{\partial M}\right). \qquad [14.30]$$

For instance:

$$\frac{\partial}{\partial M}(\det(M)) = adj(M). \qquad [14.31]$$

If M is regular, Cramer's rule [7.8] gives:

$$\frac{\partial}{\partial M}(\det(M)) = \det(M)\,M^{-1}. \qquad [14.32]$$

Also, differentiating [7.7] gives:

$$d(M^{-1}) = -M^{-1}dM\,M^{-1}. \qquad [14.33]$$

14.5. Tensor analysis

14.5.1. *Differential manifold*

A manifold is an object which, locally, just looks like an open subset of Euclidean space, but of which global topology can be quite different. Although many manifolds are realized as subsets of Euclidean spaces, the general definition is worth reviewing. More precisely, a *manifold* $\mathcal{M}$ of dimension n and class C^p is a topological space equipped with a collection of regular maps ϕ called *coordinate charts* of which (Figure 14.1):

– the definition sets are connected open subsets of $\mathbb{R}^n$,

– the value sets are open subsets of $\mathcal{M}$ covering it,

– the composite overlap maps $h = \phi^{-1} \circ \phi'$, called *coordinate changes*, are of class C^p.

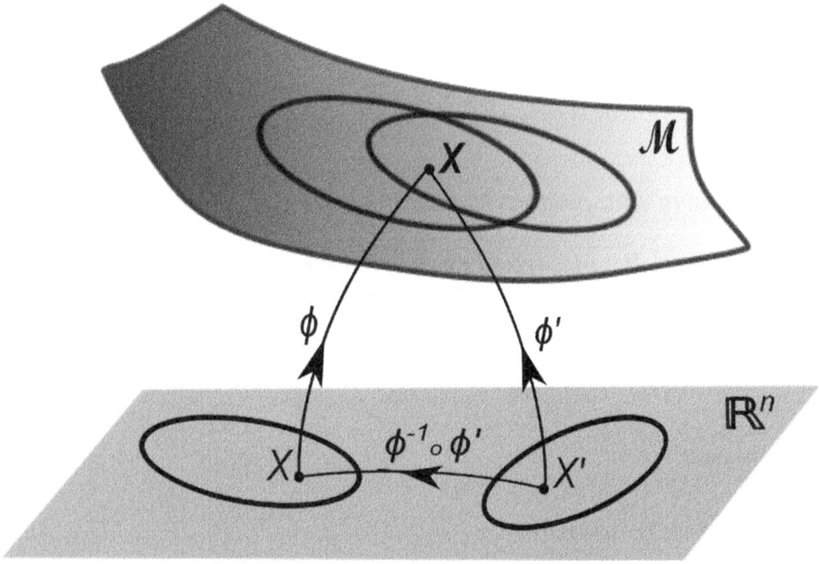

Figure 14.1. *coordinate charts of a manifold*

This allows us to define on $\mathcal{M}$ local coordinate systems:

$$X = \phi^{-1}(\boldsymbol{X}) = \begin{pmatrix} X^1 \\ X^2 \\ \vdots \\ X^n \end{pmatrix} \in \mathbb{R}^n.$$

One often expands the collection of coordinate charts to include all possible compatible charts (in the sense that the coordinate changes $X = h(X')$ are of class C^p), the resulting maximal collection defining an *atlas* on the manifold $\mathcal{M}$. In practice, it is convenient to omit explicit reference to the coordinate map and to identify the points $\boldsymbol{X}$ with their local coordinate expressions X. Nevertheless, objects living on manifolds must be defined intrinsically, independent of any choice of local coordinates. Consequently, manifolds are suitable tools to develop a coordinate-free approach to the study of their intrinsic geometry.

The basic examples of manifolds are obviously $\mathbb{R}^n$ or any open subset thereof, which are covered by a single chart. Another example is provided by the unit sphere $S^{n-1} = \{X \in \mathbb{R}^n \quad s.t. \quad \| X \| = 1\}$, which is a manifold of dimension $(n-1)$. It can be covered by two coordinate charts, obtained by omitting the north and south poles respectively. The local coordinates are provided by a stereographic projection to $\mathbb{R}^{n-1}$. Alternatively, we can use spherical coordinates on S^{n-1} which are valid away from the poles. A *submanifold* of $\mathcal{M}$ is a subset which, equipped with the topology induced by the one of $\mathcal{M}$, is a manifold in its own right. The simplest examples of submanifolds are curves (of dimension 1) and surfaces (of dimension 2).

A tangent vector to $\mathcal{M}$ at a point $\boldsymbol{X}_0 \in \mathcal{M}$ is geometrically defined by the tangent to a parameterized curve $\lambda \mapsto \boldsymbol{X} = \gamma(\lambda)$ passing through $\boldsymbol{X}_0 = \gamma(0)$. In a coordinate chart ϕ, the curve is represented by $\gamma_\phi = \phi^{-1} \circ \gamma$. We say that two curves γ and $\bar{\gamma}$ are in first order contact at $\boldsymbol{X}_0$ if

$$V = D\gamma_\phi(0) = D\bar{\gamma}_\phi(0) \in \mathbb{R}^n.$$

This equivalence relation does not depend on the choice of the coordinate chart and defines an equivalence class denoted $[\gamma]$. As the map $S_\phi(\boldsymbol{X}_0) : V = D\gamma_\phi(0) \mapsto [\gamma]$ is regular, the set of equivalence classes is, by structure transport, a vector space of dimension n called the *tangent space* to $\mathcal{M}$ at $\boldsymbol{X}_0$ and is denoted $T_{\boldsymbol{X}_0}\mathcal{M}$. Its elements are called *tangent vectors* at $\boldsymbol{X}_0$. Hence $S_\phi(\boldsymbol{X}_0)$ is a basis of the tangent space associated to the coordinate chart ϕ. The assigment $\boldsymbol{X} \mapsto S_\phi(\boldsymbol{X})$ defined on the definition set of ϕ is called a *natural frame*. In contrast, an assignment of arbitrary basis which are not related to a coordinate systems is called a *moving frame*. If ϕ' is another coordinate chart, differentiating $\gamma_\phi = h \circ \gamma_{\phi'}$ with the chain rule provides the transformation law [7.24] of vectors with:

$$P = \frac{\partial h}{\partial X'}(X'_0),$$

that is often simply denoted $\partial X / \partial X'$.

The set of linear forms on the tangent space is called the *cotangent space* and is denoted $T^*_{\boldsymbol{X}_0}\mathcal{M}$. The tangent space equipped with a structure of affine space is called the *affine tangent space* and is denoted $AT_{\boldsymbol{X}_0}\mathcal{M}$. Its elements are called *tangent points* at $\boldsymbol{X}_0$. The set of affine forms on the affine tangent space is denoted $A^*T_{\boldsymbol{X}_0}\mathcal{M}$. Of course, if the manifold $\mathcal{M}$ is a vector space (respectively an affine space), it can be identified with the tangent space (respectively affine tangent space) at any of its points.

Let $\mathcal{M}$ and $\mathcal{M}$ be manifolds of respective dimensions n and n'. A map $\boldsymbol{F} : \mathcal{M} \to \mathcal{M}'$ is of class C^p if for any coordinate charts ϕ of $\mathcal{M}$ and ϕ' of $\mathcal{M}'$, the map $F = \phi'^{-1} \circ \boldsymbol{F} \circ \phi$ is of class C^p. The *tangent map* to $\boldsymbol{F}$ at $\boldsymbol{X} \in \mathcal{M}$ is the linear map:

$$\frac{\partial \boldsymbol{F}}{\partial \mathbf{X}} : T_{\mathbf{X}}\mathcal{M} \to T_{\boldsymbol{F}(\mathbf{X})}\mathcal{M}',$$

represented in the basis S_ϕ and $S_{\phi'}$ by the $n' \times n$ matrix:

$$\frac{\partial F}{\partial X}(X) = S_{\phi'}^{-1} \circ \frac{\partial \boldsymbol{F}}{\partial \boldsymbol{X}} \circ S_\phi.$$

Let $\boldsymbol{T}$ be a field of p-covariant tensors on $\mathcal{M}'$. Its *pull-back* by $\boldsymbol{F}$ is the field of p-covariant tensor $\boldsymbol{F}^*\boldsymbol{T}$ on $\mathcal{M}$ defined by:

$$(\boldsymbol{F}^*\boldsymbol{T})(\boldsymbol{X})(\overrightarrow{d_1\mathbf{X}}, \overrightarrow{d_2\mathbf{X}}, \cdots, \overrightarrow{d_p\mathbf{X}}) = \boldsymbol{T}(\boldsymbol{F}(\boldsymbol{X}))\left(\frac{\partial \boldsymbol{F}}{\partial \boldsymbol{X}}\overrightarrow{d_1\mathbf{X}}, \frac{\partial \boldsymbol{F}}{\partial \boldsymbol{X}}\overrightarrow{d_2\mathbf{X}}, \cdots, \frac{\partial \boldsymbol{F}}{\partial \boldsymbol{X}}\overrightarrow{d_p\mathbf{X}}\right).$$

14.5.2. *Covariant differential of linear tensors*

Let $\mathcal{M}$ be a manifold of dimension n and class C^2. We call (*linear*) *covariant differential* at $\boldsymbol{X} \in \mathcal{M}$ a map ∇ such that:

– if $\boldsymbol{X} \mapsto \vec{\boldsymbol{T}}(\boldsymbol{X})$ is a tangent vector field and $\overrightarrow{d\boldsymbol{X}}$ is a tangent vector at $\boldsymbol{X}$, $\nabla_{\overrightarrow{d\boldsymbol{X}}}\vec{\boldsymbol{T}}$ is a tangent vector at $\boldsymbol{X}$;

– ϕ being a coordinate chart such that $\boldsymbol{X} = \phi(X)$, there exists a linear map $dX \mapsto \Gamma(dX) \in \mathbb{M}_{nn}$ such that:

$$\nabla_{\overrightarrow{d\boldsymbol{X}}}\vec{\boldsymbol{T}} = S_\phi \nabla_{dX} T \qquad with \qquad \nabla_{dX} T = dT + \Gamma(dX)\, T. \tag{14.34}$$

Covariant differentials are also called *connections*. The covariant differential being given by Γ' in another coordinate chart ϕ' such that $\boldsymbol{X} = \phi'(X')$, this entails:

$$\Gamma'(dX') = P^{-1}(\Gamma(P\, dX')\, P + dP), \tag{14.35}$$

with $P = \partial X/\partial X'$. By convention, the covariant differential of a scalar field is its usual one. The covariant differential is additive:

$$\nabla_{\overrightarrow{d\boldsymbol{X}}}(\vec{\boldsymbol{T}}_1 + \vec{\boldsymbol{T}}_2) = \nabla_{\overrightarrow{d\boldsymbol{X}}}\vec{\boldsymbol{T}}_1 + \nabla_{\overrightarrow{d\boldsymbol{X}}}\vec{\boldsymbol{T}}_2,$$

and, $\boldsymbol{X} \mapsto \lambda(\boldsymbol{X})$ being a scalar field:

$$\nabla_{\overrightarrow{dX}}(\lambda \vec{\boldsymbol{T}}) = d\lambda \vec{\boldsymbol{T}} + \lambda \nabla_{\overrightarrow{dX}} \vec{\boldsymbol{T}}.$$

When moving from $\boldsymbol{X}$ to the neighbour point $\boldsymbol{X} + \overrightarrow{d\boldsymbol{X}}$, the infinitesimal motion of vectors $\vec{e}_\alpha$ of the basis S_ϕ is given by:

$$\nabla_{\overrightarrow{dX}} \vec{e}_\alpha = S_\phi \nabla_{dX} e_\alpha = S_\phi(\Gamma_\alpha(dX)) = \Gamma^\beta_\alpha \vec{e}_\beta. \quad [14.36]$$

As Γ and dT , $\nabla_{dX} T$ is linear with respect to dX. Hence there exists a field $\nabla \vec{\boldsymbol{T}}$ of mixed 1-covariant and 1-contravariant tensors such that:

$$\nabla_{\overrightarrow{dX}} \vec{\boldsymbol{T}} = (\nabla \vec{\boldsymbol{T}}) \cdot \overrightarrow{d\boldsymbol{X}},$$

and represented in the basis S_ϕ by the matrix:

$$\nabla T = \frac{\partial T}{\partial X} + \Gamma(T). \quad [14.37]$$

As Γ linearly depends on dX, we introduce *Christoffel's symbols* $\Gamma^\alpha_{\mu\beta}$ such that:

$$\Gamma^\alpha_\beta(dX) = \Gamma^\alpha_{\mu\beta}\, dX^\mu. \quad [14.38]$$

A covariant differential is *symmetric* if:

$$\forall dX', \delta X', \qquad \Gamma(dX')\,\delta X' - \Gamma(\delta X')\,dX' = 0,$$

or, equivalently:

$$\Gamma^\alpha_{\mu\beta} = \Gamma^\alpha_{\beta\mu}. \quad [14.39]$$

In the following, the considered covariant differential are assumed symmetric. Putting:

$$\nabla \vec{\boldsymbol{T}} = \nabla_\beta T^\alpha\, \vec{e}_\alpha \otimes \boldsymbol{e}^\beta,$$

one has:

$$\nabla_\beta T^\alpha = \frac{\partial T^\alpha}{\partial X^\beta} + \Gamma^\alpha_{\mu\beta} T^\mu.$$

The *covariant divergence* of the vector field $\boldsymbol{T}$ is the scalar field obtained by contraction:

$$\boldsymbol{Div}\,\vec{\boldsymbol{T}} = Tr(\nabla\vec{\boldsymbol{T}}) = \nabla_\alpha T^\alpha = \frac{\partial T^\alpha}{\partial X^\alpha} + \Gamma^\alpha_{\mu\alpha} T^\mu. \qquad [14.40]$$

The extension of the covariant differential to tensors of higher order is straightforward thanks to the rule:

$$\nabla_{\overrightarrow{d\boldsymbol{X}}} (\boldsymbol{T} \otimes \boldsymbol{T}') = \nabla_{\overrightarrow{d\boldsymbol{X}}} \boldsymbol{T} \otimes \boldsymbol{T}' + \boldsymbol{T} \otimes \nabla_{\overrightarrow{d\boldsymbol{X}}} \boldsymbol{T}',$$

and by contraction:

$$\nabla_{\overrightarrow{d\boldsymbol{X}}} (\boldsymbol{T} \cdot \boldsymbol{T}') = (\nabla_{\overrightarrow{d\boldsymbol{X}}} \boldsymbol{T}) \cdot \boldsymbol{T}' + \boldsymbol{T} \cdot (\nabla_{\overrightarrow{d\boldsymbol{X}}} \boldsymbol{T}'). \qquad [14.41]$$

For instance, the covariant differential of a 2-contravariant tensor $\boldsymbol{T} = T^{\alpha\beta}\, \vec{\boldsymbol{e}}_\alpha \otimes \vec{\boldsymbol{e}}_\beta$ results from the infinitesimal variation of its components and of the infinitesimal motion of basis vectors when moving from $\boldsymbol{X}$ to the neighbour point $\boldsymbol{X} + \overrightarrow{d\boldsymbol{X}}$:

$$\nabla_{\overrightarrow{d\boldsymbol{X}}} \boldsymbol{T} = dT^{\alpha\beta}\, \vec{\boldsymbol{e}}_\alpha \otimes \vec{\boldsymbol{e}}_\beta + T^{\alpha\beta}\, \nabla_{\overrightarrow{d\boldsymbol{X}}} \vec{\boldsymbol{e}}_\alpha \otimes \vec{\boldsymbol{e}}_\beta + T^{\alpha\beta}\, \vec{\boldsymbol{e}}_\alpha \otimes \nabla_{\overrightarrow{d\boldsymbol{X}}} \vec{\boldsymbol{e}}_\beta,$$

$$\nabla_{\overrightarrow{d\boldsymbol{X}}} \boldsymbol{T} = dT^{\alpha\beta}\, \vec{\boldsymbol{e}}_\alpha \otimes \vec{\boldsymbol{e}}_\beta + T^{\alpha\beta}\Gamma^\mu_\alpha\, \vec{\boldsymbol{e}}_\mu \otimes \vec{\boldsymbol{e}}_\beta + T^{\alpha\beta}\Gamma^\mu_\beta\, \vec{\boldsymbol{e}}_\alpha \otimes \vec{\boldsymbol{e}}_\mu,$$

and, by renaming the dummy indices, we have:

$$\nabla_{\overrightarrow{d\boldsymbol{X}}} \boldsymbol{T} = \nabla_{dX} T^{\alpha\beta}\, \vec{\boldsymbol{e}}_\alpha \otimes \vec{\boldsymbol{e}}_\beta \quad with \quad \nabla_{dX} T^{\alpha\beta} = dT^{\alpha\beta} + \Gamma^\alpha_\mu T^{\mu\beta} + T^{\alpha\mu}\Gamma^\beta_\mu,$$

or in a matrix form:

$$\nabla_{dX} T = dT + \Gamma(dX)T + T\,(\Gamma(dX))^T.$$

if $\boldsymbol{X} \mapsto \boldsymbol{\Phi}(\boldsymbol{X})$ is a linear form field and ϕ is a coordinate chart such that $\boldsymbol{X} = \phi(X)$, $\nabla_{\overrightarrow{d\boldsymbol{X}}} \boldsymbol{\Phi}$ is a linear form at $\boldsymbol{X}$ such that:

$$\nabla_{\overrightarrow{d\boldsymbol{X}}} \boldsymbol{\Phi} = (\nabla_{dX}\Phi)\, S_\phi^{-1} \qquad with \qquad \nabla_{dX}\Phi = d\Phi - \Phi\,\Gamma(dX).$$

The reason of this definition is to be consistent with the rule [14.41] because the contracted product of a linear form and a vector is a scalar field then its covariant differential is the usual one:

$$(\nabla_{dX}\Phi)V + \Phi\,(\nabla_{dX}V) = \nabla_{dX}(\Phi\, V) = d\,(\Phi\, V).$$

When moving from $\boldsymbol{X}$ to the neighbour point $\boldsymbol{X} + \overrightarrow{d\boldsymbol{X}}$, the infinitesimal motion of covectors $\boldsymbol{e}^\alpha$ of the cobasis S_ϕ^{-1} is given by:

$$\nabla_{\overrightarrow{d\boldsymbol{X}}}\, \boldsymbol{e}^\alpha = \left(\nabla_{dX} e^\alpha\right) S_\phi^{-1} = -\left(\Gamma^\alpha(dX)\right) S_\phi^{-1} = -\Gamma^\alpha_\beta \boldsymbol{e}^\beta. \quad [14.42]$$

A *Riemannian metric* on a manifold $\mathcal{M}$ is a field $\boldsymbol{X} \mapsto \boldsymbol{G}(\boldsymbol{X})$ of covariant metric tensor on the tangent spaces. A manifold equipped with a Riemannian metric is called a *Riemannian manifold.* On a Riemannian manifold, there exists one and only one symmetric covariant differential such that the covariant differential of the metric vanishes. It is called *Levi-Civita covariant differential* or *connection.* Indeed the latter condition reads in any coordinate system:

$$\nabla_{dX} G_{\alpha\beta} = dG_{\alpha\beta} - \Gamma^\mu_\alpha G_{\mu\beta} - G_{\alpha\mu}\Gamma^\mu_\beta = 0, \quad [14.43]$$

or, considering the components of these differentials:

$$\frac{\partial G_{\alpha\beta}}{\partial X^\rho} = \Gamma^\mu_{\rho\alpha} G_{\mu\beta} + G_{\alpha\mu}\Gamma^\mu_{\rho\beta}.$$

Writing the equations deduced from this latter relation by circular permutation of indices, it holds, owing to [14.39] and the symmetry of the metric tensor:

$$[\rho\alpha, \beta] = \Gamma^\mu_{\rho\alpha} G_{\mu\beta},$$

where we put:

$$[\rho\alpha, \beta] = \frac{1}{2}\left(\frac{\partial G_{\alpha\beta}}{\partial X^\rho} + \frac{\partial G_{\beta\rho}}{\partial X^\alpha} - \frac{\partial G_{\rho\alpha}}{\partial X^\beta}\right). \quad [14.44]$$

As Gram's matrix is regular, we obtain:

$$\Gamma^\mu_{\alpha\beta} = G^{\mu\rho}\left[\alpha\beta, \rho\right]. \quad [14.45]$$

14.5.3. *Covariant differential of affine tensors*

The previous tensor analysis can be generalized to the affine tensors. An (*affine*) *covariant differential* at $\boldsymbol{X} \in \mathcal{M}$ is a map $\tilde{\nabla}$ such that:

– if $\boldsymbol{X} \mapsto \boldsymbol{a}(\boldsymbol{X})$ is a tangent point field and $\overrightarrow{d\boldsymbol{X}}$ is a tangent vector at $\boldsymbol{X}$, $\tilde{\nabla}_{\overrightarrow{d\boldsymbol{X}}}\, \boldsymbol{a}$ is a tangent vector at $\boldsymbol{X}$;

– V being the components of $\boldsymbol{a}$ in an affine frame f_ϕ of which the basis S_ϕ is its linear part, there exist a linear covariant differential ∇ and a linear map $dX \mapsto \Gamma_A(dX) \in \mathbb{R}^n$ such that:

$$\tilde{\nabla}_{\overrightarrow{dX}}\, \boldsymbol{a} = S_\phi \tilde{\nabla}_{dX} V \qquad with \qquad \tilde{\nabla}_{dX} V = \nabla_{dX} V + \Gamma_A(dX).$$

The covariant differential being given by Γ' and Γ'_A in another affine frame $f_{\phi'}$ obtained from f_ϕ through the affine transformation $a = (C, P)$, this entails (14.35) and:

$$\Gamma'_A(dX') = P^{-1}(\Gamma_A(P\,dX') + dC + \Gamma(P\,dX')C). \qquad [14.46]$$

Considering only linear transformations $a = (0, P)$ and taking into account the linearity of the map Γ_A, one has:

$$\Gamma'_A = P^{-1}\Gamma_A P,$$

Hence, there exists a mixed 1-covariant and 1-contravariant tensor field $\boldsymbol{A}$ represented in S_ϕ by the matrix $A = \Gamma_A$. Next we can deduce the general expression of Γ_A in any other affine frame obtained through an affine transformation $a = (C, P)$ thanks to its transformation law [14.46] which reads by inversion:

$$\Gamma_A(dX) = P\,\Gamma'_A(dX') - (dC + \Gamma(dX)\,C) = P\,A'\,dX' - (dC + \Gamma(dX)\,C),$$

or, in short:

$$\Gamma_A(dX) = A\,dX - \nabla_{dX} C. \qquad [14.47]$$

To be consistent, the symbol $\tilde{\nabla}$ is identified to ∇ when applied to a linear tensor. Moreover, the covariant differential is compatible with the action of the tangent vectors onto the tangent points, according to:

$$\tilde{\nabla}_{\overrightarrow{dX}}\,(\boldsymbol{a} + \vec{\boldsymbol{U}}) = \tilde{\nabla}_{\overrightarrow{dX}}\,\boldsymbol{a} + \nabla_{\overrightarrow{dX}}\,\vec{\boldsymbol{U}},$$

or, equivalently:

$$\tilde{\nabla}_{\overrightarrow{dX}}\,\boldsymbol{a}' = \tilde{\nabla}_{\overrightarrow{dX}}\,\boldsymbol{a} + \nabla_{\overrightarrow{dX}}\,\overrightarrow{\boldsymbol{a}\boldsymbol{a}'}.$$

The components of the origin $\mathbf{a}_0$ of the frame f_ϕ vanishing, its infinitesimal motion when moving from $\boldsymbol{X}$ to the neighbour point $\boldsymbol{X} + \overrightarrow{d\boldsymbol{X}}$ is given by:

$$\tilde{\nabla}_{\overrightarrow{dX}}\,\boldsymbol{a}_0 = S_\phi \Gamma_A(dX) = \Gamma_A^\alpha(dX)\vec{\boldsymbol{e}}_\alpha. \qquad [14.48]$$

Hence, the covariant differential of any tangent point field is:

$$\tilde{\nabla}_{\overrightarrow{dX}}\, \boldsymbol{a} = \tilde{\nabla}_{\overrightarrow{dX}}\, (\boldsymbol{a}_0 + V^\alpha \vec{e}_\alpha) = \tilde{\nabla}_{\overrightarrow{dX}}\, \boldsymbol{a}_0 + dV^\alpha \vec{e}_\alpha + V^\alpha \tilde{\nabla}_{\overrightarrow{dX}}\, \vec{e}_\alpha,$$

that leads to:

$$\tilde{\nabla}_{\overrightarrow{dX}}\, \boldsymbol{a} = \tilde{\nabla}_{dX}\, V^\alpha \vec{e}_\alpha, \qquad with \qquad \tilde{\nabla}_{dX}\, V^\alpha = \nabla_{dX}\, V^\alpha + \Gamma^\alpha_A(dX).$$

There exists a field $\tilde{\nabla}\boldsymbol{a}$ of mixed 1-covariant and 1-contravariant tensors such that:

$$\tilde{\nabla}_{\overrightarrow{dX}}\, \boldsymbol{a} = (\tilde{\nabla}\boldsymbol{a}) \cdot \overrightarrow{d\boldsymbol{X}},$$

As Γ_A linearly depends on dX, we introduce, by analogy with Christoffel's symbols–new symbols $\Gamma^\alpha_{A\beta}$ such that:

$$\Gamma^\alpha_A(dX) = \Gamma^\alpha_{A\beta}\, dX^\beta. \qquad [14.49]$$

Thus one obtains:

$$\tilde{\nabla}\boldsymbol{a} = \tilde{\nabla}_\beta V^\alpha\, \vec{e}_\alpha \otimes \boldsymbol{e}^\beta \qquad with \qquad \tilde{\nabla}_\beta V^\alpha = \nabla_\beta V^\alpha + \Gamma^\alpha_{A\beta},$$

where, according to [14.47]:

$$\Gamma^\alpha_{A\beta} = A^\alpha_\beta - \nabla_\beta C^\alpha. \qquad [14.50]$$

The *covariant divergence* of the point field $\boldsymbol{a}$ is scalar field:

$$\tilde{\boldsymbol{Div}}\, \boldsymbol{a} = Tr(\tilde{\nabla}\boldsymbol{a}) = \tilde{\nabla}_\alpha V^\alpha = \nabla_\alpha V^\alpha + \Gamma^\alpha_{A\alpha}.$$

The generalization of the covariant differential to affine tensors of higher order results from the rule:

$$\tilde{\nabla}_{\overrightarrow{dX}}\, (\boldsymbol{T} \otimes \boldsymbol{T}') = \tilde{\nabla}_{\overrightarrow{dX}}\, \boldsymbol{T} \otimes \boldsymbol{T}' + \boldsymbol{T} \otimes \tilde{\nabla}_{\overrightarrow{dX}}\, \boldsymbol{T}'. \qquad [14.51]$$

PART 3

Advanced Topics

15

Affine Structure on a Manifold

15.1. Introduction

One of the cornerstone concepts of differential geometry is the tangent space. Its modern definition leads to equip it with a structure of linear space by transport of the corresponding one on $\mathbb{R}^n$. Of course, it could also be equipped with the associated canonical structure of affine space, but this approach is purely algebraic and does not reveal the underlying differential structure on the manifold. It is this issue that we would like to highlight here.

The ordinary structure of differential manifolds, is based on stating an atlas of charts (or systems of coordinates) to which the tangent vector space is closely connected. Through this atlas of chart, the manifold is locally perceived as a linear space. The main features of this construction are well known but it seemed to us useful recalling some details of the reasoning, in particular the method of structure transport [BOU 70]. In section 15.2, we emphasize the method of transporting the structure of linear space. In section 15.3, we show that the underlying differential structure induced by an atlas of charts on the manifold allows equipping the tangent space of a linear space structure due to the linear transport method of the previous section.

This approach is scaled-down and we could enrich it by enhancing the concept of chart. Our starting point arises from É. Cartan's observation explained in his famous thesis of 1923–1924 about the manifolds with affine connections [CAR 23], [CAR 24]: "*The affine space at point* $\boldsymbol{m}$ *could be seen as the manifold itself that would be perceived in an affine manner by an observer located at* $\boldsymbol{m}$". It is this viewpoint that we hope to develop. In section 15.4, the transport method is presented for affine spaces. In section 15.5, we equip the manifold with a differential structure finer than the previous one, due to a set of one parameter smooth families of charts, called a film library. Next, the tangent space is endowed with an affine space structure by transport. In section 15.6, we show how the fields of points of the affine tangent space can be viewed as differential operators on the scalar fields. We recover the concept of particle derivative, usual in the mechanics of continua.

15.2. Endowing the structure of linear space by transport

Linear frame. Let $\mathcal{T}$ be a real linear space of dimension n and $(\vec{\mathbf{e}}_\alpha)_{1\leqslant\alpha\leqslant n}$ be a linear frame (or basis) of $\mathcal{T}$. Then, any vector $\vec{\mathbf{v}} \in \mathcal{T}$ can be expressed as:

$$\vec{\mathbf{v}} = \sum_{\alpha=1}^{n} V^\alpha \vec{\mathbf{e}}_\alpha,$$

where $V^\alpha \in \mathbb{R}$. This decomposition is unique. In other words, the correspondence between $\vec{\mathbf{v}} \in \mathcal{T}$ and the n-column V is one-to-one. This defines a one-to-one linear map $S : \mathbb{R}^n \to \mathcal{T} : V \mapsto \vec{\mathbf{v}} = S(V)$. We say it is a *linear frame* [SOU 08]. Conversely, let S be a given linear frame and $(e_\alpha)_{1\leqslant\alpha\leqslant n}$ be the canonical linear frame of $\mathbb{R}^n$ of which the elements are the key-columns. Thus, the set of vectors $\vec{\mathbf{e}}_\alpha = S(e_\alpha)$, with $\alpha = 1, \ldots, n$, is a linear frame of $\mathcal{T}$.

Linear transport. We define a *linear transport* from a linear space $\mathcal{T}_0$ over a field $\mathbb{K}$ into a set $\mathcal{T}$ as a set $\mathcal{S}$ of one-to-one maps S from $\mathcal{T}_0$ into $\mathcal{T}$ such that for any $S, S' \in \mathcal{S}$, the map $P = S^{-1} \circ S'$ from $\mathcal{T}_0$ into itself is linear.

THEOREM 15.1.– Let $\mathcal{T}$ be a set and $\mathcal{T}_0$ be a linear space over a field $\mathbb{K}$. If $\mathcal{S}$ is a linear transport from $\mathcal{T}_0$ into $\mathcal{T}$, then $\mathcal{T}$ is endowed with a unique structure of linear space for which the maps $S \in \mathcal{S}$ are linear. For this structure, the vector addition is:

$$\forall \vec{\mathbf{u}}, \vec{\mathbf{v}} \in \mathcal{T}, \quad \vec{\mathbf{u}} + \vec{\mathbf{v}} = S\,(S^{-1}(\vec{\mathbf{u}}) + S^{-1}(\vec{\mathbf{v}})), \qquad [15.1]$$

and the scalar multiplication is:

$$\forall \lambda \in \mathbb{K}, \quad \forall \vec{\mathbf{u}} \in \mathcal{T}, \quad \lambda\, \vec{\mathbf{u}} = S\,(\lambda\, S^{-1}(\vec{\mathbf{u}})). \qquad [15.2]$$

The definitions of these operations are independent of the choice of S.

If $\mathbb{K} = \mathbb{R}$ and $\mathcal{T}_0 = \mathbb{R}^n$, equipped with its canonical structure of linear space, any $S \in \mathcal{S}$ is a linear frame.

PROOF.–

For the map $S \in \mathcal{S}$ being linear, its inverse S^{-1} must be linear, hence we need:

$$\forall \vec{\mathbf{u}}, \vec{\mathbf{v}} \in \mathcal{T}, \quad S^{-1}(\vec{\mathbf{u}} + \vec{\mathbf{v}}) = S^{-1}(\vec{\mathbf{u}}) + S^{-1}(\vec{\mathbf{v}}),$$

$$\forall \lambda \in K, \quad \forall \vec{\mathbf{u}} \in \mathcal{T}, \quad S^{-1}(\lambda\, \vec{\mathbf{u}}) = \lambda\, S^{-1}(\vec{\mathbf{u}}).$$

This allows defining the vector addition by [15.1] and the scalar multiplication by [15.2].

As composition of one-to-one maps, any $P = S^{-1} \circ S'$ is one-to-one and $S' = S \circ P$, thus: $S'^{-1} = P^{-1} \circ S^{-1}$. Let us prove the definition of the sum vector is independent of the choice of S:

$$S'^{-1}(\vec{\mathbf{u}}) + S'^{-1}(\vec{\mathbf{v}}) = P^{-1}(S^{-1}(\vec{\mathbf{u}})) + P^{-1}(S^{-1}(\vec{\mathbf{v}})).$$

Because P is a linear map:

$$S'^{-1}(\vec{\mathbf{u}}) + S'^{-1}(\vec{\mathbf{v}}) = P^{-1}(S^{-1}(\vec{\mathbf{u}}) + S^{-1}(\vec{\mathbf{v}})).$$

Then, it holds:

$$S'(S'^{-1}(\vec{\mathbf{u}}) + S'^{-1}(\vec{\mathbf{v}})) = S\left(P\left(P^{-1}(S^{-1}(\vec{\mathbf{u}}) + S^{-1}(\vec{\mathbf{v}}))\right)\right) = S(S^{-1}(\vec{\mathbf{u}}) + S^{-1}(\vec{\mathbf{v}})),$$

that proves the independence of the definition [15.1] with respect to the choice of S. The demonstration of [15.2] is similar.

Moreover, it is easy to verify that the zero vector of $\mathcal{T}$ is $S\,(0)$ and the opposite of $\vec{\mathbf{u}} \in \mathcal{T}$ is $S\,(-S^{-1}(\vec{\mathbf{u}}))$. The verification of the axioms of the linear space structure is straightforward. If $\mathbb{K} = \mathbb{R}$ and $\mathcal{T}_0 = \mathbb{R}^n$, any $S \in \mathcal{S}$ is a linear frame because it is one-to-one and linear. ■

COMMENT.– the structure of linear space of $\mathcal{T}$ could be defined due to a unique one-to-one map S. The idea to introduce a set $\mathcal{S}$ of transport maps is motivated by the fact that in the next section, all the maps S are simultaneously constructed. Theorem 15.1 states the conditions ensuring that any map S equips the set $\mathcal{T}$ with the same linear space structure.

15.3. Construction of the linear tangent space

Differential structure. A differential manifold $\mathcal{M}$ of dimension n is a topological space with some additional differential structure. A chart is a homomorphism from an open subset V_ϕ of $\mathbb{R}^n$ into $\mathcal{M}$. Its value set U_ϕ is called the domain of ϕ. In other words, it is a one-to-one continuous map $\phi : V_\phi \to U_\phi : x \mapsto \boldsymbol{m} = \phi\,(x)$ and the inverse map ϕ^{-1} is continuous from U_ϕ into V_ϕ. If $\boldsymbol{m} \in U_\phi$, we call it a *chart around* $\boldsymbol{m}$. Let ϕ and ϕ' be two charts with overlapping domains. The homomorphism $h = \phi^{-1} \circ \phi'$ from $\phi'^{-1}(U_\phi \cap U_{\phi'})$ into $\phi^{-1}(U_\phi \cap U_{\phi'})$ is called a transition function or a coordinate change and the charts are said to be compatible. Thus, for any $\boldsymbol{m} \in U_\phi \cap U_{\phi'}$, $P = D\,h\,(x')$, where $x' = \phi'^{-1}(\boldsymbol{m})$, is a one-to-one linear map from $\mathbb{R}^n$ into itself. An atlas $\mathcal{A}$ is a maximal set of compatible charts $\phi : V_\phi \to \mathbb{R}^n$ whose domains cover $\mathcal{M}$. For $k = 1, 2, \ldots, +\infty$, if the transition functions are of class C^k, the atlas defines a C^k differential structure and $\mathcal{M}$ is of class C^k.

Local expression of paths through charts. Let $\boldsymbol{m}_0 \in \mathcal{M}$ and Λ be an open interval of $\mathbb{R}$ containing zero, $\gamma : \Lambda \to \mathcal{M} : \lambda \mapsto \boldsymbol{m} = \gamma(\lambda)$ be a map of class C^1 such that $\gamma(0) = \boldsymbol{m}_0$, defining a smooth path passing through $\boldsymbol{m}_0$. If ϕ is a chart around $\boldsymbol{m}_0$, the map $\gamma_\phi = \phi^{-1} \circ \gamma : \Lambda \to \mathbb{R}^n$, suitably restricted, is the local expression of the path γ with respect to the chart ϕ. Let ϕ' be another chart around $\boldsymbol{m}_0$ and $\gamma_{\phi'}$ be its local expression with respect to the chart ϕ'. Then:

$$\gamma_\phi = \phi^{-1} \circ \gamma = (\phi^{-1} \circ \phi') \circ (\phi'^{-1} \circ \gamma) = h \circ \gamma_{\phi'}.$$

Hence, it holds that:

$$D\,\gamma_\phi(0) = (D\,h\,(x'_0)) \circ (D\,\gamma_{\phi'}(0)), \qquad [15.3]$$

with $x'_0 = \phi'^{-1}(\boldsymbol{m}_0)$. In this relation, $D\,h\,(x'_0)$ is a linear map from $\mathbb{R}^n$ into itself (in other words, a $n \times n$ matrix) while $D\,\gamma_\phi(0)$ and $D\,\gamma_{\phi'}(0)$ are linear maps from $\mathbb{R}$ into $\mathbb{R}^n$ and can be identified to vectors of $\mathbb{R}^n$. In the following, we will conserve the notation $\circ$ of the composition product although, through this identification, we could remove it in matrix calculus style.

First-order contact. For $i = 1, 2$, let $\gamma_i : \Lambda_i \to \mathcal{M}$ be two smooth paths passing through $\boldsymbol{m}_0$. We say these paths are in first-order contact at $\boldsymbol{m}_0$ if for a chart ϕ around $\boldsymbol{m}_0$:

$$D\,\gamma_{1,\phi}\,(0) = D\,\gamma_{2,\phi}\,(0), \qquad [15.4]$$

where $\gamma_{i,\phi} = \phi^{-1} \circ \gamma_i$. This definition does not depend on the choice of the chart, owing to [15.3] and because the map $D\,h\,(x'_0)$ is one-to-one. First-order contact is an equivalence relation that we denote by $\gamma_1 \sim \gamma_2$. Let $\mathcal{T}$ be the set of equivalence classes $[\gamma]$ modulo first-order contact.

Structure transport. Let us arbitrarily choose a chart ϕ. For any $V \in \mathbb{R}^n$, let γ_V be the path defined by:

$$\gamma_V(\lambda) = \phi\,(\phi^{-1}(\boldsymbol{m}_0) + \lambda\,V).$$

Then, to each chart γ around $\boldsymbol{m}_0$, we can associate:

$$S_\phi(V) = [\gamma_V]\,,$$

which defines a map $S_\phi : \mathbb{R}^n \to \mathcal{T}$. Of course:

$$V = D\,(\phi^{-1} \circ \gamma_V)\,(0).$$

Conversely, because of [15.4], $V = D\,(\phi^{-1} \circ \gamma)\,(0)$ does not depend on the choice of the path γ inside the equivalence class $[\gamma]$. This defines the inverse mapping S_ϕ^{-1} by:

$$V = S_\phi^{-1}(\,[\gamma]\,) = D\,(\phi^{-1} \circ \gamma)\,(0) = D\,\gamma_\phi(0). \qquad [15.5]$$

Thus, the map S_ϕ is one-to-one. Moreover, let γ be a smooth path passing through m_0 and $V = D\,\gamma_\phi\,(0)$ (respectively, $V' = D\,\gamma_{\phi'}(0)$) be its local expression with respect to the chart ϕ (respectively, ϕ'). Owing to [15.3] and [15.5], we have:

$$S_\phi^{-1}(\,[\gamma]\,) = D\,h\,(x'_0) \circ S_{\phi'}^{-1}(\,[\gamma]\,).$$

This entails:

$$P = S_\phi^{-1} \circ S_{\phi'} = D\,h\,(x'_0),$$

which is a one-to-one linear map from $\mathbb{R}^n$ into itself. In other words, the collection $(S_\phi)_{\phi \in \mathcal{A}}$ of maps indexed by the charts of the atlas is a linear transport from $\mathbb{R}^n$ into $\mathcal{T}$. Then, owing to theorem 15.1, the set $\mathcal{T}$ is endowed with a structure of linear space. It is called the linear tangent space to $\mathcal{M}$ at m_0 and is denoted by $T_{m_0}\mathcal{M}$. Its elements are called tangent vectors.

15.4. Endowing the structure of affine space by transport

Affine space. Let $\mathcal{T}$ be a linear space over a field $\mathbb{K}$. An affine space $A\mathcal{T}$ modeled on $\mathcal{T}$ is a set with a free and transitive action of $\mathcal{T}$, viewed as an abelian group with respect to the addition:

$$\mathcal{T} \times A\mathcal{T} \to A\mathcal{T} : (\vec{\mathbf{u}}, \boldsymbol{q}) \mapsto \boldsymbol{p} = \boldsymbol{q} + \vec{\mathbf{u}}.$$

As the group $\mathcal{T}$ is abelian, the action is both right and left. We write $A\mathcal{T} = \mathbb{A}\,(\mathcal{T})$ and $\mathcal{T} = \mathbb{L}\,(A\mathcal{T})$. The elements of $A\mathcal{T}$ are called points. The unique vector $\vec{\mathbf{u}}$ such that $\boldsymbol{p} = \boldsymbol{q} + \vec{\mathbf{u}}$ is denoted by $\vec{\mathbf{u}} = \boldsymbol{p} - \boldsymbol{q}$. Of course, every linear space is canonically an affine space modeled on itself with the action:

$$\mathcal{T} \times \mathcal{T} \to \mathcal{T} : (\vec{\mathbf{u}}, \vec{\mathbf{v}}) \mapsto \vec{\mathbf{w}} = \vec{\mathbf{v}} + \vec{\mathbf{u}}.$$

If $\mathcal{T}$ is of finite dimension n, we say that $A\mathcal{T}$ has the same dimension.

Affine frame. Let $(\boldsymbol{q}, \boldsymbol{p}_1, \ldots, \boldsymbol{p}_n)$ be a set of $(n+1)$ points of an affine space $A\mathcal{T}$ of finite dimension n such that the set of the vectors $\vec{\mathbf{e}}_\alpha = \boldsymbol{p}_\alpha - \boldsymbol{q}$ is a linear frame. We say that $(\boldsymbol{q}, (\vec{\mathbf{e}}_\alpha))$ is an affine frame of $A\mathcal{T}$ of origin $\boldsymbol{q}$. For any $\boldsymbol{p} \in A\mathcal{T}$, the decomposition:

$$\boldsymbol{p} = \boldsymbol{q} + \sum_{\alpha=1}^{n} V^\alpha \vec{\mathbf{e}}_\alpha,$$

is unique. We call V^α the *components of* $\boldsymbol{p}$. In other words, the correspondence between $\boldsymbol{p} \in A\mathcal{T}$ and the column V collecting the components V^α is one-to-one. This defines a one-to-one affine map $f : \mathbb{R}^n \rightarrow A\mathcal{T} : V \mapsto \boldsymbol{p} = f(V)$. We say it is an *affine frame*. Conversely, let f be a given affine frame. It defines an affine frame of $\mathcal{T}$ by:

$$\boldsymbol{q} = f(0), \quad \boldsymbol{p}_\alpha = f(e_\alpha), \quad \vec{\mathbf{e}}_\alpha = f(e_\alpha) - \boldsymbol{q}.$$

Affine transport. We define an *affine transport* from an affine space $\mathbb{A}(\mathcal{T}_0)$ modeled on a linear space $\mathcal{T}_0$ over a field $\mathbb{K}$ into a set $A\mathcal{T}$ as a set $\mathcal{F}$ of one-to-one maps f from $\mathbb{A}(\mathcal{T}_0)$ into $A\mathcal{T}$ such that:

1) there is a map $lin : f \mapsto S_f = lin(f)$ from $\mathcal{F}$ into a linear transport $\mathcal{S}$ from $\mathcal{T}_0$ into a set $\mathcal{T}$;

2) for any $f, f' \in \mathcal{F}$, the map $a = f^{-1} \circ f'$ from $\mathbb{A}(\mathcal{T}_0)$ into itself is affine and $P = S_f^{-1} \circ S_{f'}$ is its linear part.

THEOREM 15.2.– Let $A\mathcal{T}$ be a set and $\mathcal{T}_0$ be a linear space over a field $\mathbb{K}$. If $\mathcal{F}$ is an affine transport from $\mathbb{A}(\mathcal{T}_0)$ into $A\mathcal{T}$ and $\mathcal{S}$ is its linear part, then $A\mathcal{T}$ is endowed with a unique structure of affine space for which any map $f \in \mathcal{F}$ is affine and $S_f = lin(f)$ is its linear part. This structure is defined by the action:

$$\forall \boldsymbol{q} \in A\mathcal{T}, \quad \forall \vec{\mathbf{u}} \in \mathcal{T}, \quad \boldsymbol{q} + \vec{\mathbf{u}} = f\left(f^{-1}(\boldsymbol{q}) + S_f^{-1}(\vec{\mathbf{u}})\right), \qquad [15.6]$$

The definition of this action does not depend on the choice of f.

If $\mathbb{K} = \mathbb{R}$ and $\mathcal{T}_0 = \mathbb{R}^n$, any $f \in \mathcal{F}$ is an affine frame.

PROOF.– For the map $f \in \mathcal{F}$ being affine and S_f its linear part, its inverse f^{-1} must be affine and S_f^{-1} its linear part, hence we need:

$$f^{-1}(\boldsymbol{q} + \vec{\mathbf{u}}) = f^{-1}(\boldsymbol{q}) + S_f^{-1}(\vec{\mathbf{u}}).$$

This allows defining the action by [15.6].

Let $q \in A\mathcal{T}$ and $\vec{\mathbf{u}}, \vec{\mathbf{v}} \in \mathcal{T}$. Let us verify that [15.6] defines a right action:

$$(q + \vec{\mathbf{u}}) + \vec{\mathbf{v}} = f\,(f^{-1}(q + \vec{\mathbf{u}}) + S_f^{-1}(\vec{\mathbf{v}})) = f\,((f^{-1}(q) + S_f^{-1}(\vec{\mathbf{u}})) + S_f^{-1}(\vec{\mathbf{v}})).$$

As the action of $\mathcal{T}_0$ onto $\mathbb{A}\,(\mathcal{T}_0)$ is a right action, we have:

$$(q + \vec{\mathbf{u}}) + \vec{\mathbf{v}} = f\,(f^{-1}(q) + (S_f^{-1}(\vec{\mathbf{u}}) + S_f^{-1}(\vec{\mathbf{v}}))).$$

Owing to theorem 15.1 and $\mathcal{S}$ being a linear transport, $\mathcal{T}$ is a linear space and S_f is a linear map. Thus, it holds:

$$(q + \vec{\mathbf{u}}) + \vec{\mathbf{v}} = f\,(f^{-1}(q) + S_f^{-1}(\vec{\mathbf{u}} + \vec{\mathbf{v}})) = q + (\vec{\mathbf{u}} + \vec{\mathbf{v}}).$$

Moreover, it holds:

$$q + \vec{\mathbf{0}} = f\,(f^{-1}(q) + S_f^{-1}(\vec{\mathbf{0}})) = f\,(f^{-1}(q) + 0) = f\,(f^{-1}(q)) = q.$$

Let us verify that the action is transitive. If $p, q \in A\mathcal{T}$, then $f^{-1}(p), f^{-1}(q) \in \mathbb{A}\,(\mathcal{T}_0)$. As the action of $\mathcal{T}_0$ onto $\mathbb{A}\,(\mathcal{T}_0)$ is transitive, thus there exists $x \in \mathcal{T}_0$ such that:

$$f^{-1}(p) = f^{-1}(q) + x.$$

Putting: $\vec{\mathbf{u}} = S_f(x)$, we have:

$$f^{-1}(p) = f^{-1}(q) + S_f^{-1}(\vec{\mathbf{u}}).$$

Thus, there exists $\vec{\mathbf{u}} \in \mathcal{T}$ such that:

$$p = f\,(f^{-1}(q) + S_f^{-1}(\vec{\mathbf{u}})) = q + \vec{\mathbf{u}}.$$

Let us verify that the action is free. Let $q \in A\mathcal{T}$ and $\vec{\mathbf{u}}, \vec{\mathbf{v}} \in \mathcal{T}$ such that: $q + \vec{\mathbf{u}} = q + \vec{\mathbf{v}}$. Thus:

$$f\,(f^{-1}(q) + S_f^{-1}(\vec{\mathbf{u}})) = f\,(f^{-1}(q) + S_f^{-1}(\vec{\mathbf{v}})).$$

As f is one-to-one, we have:

$$f^{-1}(q) + S_f^{-1}(\vec{\mathbf{u}}) = f^{-1}(q) + S_f^{-1}(\vec{\mathbf{v}}).$$

As the action of $\mathcal{T}_0$ onto $\mathbb{A}\,(\mathcal{T}_0)$ is free, it holds:

$$S_f^{-1}(\vec{\mathbf{u}}) = S_f^{-1}(\vec{\mathbf{v}}).$$

Because S_f^{-1} is one-to-one, $\vec{\mathbf{u}} = \vec{\mathbf{v}}$. Thus, we proved that [15.6] defines a transitive and free action of the linear space $\mathcal{T}$ onto $A\mathcal{T}$. In other words, $A\mathcal{T}$ is an affine space modeled on $\mathcal{T}$.

As composition of one-to-one maps, any $a = f^{-1} \circ f'$ is one-to-one and $f' = f \circ a$, thus: $f'^{-1} = a^{-1} \circ f^{-1}$. Let us prove the definition [15.6] is independent of the choice of f:

$$f'^{-1}(q) + S'^{-1}_f(\vec{\mathbf{u}})) = a^{-1}(f^{-1}(q)) + P^{-1}(S_f^{-1}(\vec{\mathbf{u}})).$$

Because a^{-1} is an affine map and P^{-1} is its linear part:

$$f'^{-1}(q) + S'^{-1}_f(\vec{\mathbf{u}})) = a^{-1}(f^{-1}(q) + S_f^{-1}(\vec{\mathbf{u}})).$$

Then, it holds:

$$f'(f'^{-1}(q) + S'^{-1}_f(\vec{\mathbf{u}})) = f(a\,(a^{-1}(f^{-1}(q) + S_f^{-1}(\vec{\mathbf{u}})))) = f(f^{-1}(q) + S_f^{-1}(\vec{\mathbf{u}})),$$

that proves the independence of the definition [15.6] with respect to the choice of f.

If $\mathbb{K} = \mathbb{R}$ and $\mathcal{T}_0 = \mathbb{R}^n$, any $f \in \mathcal{F}$ is an affine frame because it is affine and one-to-one. ■

REMARK.– As for any affine map $f \in \mathcal{F}$, $S_f = lin\,(f)$ is its linear part, we say the transport $\mathcal{S} = lin\,(\mathcal{F})$ is the *linear part of* $\mathcal{F}$.

15.5. Construction of the affine tangent space

Film. Let $\mathcal{M}$ be a topological space. We call *film* a one parameter smooth family ϕ of charts. More precisely, let Λ be an open interval of $\mathbb{R}$ containing zero and V be an open subset of $\mathbb{R}^n$. The map:

$$\phi : \Lambda \times V \to \mathcal{M} : (\lambda, x) \mapsto \boldsymbol{m} = \phi\,(\lambda, x),$$

is continuous and for any $\lambda \in \Lambda$, the map $\phi_\lambda(\bullet) = \phi\,(\lambda, \bullet)$ is a homeomorphism. Thus, ϕ_λ is a chart. If $\boldsymbol{m} \in U_{\phi_0}$, we say that ϕ is a *film around* $\boldsymbol{m}$. We call *film library* a set L of films ϕ such that:

1) for all $\lambda \in \Lambda$, the charts ϕ_λ cover $\mathcal{M}$;

2) the set L contains the *motionless films* ϕ for which:

$$\forall \lambda \in \Lambda, \quad \forall x \in V, \quad \phi_\lambda(x) = \phi_0(x).$$

If for all $\lambda \in \Lambda$ the transition functions between the charts ϕ_λ are of class C^k and for all $x \in V_\phi$, the maps $\phi\,(\bullet\,, x)$ are of class C^k, we say the film library defines a C^k *affine differential structure*. Of course, the sublibrary of motionless films can be identified to the atlas $\mathcal{A}$ of the charts ϕ_0. In this sense, the underlying affine differential structure on a manifold is finer than the usual one. We propose to say that a set endowed with such a structure is an *affine differential manifold*.

Local expression of paths through films. Let $m_0 \in \mathcal{M}$ and $\gamma : \Lambda \to \mathcal{M}$ be a path of class C^1 passing through m_0. If ϕ is a film around m_0, the map $\gamma_\phi : \Lambda \to \mathbb{R}^n : \lambda \mapsto x = (\phi_\lambda^{-1} \circ \gamma)\,(\lambda)$, suitably restricted, is the local expression of the path γ with respect to the film ϕ. Let ϕ' be another film around m_0. Let $\gamma_{\phi'} : \Lambda \to \mathbb{R}^n : \lambda \mapsto x = ({\phi'}_\lambda^{-1} \circ \gamma)\,(\lambda)$ be the local expression of γ with respect to the film ϕ'. For convenience, we introduce the one parameter family of transition functions, suitably restricted and defined by:

$$H\,(\lambda, x') = (\phi_\lambda^{-1} \circ \phi'_\lambda)\,(x').$$

Then, we have:

$$\gamma_\phi(\lambda) = (\phi_\lambda^{-1} \circ \gamma)\,(\lambda) = ((\phi_\lambda^{-1} \circ \phi'_\lambda)\,(({\phi'}_\lambda^{-1} \circ \gamma))\,(\lambda)) = H\,(\lambda, \gamma_{\phi'}(\lambda)).$$

Hence, it holds that:

$$D\,\gamma_\phi(0) = (D\,H\,(0, \bullet)\,(x'_0)) \circ (D\,\gamma_{\phi'}(0)) + D\,H\,(\bullet\,, x'_0)\,(0), \qquad [15.7]$$

with $x'_0 = {\phi'}_0^{-1}(m_0) = \gamma_{\phi'}(0)$. The essential point to underline is that $D\,\gamma_\phi(0)$ is an affine function of $D\,\gamma_{\phi'}(0)$ through [15.7], while it was a linear function in [15.3].

First-order affine contact. For $i = 1, 2$ let $\gamma_i : \Lambda_i \to \mathcal{M}$ be two smooth paths passing through m_0. We say these paths are in *first-order affine contact* at m_0 if:

$$D\,\gamma_{1,\phi}\,(0) = D\,\gamma_{2,\phi}\,(0). \qquad [15.8]$$

where: $\gamma_{i,\phi}(\lambda) = (\phi_\lambda^{-1} \circ \gamma_i)\,(\lambda)$. This definition does not depend on the choice of the film, owing to [15.7] and because the map $D\,H\,(0, \bullet)\,({x'}_0)$ is one-to-one. First-order affine contact is an equivalence relation that we denote by $\gamma_1 \approx \gamma_2$. Let $A\mathcal{T}$ be the set of equivalence classes $[[\gamma]]$ modulo first-order affine contact.

It is emphasized that $\gamma_1 \approx \gamma_2$ whenever [15.8] is fulfilled for one given film. Since motionless films are identified to charts, $\gamma_1 \approx \gamma_2$ entails $\gamma_1 \sim \gamma_2$ and conversely. Hence, the set $A\mathcal{T}$ of equivalence classes $[[\gamma]]$ modulo first-order affine contact is identical to the set $\mathcal{T}$ of equivalence classes $[\gamma]$ modulo first-order contact, but it is worthwhile to distinguish these two equivalence relations because each of them leads to define different algebraic structures on this set.

Structure transport. For any $V \in \mathbb{R}^n$, let γ_V be the path defined by:

$$\gamma_V(\lambda) = \phi_\lambda\left(\phi_0^{-1}(\boldsymbol{m}_0) + \lambda V\right).$$

Then, to each film ϕ around $\boldsymbol{m}_0$, we can define the map $f_\phi : \mathbb{R}^n \mapsto A\mathcal{T}$ by:

$$f_\phi(V) = [[\gamma_V]],$$

We can easily check:

$$V = D\left(\phi_\lambda^{-1} \circ \gamma_V\right)(0).$$

Conversely, because of [15.8], $V = D\,\gamma_\phi(0)$ does not depend on the choice of γ inside the equivalence class $[[\gamma]]$. This defines the inverse mapping f_ϕ^{-1} by:

$$V = f_\phi^{-1}(\,[[\gamma]]\,) = D\,\gamma_\phi(0). \qquad [15.9]$$

Thus, the map f_ϕ is one-to-one. Moreover, let γ be a smooth path passing through $\boldsymbol{m}_0$ and $V = D\,\gamma_\phi(0)$ (respectively, $V' = D\,\gamma_{\phi'}(0)$) be its local expression with respect to the film ϕ (respectively, ϕ'). Owing to [15.7] and [15.9], we have:

$$f_\phi^{-1}(\,[[\gamma]]\,) = \left(D\,H\,(0,\,\bullet)\,(x_0')\right) \circ \left(f_{\phi'}^{-1}(\,[[\gamma]]\,)\right) + D\,H\,(\bullet\,,x_0')\,(0).$$

This entails:

$$a\,(V') = \left(f_\phi^{-1} \circ f_{\phi'}\right)(V') = P\,V' + C, \qquad [15.10]$$

with $P = D\,H\,(0,\,\bullet)\,(x_0')$ and $C = D\,H\,(\bullet\,,x_0')\,(0)$. Thus, $a = f_\phi^{-1} \circ f_{\phi'}$ is a one-to-one affine map from $\mathbb{A}\,(\mathbb{R}^n)$ into itself. Moreover, its linear part is $P = D\,h_0\,(x_0')$ where $h_0 = \phi_0^{-1} \circ \phi_0'$ is a transition function, then $P = S_{\phi_0}^{-1} \circ S_{\phi_0'}$, which suggests, with the notations of section 15.3, to define the map $lin\,(f_\phi) = S_{\phi_0}$. In other words, the collection $(f_\phi)_{\phi \in L}$ of maps indexed by the films of the library is an affine transport from $\mathbb{A}\,(\mathbb{R}^n)$ into $A\mathcal{T}$. Its linear part is the linear transport from $\mathbb{R}^n$ into $T_{\boldsymbol{m}_0}\mathcal{M}$ defined in section 15.3.

Then, owing to theorem 15.2, the set $A\mathcal{T}$ is endowed with a structure of affine space modeled on the linear tangent space $T_{\boldsymbol{m}_0}\mathcal{M}$. The set $A\mathcal{T}$ of the equivalence classes modulo first-order affine contact is called the affine tangent space to $\mathcal{M}$ at $\boldsymbol{m}_0$ and is denoted by $AT_{\boldsymbol{m}_0}\mathcal{M}$. We say its elements are *tangent points* or more simply *points*. According to the terminology of section 15.4, the elements V^α of the column $V = D\,\gamma_\phi(0)$ are called the *components of* $M = [[\gamma]]$ *in the film* ϕ.

15.6. Particle derivative and affine functions

Scalar fields. Let $\mathcal{M}$ be an affine differential manifold. Any map $g : \mathcal{M} \mapsto \mathbb{R}$ of class C^k is called a *scalar field*. The restriction of g to a path γ is the composed map $g \circ \gamma$. Its local expression with respect to a film ϕ is given by a family of maps $g_\phi = g \circ \phi_\lambda$. With the notations of the previous section, we have:

$$g \circ \gamma = g_\phi \circ \gamma_\phi.$$

For convenience, we use the following notation:

$$\bar{g}_\phi(\lambda, x) = g\left(\phi_\lambda(x)\right).$$

Point fields as differential operators. Any map $\boldsymbol{m} \mapsto \boldsymbol{M} \in AT_{\boldsymbol{m}}\mathcal{M}$ of class C^k on $\mathcal{M}$ is called a *point field*. For $\boldsymbol{m} \in \mathcal{M}$, let γ be a path film passing through $\boldsymbol{m}$ such that $\boldsymbol{M} = [[\gamma]]$. The column V collects the components of $\boldsymbol{M}$ in the film ϕ around $\boldsymbol{m}$.

To each point field $\boldsymbol{m} \mapsto \boldsymbol{M}$, we can associate a map $D_{\boldsymbol{M}}$ defined on the set of the scalar fields with real values:

$$g \mapsto D_{\boldsymbol{M}} g = D\left(g \circ \gamma\right)(0) = D\left(g_\phi \circ \gamma_\phi\right)(0).$$

It does not depend on the choice of the film. Differentiating the composed function gives:

$$D_{\boldsymbol{M}} g = \left(D\,\bar{g}_\phi(0, \bullet)\,(x_0)\right) \circ D\,\gamma_\phi(0) + D\,\bar{g}_\phi\left(\bullet, x_0\right)(0).$$

We introduce simplified notations for the n-row:

$$\Phi = D\,\bar{g}_\phi(0, \bullet)\,(x_0), \qquad [15.11]$$

and the scalar:

$$\chi = D\,\bar{g}_\phi(\bullet, x_0)\,(0), \qquad [15.12]$$

Omitting the symbol $\circ$ as in the matrix calculus, we see that $D_{\boldsymbol{M}} g$ is an affine function with respect to V, denoted by $\Psi_{g,\phi}$:

$$D_{\boldsymbol{M}} g = \Phi\, V + \chi = \Psi_{g,\phi}(V).$$

The function $\boldsymbol{\Psi}_g = \Psi_{g,\phi} \circ f_\phi^{-1}$ from $AT_{\boldsymbol{m}}\mathcal{M}$ into $\mathbb{R}$ is independent of the choice of the film and is an affine function, as composed of two affine maps. Then as a function of $\boldsymbol{M}$, $D_{\boldsymbol{M}}g$ is an affine form. It can be easily verified that $D_{\boldsymbol{M}}$ is a *derivation* on the real algebra of the scalar fields:

$$D_{\boldsymbol{M}}(gh) = g\,D_{\boldsymbol{M}}h + h\,D_{\boldsymbol{M}}g,$$

$D_{\boldsymbol{M}}g$ is *linear* with respect to g (but *affine* with respect to $\boldsymbol{M}$).

This derivative operator with respect to a point field is the counterpart of the derivative operator of a scalar field with respect to a vector field (when the manifold is equipped with the ordinary differential structure). From this point of view, point fields can be considered as differential operators.

Besides, to each scalar field g, we can associate an affine function Ψ_g defined at $\boldsymbol{m} \in \mathcal{M}$. The set of such functions is naturally equipped with a structure of linear space. In [DES 03], it is denoted by $A^*T_{\boldsymbol{m}}\mathcal{M}$. In [GRA 04], it is called the *dual vector* of the affine space.

Mechanical viewpoint. If the manifold is the physical space and the path parameter λ is the time, we can easily recognize in $D_{\boldsymbol{M}}g$ the *material derivative* or *total derivative* commonly used in the mechanics of continua, especially in fluid mechanics. In the literature, this object has a mechanical status, curiously out of the scope of the differential geometry.

16

Galilean, Bargmannian and Poincarean Structures on a Manifold

16.1. Toupinian structure

The set $\mathbb{GAL}$ of the affine Galilean transformations is a Lie subgroup of dimension 10 of $\mathbb{A}ff\ (4)$ called Galileo's group. The subset $\mathbb{GAL}_0 \subset \mathbb{GAL}$ of the linear Galilean transformations P given by [1.4] is a Lie subgroup of dimension 6 of $\mathbb{GL}(4)$.

The group $\mathbb{GAL}_0$ of linear Galilean transformations (or equivalently Galileo's group) equips the space–time with the structure proposed by Toupin [TOU 58], taken up later on by Noll [NOL 73] and Künzle [KUN 72]. This modeling offers a theoretical framework for the universal or absolute time and space.

DEFINITION 16.1.– We call a *Galilean basis* (respectively, *frame*) a $\mathbb{GAL}$-basis $(\vec{e}_\alpha)$ (respectively, $\mathbb{GAL}$-frame) in the meaning of section 14.2.3.

The *Toupinian structure* of the space–time is based on two canonical tensors:

1) A linear form, the *time arrow* $\mathbf{e}^0$ represented in any Galilean basis by the same row [10.15]:

$$e^0 = \left(1 \ 0^T \right).$$

It could also be denoted by dt, hence the name of time arrow.

2) A symmetric 2-contravariant linear tensor γ represented in any Galilean basis by the same matrix [10.16]:

$$\gamma = \begin{pmatrix} 0 & 0 \\ 0 & 1_{\mathbb{R}^3} \end{pmatrix}.$$

It can read:

$$\gamma = \delta^{ij} \vec{e}_i \otimes \vec{e}_j,$$

the subcobasis $(\boldsymbol{e}^i)$ being orthonormal:

$$\gamma(\boldsymbol{e}^i, \boldsymbol{e}^j) = \delta^{ij},$$

and the time arrow being orthogonal to it:

$$\boldsymbol{e}^0 \cdot \gamma = \mathbf{0}.$$

Then, it is not a contravariant metric but it is semi-positive in the sense that:

$$\forall \boldsymbol{\Phi} \in T^*_{\mathbf{X}}\mathcal{M}, \qquad \gamma(\boldsymbol{\Phi}, \boldsymbol{\Phi}) \geq 0,$$

and for any linear form $\boldsymbol{\Phi} \neq \mathbf{0}$, we have:

$$\gamma(\boldsymbol{\Phi}, \boldsymbol{\Phi}) = 0 \quad \Leftrightarrow \quad \boldsymbol{\Phi} \ \ is \ \ colinear \ \ to \ \ \mathbf{e}^0.$$

3) A volume 4-form $\boldsymbol{vol} = \mathbf{e}^0 \wedge \mathbf{e}^1 \wedge \boldsymbol{e}^2 \wedge \boldsymbol{e}^3$.

Conversely, if a Toupinian structure $(\mathbf{e}^0, \gamma, \boldsymbol{vol})$ is given, the linear transformations preserving it are Galilean. Indeed, let us consider the transformation matrix:

$$P = \begin{pmatrix} \alpha \ w^T \\ u \ F \end{pmatrix}.$$

where $\alpha \in \mathbb{R}$, $u, w \in \mathbb{R}^3$ and F is a 3×3 matrix. Because of the transformation law [7.28], $\mathrm{e}^0 = \mathrm{e}^0 P$ that leads to $\alpha = 1$ and $w = 0$. Owing to the transformation law [14.3], $\gamma = P^{-1} \gamma P^{-T}$ that shows F is an orthogonal matrix. According to the transformation law of 4 forms, $\det(P) = 1$ hence F is a rotation. This is just what we did in theorem 1.1 but rephrased now in the tensor language.

THEOREM 16.1.– If a Toupinian structure is given, there exists a Galilean basis.

PROOF.– Equivalently, we prove the existence of the corresponding cobasis. We take the time arrow $\mathbf{e}^0$ as first cobasis form and we can complete it by three other linear forms $\boldsymbol{e}^i$ to build a cobasis of the tangent space. As the $\boldsymbol{e}^i$ are linearly independent of e^0, then:

$$\gamma(\boldsymbol{e}^i, \boldsymbol{e}^i) > 0,$$

and for any non-vanishing linear combinations of them:

$$\gamma(\mathbf{\Phi}, \mathbf{\Phi}) > 0, \qquad [16.1]$$

in which case, we introduce the notation:

$$N(\mathbf{\Phi}) = \mid \gamma(\mathbf{\Phi}, \mathbf{\Phi}) \mid^{1/2},$$

but the subcobasis (e^i) is not necessarily orthonormal, hence we apply Gram–Schmidt process. The tensor γ is not exactly a metric but [16.1] is just we need to apply it. The cobasis (e'^α) with:

$$e'^0 = e^0, \qquad e'^1 = (N(e^1))^{-1}\, e^1,$$
$$e'^2 = (N(e^2 - \gamma(e^2, e'^1)\, e'^1))^{-1}\,(e^2 - \gamma(e^2, e'^1)\, e'^1),$$
$$e'^3 = (N(e^3 - \gamma(e^3, e'^1)\, e'^1 - \gamma$$
$$(e^3, e'^2)\, e'^2))^{-1}\,(e^3 - \gamma(e^3, e'^1)\, e'^1 - \gamma(e^3, e'^2)\, e'^2).$$

is Galilean. ■

16.2. Normalizer of Galileo's group in the affine group

Once we have a Galilean basis and a corresponding Galilean affine frame, we can deduce a family of Galilean frames by Galilean transformations. Now, we would like to address the following issue: are there other similar families and how do we make a change of family? To be more precise, let $a = (C, P)$ be an affine Galilean transformation transforming the Galilean frame f' into another one f. Which are the regular affine transformations a_u transforming the Galilean frame $\bar{f}'$ into f' and the Galilean frame $\bar{f}$ into f such that the resulting transformation from $\bar{f}'$ to $\bar{f}$ is Galilean? Hence, which are the $a_u \in \mathbb{A}ff(4)$ such that:

$$\tilde{a} = a_u a\, a_u^{-1}, \qquad [16.2]$$

is a Galilean affine transformation? In other words, we aim to determine the normalizer of Galileo's group in the affine group:

$$N_{\mathbb{A}ff(4)}(\mathbb{GAL}) = \{a_u \in \mathbb{A}ff(4) \quad s.t. \quad a_u\, \mathbb{GAL} = \mathbb{GAL}\, a_u\}\,.$$

The answer is given by the following proposition:

THEOREM 16.2.– The normalizer of Galileo's group in the affine group is the set generated by the Galilean transformations and the transformations $a_u = (C_u, P_u)$ of which the linear part is a scaling:

$$P_u = \begin{pmatrix} T & 0 \\ 0 & L1_{\mathbb{R}^3} \end{pmatrix}.$$

PROOF.– Of course, the normalizer of a group contains the group itself then we have to find the other transformations. If $a = (C, P)$ and $a_u = (C_u, P_u)$, calculating $\tilde{a}_u = (\tilde{C}_u, \tilde{P}_u)$ by [16.2], we have to find $C_u, \tilde{C} \in \mathbb{R}^4$, $P_u \in \mathbb{GL}(4)$ and $\tilde{P} \in \mathbb{GAL}_0$ such that:

$$\tilde{C} + \tilde{P}\, C_u = C_u + P_u C, \qquad \tilde{P}\, P_u = P_u P,$$

for all $C \in \mathbb{R}^4$ and $P \in \mathbb{GAL}_0$. The first equation is satisfied anyway. Moreover, if:

$$P = \begin{pmatrix} 1 & 0 \\ u & R \end{pmatrix}, \qquad \tilde{P} = \begin{pmatrix} 1 & 0 \\ \tilde{u} & \tilde{R} \end{pmatrix}, \qquad P_u = \begin{pmatrix} T & w^T \\ v & M \end{pmatrix}.$$

we have to find $T \in \mathbb{R}$, $v, w, \tilde{u} \in \mathbb{R}^3$, $M \in \mathfrak{gl}(3)$ and $\tilde{R} \in \mathbb{SO}(3)$ such that:

$$w \cdot u = 0, w = R^T w, T\,\tilde{u} + \tilde{R}\,v = v + M\,u, \tilde{u}\,w^T + \tilde{R}\,M = M\,R,$$

for all $u \in \mathbb{R}^3$ and $R \in \mathbb{SO}(3)$. The first two conditions are satisfied if $w = 0$. Thus, the condition $\det(P_u) = T\,\det(M) \neq 0$ enforces $T \neq 0$ and $\det(M) \neq 0$. The next condition can be satisfied anyway and the last condition becomes:

$$\tilde{R}\,M = M\,R.$$

As M is regular, we have to find M and $\tilde{R} = M\,R\,M^{-1} \in \mathbb{SO}(3)$ for any $R \in \mathbb{SO}(3)$. The condition $\tilde{R}\,\tilde{R}^T = 1_{\mathbb{R}^3}$ leads to:

$$\forall R \in \mathbb{SO}(3), \qquad R\,(M^T M)^{-1} R^T = (M^T M)^{-1},$$

then the matrix $(M^T M)^{-1}$ is isotropic and $(M^T M)$ so is. Because it is positive definite, there exists $L \neq 0$ such that $M^T M = L^2 1_{\mathbb{R}^3}$ and it is possible to choose the sign of L in order that $M = L\,Q$ where Q is a rotation, hence:

$$P_u = \begin{pmatrix} T & 0 \\ v & LQ \end{pmatrix} = \begin{pmatrix} 1 & 0 \\ v/T & Q \end{pmatrix} \begin{pmatrix} T & 0 \\ 0 & L1_{\mathbb{R}^3} \end{pmatrix},$$

that achieves the proof. ■

We recover the scaling of section 1.3.5. This Theorem sheds a light on the existence of different families of Galilean frames. Every one corresponds to a particular choice of units of measuments of time and length.

16.3. Momentum tensors

If you need to brush up on Lie groups, this would be a good time to consult [CHE 06], for instance. All throughout this book, we used the concept of torsor, a skew-symmetric 2-contravariant affine tensor but there is another one very relevant for the mechanics that we call momentum. Let us consider a linear map $\bar{\mu}$ from the space $A^*T_{\mathbf{X}}\mathcal{M}$ of affine forms into the one $T^*_{\mathbf{X}}\mathcal{M}$ of linear forms. It is a vector valued affine tensor which can be identified with the scalar valued mixed 1-covariant and 1-contravariant affine tensor μ defined by:

$$\mu(\vec{\mathbf{V}}, \Psi) = (\bar{\mu}(\Psi))\vec{\mathbf{V}}.$$

It can be decomposed with respect to the affine frame f:

$$\mu = e^\beta \otimes (F_\beta\, a_0 + L^\alpha_\beta\, \vec{e}_\alpha),$$

where the affine components of μ are defined by:

$$F_\beta = \mu(\vec{\mathbf{e}}_\beta, 1), \qquad L^\alpha_\beta = \mu(\vec{\mathbf{e}}_\beta, \mathbf{e}^\alpha).$$

Taking into account the bilinearity, its value for a vector and an affine form is:

$$\mu(\vec{\mathbf{V}}, \Psi) = \mu(V^\beta \vec{\mathbf{e}}_\beta, \chi 1 + \Phi_\alpha \mathbf{e}^\alpha) = (\chi\, F_\beta + \Phi_\alpha\, L^\alpha_\beta)\, V^\beta,$$

or, introducing the row F collecting the F_β and the $n \times n$ matrix L of elements L^α_β:

$$\mu(\vec{\mathbf{V}}, \Psi) = (\chi\, F + \Phi\, L)\, V,$$

that reads in compact form:

$$\mu(\vec{\mathbf{V}}, \Psi) = \tilde{\Psi}\, \tilde{\mu}\, V$$

by introducing the matrix:

$$\tilde{\mu} = \begin{pmatrix} F \\ L \end{pmatrix}. \qquad [16.3]$$

According to the usual rules of the tensorial calculus, the transformation laws [7.24] of the vectors and [7.37] of the affine forms induces:

$$\tilde{\mu} = \tilde{P}\, \tilde{\mu}'\, P^{-1}.$$

Owing to [7.35] and [16.3], it is equivalent to:

$$F = F' P^{-1}, \qquad L = (P L' + C F') P^{-1}. \tag{16.4}$$

It is a left action of the affine group onto the space of momentum components. If the action is restricted to the Lie subgroup G, the momentum is a G-tensor.

On the other hand, have a look to the Lie algebra $\mathfrak{g}$ of G, that is the set of infinitesimal generators $Z = da = (dC, dP)$ with $a \in G$. Let us identify the space of the momentum components $\mu = (F, L)$ to the dual $\mathfrak{g}^*$ of the Lie algebra thanks to the dual pairing :

$$\mu Z = \mu\, da = (F, L)\,(dC, dP) = F\, dC + Tr(L\, dP) \tag{16.5}$$

We know that the group acts on its Lie algebra by the adjoint representation (18.1):

$$Ad(a) : \mathfrak{g} \to \mathfrak{g} : Z' \mapsto Z = Ad(a)\, Z' = a\, Z'\, a^{-1}.$$

As G is a group of affine transformations, any infinitesimal generator Z is represented by:

$$\tilde{Z} = d\tilde{P} = d \begin{pmatrix} 1 & 0 \\ C & P \end{pmatrix} = \begin{pmatrix} 0 & 0 \\ dC & dP \end{pmatrix}.$$

Then $\tilde{Z} = \tilde{P}\, \tilde{Z}'\, \tilde{P}^{-1}$ leads to:

$$dC = P\,(dC' - dP'\, P^{-1} C), \qquad dP = P\, dP'\, P^{-1}. \tag{16.6}$$

This adjoint representation induces the coadjoint representation of G in $\mathfrak{g}^*$ defined by:

$$(Ad^*(a)\, \mu')\, Z = \mu'\, (Ad(a^{-1})\, Z).$$

Taking into account [16.5], one finds that the coadjoint representation [18.2]:

$$Ad^*(a) : \mathfrak{g}^* \to \mathfrak{g}^* : \mu' \mapsto \mu = Ad^*(a)\, \mu'$$

is given by:

$$F = F' P^{-1}, \qquad L = (P L' + C F') P^{-1}.$$

It is noteworthy to observe that *the transformation law* [16.4] *of momentum tensors is nothing else but the coadjoint action.*

Although elegant, this mathematical construction is, however, not relevant for the addressed physical applications and this brings us to consider a map θ from G into $\mathfrak{g}^*$ and a more general transformation law:

$$\mu = a \cdot \mu' = Ad^*(a)\,\mu' + \theta(a), \qquad [16.7]$$

which is an affine representation of G in $\mathfrak{g}^*$:

$$\forall a', a \in G, \qquad \theta(a'a) = \theta(a') + Ad^*(a')\,\theta(a). \qquad [16.8]$$

Hence, the affine tensor nature of the map $\boldsymbol{\mu}$ is preserved. The only change is that the space $\mathfrak{g}^*$ is considered now with its affine structure. By the way, we should denote it by $A\mathfrak{g}^*$ but we will follow to forget the A out of habit. It is worth observing that there are obvious solutions of equation [16.8] in θ:

$$\theta_{\mu_0}(a) = Ad^*(a)\,\mu_0 - \mu_0, \qquad [16.9]$$

with constant $\mu_0 \in \mathfrak{g}^*$.

DEFINITION 16.2.– The *momentum G-tensor* $\boldsymbol{\mu}$ of a particle is a mixed 1-covariant and 1-contravariant affine tensor:

$$\boldsymbol{\mu} : T_{\mathbf{X}}\mathcal{M} \times A^*T_{\mathbf{X}}\mathcal{M} \rightarrow \mathbb{R} : (\vec{\boldsymbol{V}}, \boldsymbol{\Psi}) \mapsto \boldsymbol{\mu}(\vec{\boldsymbol{V}}, \boldsymbol{\Psi}),$$

of which the components $\mu = (F, L) \in \mathfrak{g}^*$ are modified according to an affine representation of which the linear part is the coadjoint representation:

$$\boxed{\mu = a \cdot \mu' = Ad^*(a)\,\mu' + \theta(a).} \qquad [16.10]$$

This action induces a *structure of affine space* on the set of components of momentum tensors. Let $\pi : \mathcal{F} \rightarrow \mathcal{M}$ be a G-principal bundle of affine frames with the free action $(a, f) \mapsto f' = a \cdot f$ on each fiber. Then, we can build the associated G-principal bundle:

$$\hat{\pi} : \mathfrak{g}^* \times \mathcal{F} \rightarrow (\mathfrak{g}^* \times \mathcal{F})/G : (\mu, f) \mapsto \boldsymbol{\mu} = orb(\mu, f)$$

for the free action:

$$(a, (\mu, f)) \mapsto (\mu', f') = a \cdot (\mu, f) = (a \cdot \mu, a \cdot f)$$

where the action on $\mathfrak{g}^*$ is [16.7]. Clearly, the orbit $\boldsymbol{\mu} = orb(\mu, f)$ can be identified to the momentum G-tensor $\boldsymbol{\mu}$ of components μ in the G-frame f.

16.4. Galilean momentum tensors

16.4.1. *Coadjoint representation of Galileo's group*

Let us put:

$$F = \begin{pmatrix} -e & p^T \end{pmatrix},$$

$$L = \begin{pmatrix} \zeta & -q^T \\ -w & -E \end{pmatrix}$$

where $e, \zeta \in \mathbb{R}, p, q, w \in \mathbb{R}^3$ and $E \in \mathbb{M}_{33}$. Then, for a Galilean transformation [1.9], the coadjoint representation [16.4] reads:

$$e = e' + u \cdot (R\,p'), \qquad p = R\,p', \qquad q = R\,(q' - \tau_0\,p'), \tag{16.11}$$

$$\zeta = \zeta' - e'\,\tau_0 + u \cdot (R\,(q' - \tau_0 p')),$$

$$E = R\,E'\,R^T - k\,(R\,p')^T + u\,(R\,q')^T,$$

$$w = R\,w' - \zeta'\,u + e'\,k - \left[R\,E'\,R^T - k\,(R\,p')^T + u\,(R\,q')^T\right]\,u.$$

Moreover, considering an infinitesimal generator of Galileo's group:

$$dC = \begin{pmatrix} d\tau_0 \\ dk \end{pmatrix}, \qquad dP = \begin{pmatrix} 0 & 0 \\ du & j(d\varpi) \end{pmatrix},$$

Introducing:

$$l = j^{-1}(E - E^T),$$

and using [7.3] and [7.13], the dual pairing [16.5] reads:

$$\mu\,Z = l \cdot d\varpi - q \cdot du + p \cdot dk - e\,d\tau_0. \tag{16.12}$$

As ζ, w and the symmetric part of E are masked by the dual pairing, we will not consider it in the sequel. To obtain the coadjoint representation of Galileo's group, we only need to remember [16.11] and the derived relation:

$$l = R\,l' - u \times (R\,q') + k \times (R\,p').$$

16.4.2. *Galilean momentum transformation law*

We are now able to determine the Galilean momentum transformation law of the form [16.7].

THEOREM 16.3.– The most general affine representation of Galileo's group into the space of momentum components of which the linear part is the coadjoint representation is given by:

$$\theta(a)\, Z = l_\theta(a) \cdot d\varpi - q_\theta(a) \cdot du + p_\theta(a) \cdot dk - e_\theta(a)\, d\tau_0,$$

with:

- $\diamondsuit$ $p_\theta(a) = m\, u \,+ R\, p_0 - p_0,$
- $\heartsuit$ $e_\theta(a) = \frac{1}{2}\, m \parallel u \parallel^2 + e_1\, \tau_0 + u \cdot (R\, p_0),$
- $\spadesuit$ $q_\theta(a) = m\,(k - \tau_0\, u) + R\, q_0 - q_0 - \tau_0 u\, R\, p_0,$
- $\clubsuit$ $l_\theta(a) = m\, k \times u + s\, u + R\, l_0 - l_0 - u \times (R\, q_0) + k \times (R\, p_0),$

where $m, s, e_1 \in \mathbb{R}$ and $p_0, q_0, l_0 \in \mathbb{R}^3$ are constants.

PROOF.– We have to solve equation [16.8]. Some calculations are rather lengthy and we give only the sketch of the demonstration.

– *Step 1: demonstrating* $\diamondsuit$. We start with p because its transformation law by the coadjoint representation does not involve other components. Condition [16.8] together with [16.11] provides:

$$p_\theta(a'a) = p_\theta(a') + R\, p_\theta(a). \qquad [16.13]$$

The difficulty lays in the nonlinear feature of this equation. To skirt the pitfall, we differentiate with respect to u', which gives, taking into account Galileo's group law [1.14]:

$$\frac{\partial p_\theta(a'a)}{\partial u} + \tau_0\, \frac{\partial p_\theta(a'a)}{\partial k} = \frac{\partial p_\theta(a')}{\partial u},$$

next, we consider the limit as a' approaches e:

$$\frac{\partial p_\theta(a)}{\partial u} + \tau_0\, \frac{\partial p_\theta(a)}{\partial k} = \frac{\partial p_\theta(e)}{\partial u}, \qquad [16.14]$$

where the right-hand member is a constant matrix that we denote by M_0. Repeating this procedure but differentiating with respect to k:

$$\frac{\partial p_\theta(a)}{\partial k} = \frac{\partial p_\theta(e)}{\partial k}, \qquad [16.15]$$

where the right-hand member is a constant matrix M_1, and differentiating with respect to τ_0:

$$\frac{\partial p_\theta(a)}{\partial \tau_0} = \frac{\partial p_\theta(e)}{\partial \tau_0}, \qquad [16.16]$$

where the right-hand member is a constant column p_1. Integrating [16.14] gives:

$$p_\theta(a) = (M_0 - \tau_0 M_1)\, u + \tilde{p}(k, \tau_0, R),$$

where $\tilde{p}$ is an arbitrary function. Introducing it in [16.15] gives:

$$p_\theta(a) = (M_0 - \tau_0 M_1)\, u + M_1\, k + \hat{p}(\tau_0, R).$$

Putting it into [16.16] leads to:

$$p_\theta(a) = M_0\, u + M_1\, k + p_1\, \tau_0 + \bar{p}(R).$$

Introducing this expression into [16.13] gives, after simplification:

$$M_1\, \tau_0 u' + (M_0 R' - R' M_0)\, u + (M_1 R' - R' M_1)\, k + \tau_0 (p_1 - R'\, p_1)$$
$$+\, \bar{p}(R' R) - \bar{p}(R') - R'\, \bar{p}(R) = 0,$$

which is satisfied if and only if:

- $M_0 R' = R' M_0$, then the matrix M_0 is isotropic: $M_0 = m\, 1^3_{\mathbb{R}}$,
- $M_1 = 0$,
- $p_1 = R'\, p_1$, then $p_1 = 0$,
- and :

$$\bar{p}(R' R) = \bar{p}(R') + R'\, \bar{p}(R). \qquad [16.17]$$

The latter condition shows that the map $f : \mathbb{SO}(3) \to \mathbb{SE}(3)$ represented in $\mathbb{GL}(4)$ by:

$$\tilde{f}(R) = \begin{pmatrix} 1 & 0 \\ \bar{p}(R) & R \end{pmatrix},$$

is a group homomorphism. Clearly $Ker(f) = \{e\}$ and the map f is injective. If H is the value set of f, $dim(H) = dim(\mathbb{SO}(3)) = 3$. As $\mathbb{SO}(3)$ is compact, there exists a Haar's measure on it and we define for $x \in \mathbb{R}^3$:

$$x_0 = \int_{\mathbb{SO}(3)} (f(R))(x)\, dR.$$

For any $a \in H$, there exists $Q \in \mathbb{SO}(3)$ such that $a = f(Q)$ and because Haar's measure is left invariant:

$$\begin{aligned} a(x_0) &= \int_{\mathbb{SO}(3)} (f(Q) f(R))(x)\, dR \\ &= \int_{\mathbb{SO}(3)} (f(Q\,R))(x)\, dR \\ &= \int_{\mathbb{SO}(3)} (f(R))(x)\, dR \\ &= x_0, \end{aligned}$$

and a belongs to the isotropy group of x_0, then $H \subset iso(x_0)$. But $a = (k, R) \in \mathbb{SE}(3)$ leaves x_0 invariant if $k = x_0 - R\,x_0$ then $dim(iso(x_0)) = 3 = dim(H)$ that proves $H = iso(x_0)$ and H is conjugate to $\mathbb{SO}(3)$ in $\mathbb{SE}(3)$:

$$\begin{pmatrix} 1 & 0 \\ \bar{p}(R) & R \end{pmatrix} = \begin{pmatrix} 1 & 0 \\ -p_0 & Q_0 \end{pmatrix} \begin{pmatrix} 1 & 0 \\ 0 & R' \end{pmatrix} \begin{pmatrix} 1 & 0 \\ -p_0 & Q_0 \end{pmatrix}^{-1},$$

then:

$$R = Q\,R'\,Q^T, \qquad \bar{p}(R) = Q\,R'\,Q^T p_0 - p_0,$$

that gives the general solution of equation [16.13]:

$$\bar{p}(R) = R\,p_0 - p_0, \qquad p_0 \in \mathbb{R}^3,$$

and leads to $\diamondsuit$:

$$p_\theta(a) = m\,u + R\,p_0 - p_0.$$

– *Step 2: demonstrating* $\heartsuit$. Condition [16.18] together with [16.11] provides:

$$e_\theta(a'a) = e_\theta(a') + e_\theta(a) + u' \cdot (R'\,p_\theta(a)). \qquad [16.18]$$

The resolution method is similar to the one used for p_θ. Taking into account $\diamondsuit$, differentiating with respect to u', k' and τ_0', next considering the limit as a' approaches e, we obtain:

$$\frac{\partial e_\theta(a)}{\partial u} + \tau_0 \frac{\partial e_\theta(a)}{\partial k} = \frac{\partial e_\theta(e)}{\partial u} + (m\,u + R\,p_0 - p_0)^T,$$

$$\frac{\partial e_\theta(a)}{\partial k} = \frac{\partial e_\theta(e)}{\partial k}, \qquad \frac{\partial e_\theta(a)}{\partial \tau_0} = \frac{\partial e_\theta(e)}{\partial \tau_0}.$$

Integrating successively with respect to u, k and τ_0 leads to:

$$e_\theta(a) = \frac{1}{2} m \parallel u \parallel^2 + u \cdot (w_0 + R\, p_0) + w_1 \cdot k + e_1\, \tau_0 + \bar{e}(R)$$

where $w_0, w_1 \in \mathbb{R}^3$, $e_1 \in \mathbb{R}$ are constants. Then, [16.18] is satisfied by the previous expression if and only if $w_0 = w_1 = 0$ and $\bar{e}$ is a group homomorphism from $\mathbb{SO}(3)$ into the additive group $\mathbb{R}$:

$$\bar{e}(R' R) = \bar{e}(R') + \bar{e}(R),$$

thus its kernel is a normal subgroup of $\mathbb{SO}(3)$. As it is a simple group, the only normal subgroups are $\{e\}$ or $\mathbb{SO}(3)$ itself but this former case is absurd because $\bar{e}$ would be injective while $dim(\mathbb{SO}(3)) > dim(\mathbb{R})$. Then $\bar{e} = 0$ and we prove ♡:

$$e_\theta(a) = \frac{1}{2} m \parallel u \parallel^2 + e_1\, \tau_0 + u \cdot (R\, p_0).$$

– *Step 3: demonstrating* ♠. Condition [16.8] together with [16.11] provides:

$$q_\theta(a' a) = q_\theta(a') + R' \,(q_\theta(a) - \tau_0'\, p_\theta(a)). \qquad [16.19]$$

Similarly to the previous steps, we deduce:

$$\frac{\partial q_\theta(a)}{\partial u} + \tau_0 \frac{\partial q_\theta(a)}{\partial k} = \frac{\partial q_\theta(e)}{\partial u},$$

$$\frac{\partial q_\theta(a)}{\partial k} = \frac{\partial q_\theta(e)}{\partial k}, \qquad \frac{\partial q_\theta(a)}{\partial \tau_0} = \frac{\partial q_\theta(e)}{\partial \tau_0} - (m\, u + R\, p_0 - p_0).$$

By integration, we obtain:

$$q_\theta(a) = N_0 u + N_1 k - m\, \tau_0 u + (q_1 - R\, p_0)\, \tau_0 + \bar{q}(R),$$

depending on constant quantities $N_0, N_1 \in \mathbb{M}_{33}$ and $q_1 \in \mathbb{R}^3$, and replacing these expression into [16.19], it holds after identification:

$$N_0 = n\, 1_{\mathbb{R}^3}, \qquad N_1 = m\, 1_{\mathbb{R}^3}, \qquad q_1 = 0,$$

and:

$$\bar{q}(R' R) = \bar{q}(R') + R'\, \bar{q}(R).$$

Similarly to the first step, we conclude that $\bar{q}(R) = R\,q_0 - q_0$, that leads to:

$$q_\theta(a) = m\,(k - \tau_0 u) + n\,u + R\,q_0 - q_0 - \tau_0 R\,p_0. \qquad [16.20]$$

Latter on, it will appear that, in fact, n vanishes and ♠ will be proved.

– *Step 4: demonstrating* ♣. Condition [16.8] together with [16.11] provides:

$$l_\theta(a'a) = l_\theta(a') + R'\,l_\theta(a) + j(R'\,q_\theta(a))\,u' - j(R'\,p_\theta(a))\,k'. \qquad [16.21]$$

Similarly to the previous steps, we deduce:

$$\frac{\partial l_\theta(a)}{\partial u} + \tau_0 \frac{\partial l_\theta(a)}{\partial k} = \frac{\partial l_\theta(e)}{\partial u} + j(q_\theta(a)),$$
$$\frac{\partial l_\theta(a)}{\partial k} = \frac{\partial l_\theta(e)}{\partial k} - j(p_\theta(a)), \qquad \frac{\partial l_\theta(a)}{\partial \tau_0} = \frac{\partial_\theta(e)}{\partial \tau_0}. \qquad [16.22]$$

Combining the two former conditions, introducing the constant quantities:

$$P_0 = \frac{\partial l_\theta(e)}{\partial u}, \qquad P_1 = \frac{\partial l_\theta(e)}{\partial k}, \qquad l_1 = \frac{\partial l_\theta(e)}{\partial \tau_0},$$

and taking into account ♢ and ♡ leads to:

$$\frac{\partial l_\theta(a)}{\partial u} = P_0 - \tau_0 P_1 + j(m\,k + n\,u + R\,q_0 - q_0 - \tau_0 p_0).$$

which is a linear system of partial differential equations. First of all, let us examine the system of 9 scalar equations and 3 unknown components of l_θ:

$$\frac{\partial l_\theta(a)}{\partial u} = n\,j(u).$$

According to [14.28] and [14.26], it is integrable provided:

$$curl_u\,(grad_u\,l_\theta) = -n\,curl_u\,(j(u)) = 2\,n\,1_{\mathbb{R}^3} = 0$$

thus $n = 0$ and, owing to [16.20], we prove ♠. By integration of the general system, we obtain:

$$l_\theta(a) = (P_0 - \tau_0 P_1)\,u + j(m\,k + R\,q_0 - q_0 - \tau_0 p_0)\,u + \tilde{l}(k, \tau_0, R).$$

Introducing this expression into the second equation [16.22] and integrating leads to:

$$\tilde{l}(k, \tau_0, R) = (P_1 - j(R\,p_0 - p_0))\,k + \hat{l}(\tau_0, R).$$

Hence, by integration of the latter equation [16.22], one has:

$$l_\theta(a) = m\,k \times u + (P_0 + j(R\,q_0 - q_0))\,u + (P_1 - j(R\,p_0 - p_0))\,k + l_1\tau_0 + \bar{l}(R).$$

Replacing this expression into [16.21], it holds after identification that:

$$P_0 = s\,1_{\mathbb{R}^3} + j(q_0), \qquad P_1 = -j(p_0), \qquad l_1 = 0, \qquad \bar{l}(R) = R\,l_0 - l_0,$$

and ♣ is proved. ■

It is worth noting that the general solution θ given by theorem 16.3 contains solutions of the form [16.9] with:

$$\mu_0\,Z = l_0 \cdot d\varpi - q_0 \cdot du + p_0 \cdot dk - e_0\,d\tau_0.$$

If we forget these obvious solution, the action [16.7] reads:

$$p = R\,p' + m\,u, \quad q' = R\,(q' - \tau_0 p') + m\,(k - \tau_0 u),$$

$$l = R\,l' - u \times (R\,q') + k \times (R\,p') + m\,k \times u + s\,u,$$

$$e = e' + u \cdot (R\,p') + \frac{1}{2}\,m \parallel u \parallel^2 + e_1\tau_0.$$

Taking into account [1.19], the inverse law restitues the transformation laws [3.6], [3.7], [3.8] of the components of the dynamical torsor, provided $s = 0$ and m is the particle mass. On the other hand, if we put $p' = m\,v'$, according to [3.17], the last relation is nothing else the transformation law of the energy [14.3], provided $e_1 = 0$. Hence the most general action [16.7] useful for the physical applications we are concerned is:

$$\boxed{p = R\,p' + m\,u, \qquad q = R\,(q' - \tau_0 p') + m\,(k - \tau_0 u),} \qquad [16.23]$$

$$\boxed{l = R\,l' - u \times (R\,q') + k \times (R\,p') + m\,k \times u,} \qquad [16.24]$$

$$\boxed{e = e' + u \cdot (R\,p') + \tfrac{1}{2}\,m \parallel u \parallel^2 .} \qquad [16.25]$$

Taking into account [16.4], the transformation law [16.10] of the Galilean momentum tensor $\boldsymbol{\mu}$ reads:

$$F = F'\,P^{-1} + F_m(C, P), \qquad L = (P\,L' + C\,F')\,P^{-1} + L_m(C, P), \quad [16.26]$$

where F_m and L_m are the components of θ. In particular, one has:

$$F_m(C,P) = m\left(-\frac{1}{2}\parallel u\parallel^2, u^T\right). \qquad [16.27]$$

16.4.3. *Structure of the orbit of a Galilean momentum torsor*

We saw in section 16.3 that the momentum G-tensor $\boldsymbol{\mu}$ is identified to the orbit $\boldsymbol{\mu} = orb(\mu, f)$ and, disregarding the frames for simplification, we can identify $\boldsymbol{\mu}$ to the orbit $orb(\mu)$ in $\mathfrak{g}^*$, i.e. the set of μ' that can be obtained by applying a Galilean transformation to μ. As submanifold of $\mathfrak{g}^*$, it can be defined by either a family of equations or a local coordinates system. To reveal its structure, let us construct its representations by two ways.

– *Representation of the orbit by equations.* To obtain them, we have to determine a functional basis. The first step is to calculate their number. According to the method presented in section 18.2, we start determining the isotropy group of μ. The analysis will be restricted to massive particles: $m \neq 0$. The components p, q, l, e being given, we have to solve the following system:

$$p = Rp + m\,u, \qquad [16.28]$$

$$q = Rq - \tau_0(Rp + m\,u) + m\,k, \qquad [16.29]$$

$$l = Rl - u\times(Rq) + k\times(Rp) + m\,k\times u, \qquad [16.30]$$

$$u\cdot(Rp) + \frac{1}{2}m\parallel u\parallel^2 = 0, \qquad [16.31]$$

with respect to τ_0, k, R, u. Owing to [16.28], the boost u can be expressed with respect to the rotation R by:

$$u = \frac{1}{m}(p - Rp), \qquad [16.32]$$

that allows automatically satisfying [16.31]. Next, owing to [16.28], equation [16.29] can be simplified as follows:

$$q = Rq - \tau_0 p + m\,k,$$

that allows us to determine the spatial translation k with respect to R and the clock change τ_0:

$$k = \frac{1}{m}(q - Rq + \tau_0 p). \qquad [16.33]$$

Finally, because of [16.28], equation [16.30] is simplified as follows:

$$l = R\,l - u \times (R\,q) + k \times p.$$

Substituting [16.33] into the last relation gives:

$$l = R\,l - u \times (R\,q) + \frac{1}{m}\,q \times p - \frac{1}{m}\,(R\,q) \times p.$$

Owing to [16.28] and the definition [3.13] of the spin angular momentum l_0 leads to:

$$l_0 = R\,l_0. \qquad [16.34]$$

These quantities being given, we have to determine the rotations satisfying the previous relation. It turns out that two cases must be considered.

1) *Generic orbits: massive particle with spin or rigid body.* If l_0 does not vanish, the solutions of [16.34] are the rotations of an arbitrary angle θ about the axis l_0. We know by [16.32] and [16.33] that u and k are determined in a unique manner with respect to R and τ_0. The isotropy group of μ can be parametrized by θ and τ_0. It is a Lie group of dimension 2. According to [18.6], the dimension of the orbit of μ is $10-2=8$. Owing to [18.7], the maximum number of independent invariant functions is $10-8=2$. A possible functional basis is composed of:

$$s_0 = \parallel l_0 \parallel, \qquad [16.35]$$

$$e_0 = e - \frac{1}{2\,m} \parallel p \parallel^2, \qquad [16.36]$$

of which the values are constant on the orbit which represents a massive particle with spin or a rigid body (seen from a long way off).

2) *Singular orbits: spinless massive particle.* In the particular case $l_0 = 0$, all the rotations of $\mathbb{SO}(3)$ satisfy [16.34], then the isotropy group is of dimension 4. By similar reasoning to the case of non-vanishing l_0, we conclude that dimension of the orbit is 6 and the number of invariant functions is 4. A possible functional basis is composed of e_0 and the three null components of l_0.

For the orbits with $m = 0$, the readers are referred to [GUI 84b] (pages 440 and 441). As an exercise, we leave to the readers the calculation of the dimension of the orbits and functional basis of invariants of the torsor of a particle by Lie group method [DES 11]. The calculations are very similar to the one of the boost method (section 3.1.1) which can be seen as a heuristic method inspired by Lie group technics but easier to teach to undergraduates. The dimension of generic orbits is also equal to 8 but the invariant e_0 is replaced by m.

–*Parametrization of the orbit.* Let us consider the generic orbits of dimension 8. Comparing to the expression [5.39] given by König's second theorem, taking into account $p_{\mathcal{B}} = m_{\mathcal{B}} \dot{x}_{\mathcal{B}}$ and leaving out the index $\mathcal{B}$, formula [16.36] reads:

$$e - \frac{1}{2\,m} \parallel p \parallel^2 = e - \frac{m}{2} \parallel \dot{x} \parallel^2 = e_0,$$

with:

$$e_0 = \frac{1}{2}\,\varpi \cdot (\mathcal{J}\,\varpi).$$

Taking into account [16.35], let us put:

$$l_0 = s_0 n,$$

where n is the unit vector giving the spin angular momentum direction. As we prove in section 5.2.3 that the moment of inertia matrix $\mathcal{J}$ is definite positive, it is regular and:

$$e_0 = \frac{s_0^2}{2}\, n \cdot (\mathcal{J}^{-1}\, n).$$

Hence, each generic orbit describing a particle of mass m, spin s_0 and inertia $\mathcal{J}$ can be parametrized by 8 coordinates, the 3 components of q, the 3 components of p and the 2 independent components of n defining the spin direction, due to the map:

$$\psi : \mathbb{R}^3 \times \mathbb{R}^3 \times \mathbb{S}^2 \to \mathfrak{g}^* : (q, p, n) \mapsto \mu = \psi(q, p, n),$$

such that:

$$l = \frac{1}{m}\, q \times p + s_0 n, \qquad e = \frac{1}{2\,m} \parallel p \parallel^2 + \frac{s_0^2}{2}\, n \cdot (\mathcal{J}^{-1}\, n).$$

A singular orbit of dimension 6, representing a spinless particle of mass m, corresponds to the particular case $l_0 = 0$ then $n = 0$. It can be parametrized by 6 coordinates, the 3 components of q and the 3 components of p due to the map:

$$\psi : \mathbb{R}^3 \times \mathbb{R}^3 \to \mathfrak{g}^* : (q, p) \mapsto \mu = \psi(q, p),$$

such that:

$$l = \frac{1}{m}\, q \times p, \qquad e = \frac{1}{2\,m} \parallel p \parallel^2 .$$

16.5. Galilean coordinate systems

16.5.1. *G-structures*

Let $\mathcal{M}$ be a differentiable manifold of dimension n and the corresponding principal fiber bundle $\pi : L(\mathcal{M}) \to \mathcal{M}$ of basis with structure group $\mathbb{GL}(n)$. Let G be a Lie subgroup of $\mathbb{GL}(n)$. By a *G-structure* on $\mathcal{M}$, we mean a differentiable subbundle L_G of $L(\mathcal{M})$ with structure group G [KOB 72]. Then, the fiber over $\boldsymbol{X} \in \mathcal{M}$ is the set of G-basis at $\boldsymbol{X}$. A G-structure L_G is integrable if every point $\boldsymbol{X}$ of $\mathcal{M}$ has a chart $\phi : V_\phi \mapsto U_\phi$ around $\boldsymbol{X}$ with local coordinate system X such that the cross-section $\boldsymbol{X} \mapsto S_\phi(\boldsymbol{X})$ over U_ϕ is a cross-section of L_G over U_ϕ. In other words, $\boldsymbol{X} \mapsto S_\phi(\boldsymbol{X})$ is a natural frame and X is said to be G-admissible. If X' is a G-admissible local coordinate system over $U_{\phi'}$, then the Jacobian matrix $\partial X'/\partial X$ belongs to G at each point of $U_\phi \cap U_{\phi'}$:

$$\frac{\partial X'}{\partial X} = P^{-1} \in G. \qquad [16.37]$$

We say that $X \mapsto X'$ is a *G-morphism*. Although in general G-structures are not integrable – in particular in the important case of the Riemannian geometry, the obstruction being the curvature –, it is worth to notice that the Galilean structures are integrable, as it will be proved further on.

16.5.2. *Galilean coordinate systems*

We hope to find the $\mathbb{GAL}_0$-admissible coordinate systems called Galilean coordinate systems by determining the $\mathbb{GAL}_0$-morphisms $X \mapsto X'$ called *galileomorphisms*. The partial derivative system [16.37]:

$$\frac{\partial X'}{\partial X} = P^{-1} = \begin{pmatrix} 1 & 0 \\ -R^T u & R^T \end{pmatrix} \in \mathbb{GAL}_0, \qquad [16.38]$$

involves $(4 \times 4 = 16)$ equations for 4 unknowns (X'^α). It is overdetermined and has generally no solutions, except if the equations satisfy suitable compatibility conditions. To discover them, we apply Frobenius method. If a solution exists, we may verify, for any perturbations dX and δX, the Lie bracket of vector fields $\delta X'$ and dX' vanishes:

$$[\delta, d]\, X' = \delta(dX') - d(\delta X') = 0, \qquad [16.39]$$

As it can be easily checked:

$$\delta(dX') = \delta(P^{-1} dX) = \delta(P^{-1}) dX + P^{-1} \delta(dX).$$

By skew-symmetrization with respect to δ and d, we have:

$$[\delta, d]\, X' = \delta(P^{-1})dX - d(P^{-1})\delta X + P^{-1}\,[\delta, d]\, X, \qquad [16.40]$$

and owing to $[\delta, d]\, X = 0$, equation [16.39] becomes:

$$[\delta, d]\, X' = \delta(P^{-1})dX - d(P^{-1})\delta X = 0. \qquad [16.41]$$

THEOREM 16.4.– Any galileomorphism $X \mapsto X'$ is compound of a rigid body motion:

$$x' = (R\,(t))^T\,(x - x_0\,(t)), \qquad [16.42]$$

and a clock change:

$$t' = t + \tau_0, \qquad [16.43]$$

where $t \mapsto R\,(t) \in \mathbb{SO}(3)$ and $t \mapsto x_0(t) \in \mathbb{R}^3$ are smooth mappings, and $\tau_0 \in \mathbb{R}$. Then, the velocity of transport is given by [3.29]:

$$u = \varpi\,(t) \times (x - x_0\,(t)) + \dot{x}_0\,(t), \qquad [16.44]$$

where ϖ is Poisson's vector defined by [3.25]:

$$\dot{R}R^T = j(\varpi). \qquad [16.45]$$

PROOF.– Taking into account equation [16.38], it holds that:

$$\delta(P^{-1})dX = \begin{pmatrix} 0 & 0 \\ -\delta\, R^T u - R^T \delta\, u & \delta\, R^T \end{pmatrix} \begin{pmatrix} d\,t \\ d\,x \end{pmatrix}$$
$$= \begin{pmatrix} 0 \\ \delta\, R^T d\,x - (\delta\, R^T u + R^T \delta\, u)\, d\,t \end{pmatrix}.$$

The compatibility condition $\delta\,(d\,t') = d\,(\delta\,t')$ is automatically satisfied. On the other hand, any infinitesimal rotation has the form:

$$\delta\, R = j\,(\delta\,\psi)\, R,$$

where $\delta\,\psi \in \mathbb{R}^3$. The second compatibility condition reads:

$$\delta\,(d\,x') - d\,(\delta\,x') = -R^T\ [\,j\,(\delta\,\psi)\, d\,x - j\,(d\,\psi)\,\delta\,x$$
$$-j\,(\delta\,\psi\, d\,t - d\,\psi\,\delta\,t)\, u + \delta\,u\, d\,t - d\,u\,\delta\,t\,] = 0.$$

The matrix R being regular, we have:

$$j\,(\delta\,\psi)\,d\,x - j\,(d\,\psi)\,\delta\,x - j\,(\delta\,\psi\,d\,t - d\,\psi\,\delta\,t)\,u + \delta\,u\,d\,t - d\,u\,\delta\,t = 0. \quad [16.46]$$

The perturbations $\delta\,\psi$ being infinitesimal, they linearly depend on $\delta\,t$ and $\delta\,x$:

$$\delta\,\psi = \frac{\partial\,\psi}{\partial\,x}\,\delta\,x + \frac{\partial\,\psi}{\partial\,t}\,\delta\,t = A\,\delta\,x + \varpi\,\delta\,t, \delta\,u = \frac{\partial\,u}{\partial\,x}\,\delta\,x + \frac{\partial\,u}{\partial\,t}\,\delta\,t = B\,\delta\,x + a\,\delta\,t. \quad [16.47]$$

Introducing the expressions [16.47] into the compatibility condition [16.46] gives after simplifying:

$$(A\,\delta\,x) \times d\,x + \delta\,x \times (A\,d\,x) + j\,(\varpi)\,(d\,x\,\delta\,t - \delta\,x\,d\,t) - (A\,(\delta\,x\,d\,t - d\,x\,\delta\,t))$$
$$\times u + B\,(\delta\,x\,d\,t - d\,x\,\delta\,t) = 0.$$

After some algebraic handlings, we have:

$$(Tr\,(A)\,1_{\mathbb{R}^3} - A)^T\,\delta\,x \times d\,x + (j\,(\varpi) - j\,(u)\,A - B)\,(d\,x\,\delta\,t - \delta\,x\,d\,t) = 0.$$

The perturbations $d\,X$, $\delta\,X$ being arbitrary, it holds:

$$Tr\,(A)\,1_{\mathbb{R}^3} = A, \quad j\,(\varpi) = j\,(u)\,A + B. \quad [16.48]$$

The first condition leads to:

$$A = \frac{\partial\,\psi}{\partial\,x} = 0, \quad [16.49]$$

that entails ψ is independent of the position: $\psi = \psi\,(t)$, then integrating:

$$\varpi = \frac{\partial\,\psi}{\partial\,t} = \dot{\psi}\,(t),$$

Hence, ϖ does not depend on the position: $\varpi = \varpi\,(t)$. Owing to [16.49], the second condition [16.48] gives:

$$\frac{\partial\,u}{\partial\,x} = B = j\,(\varpi\,(t))$$

There exists an arbitrary column $u_0\,(t) \in \mathbb{R}^3$ such that the velocity of transport has the following form:

$$u = \varpi\,(t) \times x + u_0\,(t). \quad [16.50]$$

Finally, integrating the differential system:

$$d\,R = j\,(d\,\psi)\,R = j\,(\varpi\,(t))\,R\,d\,t, \qquad [16.51]$$

we see that the rotation R does not depend on the position: $R = R\,(t)$, the column ϖ being Poisson's vector defined by [16.45].

In short, we showed that the compatibility conditions [16.41] are satisfied if and only if R does not depend on the position and the velocity of transport has the form [16.50]. Under these conditions, we can integrate the equation system [16.38]:

$$\frac{\partial\,t'}{\partial\,t} = 1, \quad \frac{\partial\,t'}{\partial\,x} = 0, \quad \frac{\partial\,x'}{\partial\,x} = R^T\,(t), \quad \frac{\partial\,x'}{\partial\,t} = -R^T\,(t)\,(\varpi\,(t)\times x + u_0\,(t)). \qquad [16.52]$$

The integration of the first two equations leads to [16.43] where $\tau_0 \in R$ is an arbitrary constant representing a clock change. Integrating the third equation introduces a column $x_0\,(t) \in \mathbb{R}^3$ such that the new position takes the form [16.42]. Introducing [16.42] in the last equation [16.52] and taking into account [16.51], we have:

$$u_0 = \dot{x}_0 - \varpi \times x_0.$$

Therefore, the velocity of transport [16.50] is given by expression [16.44]. ■

In section 3.2, we introduced directly the galileomorphisms as rigid body motions. It is straightforward to verify that the Jacobian matrix of these coordinate changes is a linear Galilean transformation. The present theorem is the converse proposition.

16.6. Galilean curvature

In general relativity, it is well known that space–time is curved due to gravitation. But what happens in Galilean mechanics? Is space–time flat or curved? To answer these questions, we have to calculate the Riemann–Christoffel curvature tensor. Using [18.10] is tedious but shows the only non-vanishing components are:

$$R^i_{0j0} = -R^i_{j00} = \frac{\partial g^i}{\partial x^j} + \frac{\partial \Omega^i_j}{\partial t} + \Omega^i_k\,\Omega^k_j, \qquad [16.53]$$

$$R^i_{kj0} = -R^i_{jk0} = \frac{\partial \Omega^i_j}{\partial x^k} - \frac{\partial \Omega^i_k}{\partial x^j}, \qquad R^i_{k0j} = -R^i_{0kj} = \frac{\partial \Omega^i_j}{\partial x^k}. \qquad [16.54]$$

Another way consists of applying [18.9]. By differentiation of [3.38] and antisymmetrization, we obtain:

$$d\Gamma(\delta X) - \delta\Gamma(dX) =$$

$$\begin{pmatrix} 0 & 0 \\ \left(\frac{\partial\Omega}{\partial x}\,dx\right)\times\delta x + dx \times \left(\frac{\partial\Omega}{\partial x}\,\delta x\right) - \left(\frac{\partial g}{\partial x} + j\left(\frac{\partial\Omega}{\partial t}\right)\right)(dx\,\delta t - \delta x\,dt) & j\left(\frac{\partial\Omega}{\partial x}\,(dx\,\delta t - \delta x\,dt)\right) \end{pmatrix},$$

where, taking into account [7.19] and [14.26]:

$$\left(\frac{\partial\Omega}{\partial x}\,dx\right)\times\delta x + dx\times\left(\frac{\partial\Omega}{\partial x}\,\delta x\right) = \left(Tr\left(\frac{\partial\Omega}{\partial x}\right)1_{\mathbb{R}^3} - \frac{\partial\Omega}{\partial x}\right)^T dx\times\delta x$$
$$= curl\,(j(v))\,\delta x\times dx.$$

On the other hand, owing to [3.38] and after simplification:

$$\Gamma(dX)\,\Gamma(\delta X) - \Gamma(\delta X)\,\Gamma(dX) = \begin{pmatrix} 0 & 0 \\ (j(\Omega))^2\,(\delta x\,dt - dx\,\delta t) & 0 \end{pmatrix}.$$

Hence, the *Galilean curvature tensor* is:

$$R(dX,\delta X) =$$

$$\begin{pmatrix} 0 & 0 \\ \left(\frac{\partial g}{\partial x} + j\left(\frac{\partial\Omega}{\partial t}\right) + (j(\Omega))^2\right)(\delta x\,dt - dx\,\delta t) + curl\,(j(\Omega))\,\delta x\times dx & j\left(\frac{\partial\Omega}{\partial x}\,(dx\,\delta t - \delta x\,dt)\right) \end{pmatrix}.$$

It can be checked this fits [16.53] and [16.54].

A noticeable fact is that $R^i_{jkm} = 0$ where Latin indices run over the spacial coordinate labels, which means the slices $t = C^{te}$ representing the 3-dimensional space are flat subspaces, while *the space–time is curved, even in classical mechanics*.

THEOREM 16.5.– If the Galilean curvature is null and there are no other forces, there exists a Galilean coordinate system in which the material particles with constant mass are in uniform straight motion (USM).

PROOF.– From [16.55], it results that in any Galilean coordinate system X the Galilean curvature tensor vanishes if and only if:

$$\frac{\partial\Omega}{\partial x} = 0, \qquad curl\,(j(\Omega)) = 0, \qquad \frac{\partial g}{\partial x} + j\left(\frac{\partial\Omega}{\partial t}\right) + (j(\Omega))^2 = 0.$$

The former equation means that Ω depends only on the time and the second condition is, therefore, automatically satisfied. Integrating the latter equation leads to:

$$g = g_0(t) - \left(j\left(\dot{\Omega}(t)\right) + (j(\Omega(t)))^2 \right) (x - x_0(t)),$$

where $g_0(t)$ and $x_0(t)$ are arbitrary smooth functions. We have to find a Galilean coordinate change $X \mapsto X'$, given by smooth functions $x_0(t) \in \mathbb{R}^3$, $R(t) \in \mathbb{SO}(3)$ and corresponding Poisson's vector $\varpi(t)$, such that $g' = \Omega' = 0$. Considering the transformation law [3.51] of the spinning, we enforce $\Omega' = 0$ by choosing $\varpi = -\Omega$, which is possible because Ω is uniform and gives $R(t)$ by integrating [16.45]. The gravity becomes:

$$g = g_0 + \dot{\varpi} \times (x - x_0) - \varpi \times (\varpi \times (x - x_0)).$$

Besides, considering the transformation law [3.62] of the gravity and the expression [3.29] of the velocity of transport, we have:

$$g = \ddot{x}_0 - 2\,\varpi \times \dot{x}_0 + \dot{\varpi} \times (x - x_0) - \varpi \times (\varpi \times (x - x_0)) + R\,g'.$$

Comparing the latter two conditions shows that $g' = 0$ by picking:

$$g_0 = \ddot{x}_0 - 2\,\varpi \times \dot{x}_0,$$

which is possible because g_0 is uniform. According to law 3.4, the equation of motion [3.78] in the new coordinate system X' shows, if there are no other forces, that the linear momentum is constant and the velocity so is, which achieves the proof. ■

For the Newtonian gravitation (law 3.3), the curvature tensor reads:

$$R(dX, \delta X) = \begin{pmatrix} 0 & 0 \\ \dfrac{\partial g}{\partial x} & 0 \end{pmatrix},$$

where:

$$\frac{\partial g}{\partial x} = -\frac{k_g\, m'}{\parallel x - x' \parallel^3} \left[1_{\mathbb{R}^3} - 3\, \frac{x - x'}{\parallel x - x' \parallel}\, \frac{(x - x')^T}{\parallel x - x' \parallel} \right] \neq 0.$$

Hence, the space–time is curved and no USM is possible.

We obtained previously the transformation law of the Galilean gravitation composed of g and Ω (theorem 3.2) and the one of the Galilean gravitation potentials ϕ and A (theorem 6.1). Incidentally, we obtained a Galilean invariant of these potentials, I_0 given by [6.24]. Let us now construct a Galilean joint invariant of the gravitation and its potentials.

THEOREM 16.6.– The quantity:

$$I = div\, g - 2 \parallel \Omega \parallel^2 + 2\, A \cdot curl\, \Omega.$$

is a Galilean invariant.

PROOF.– We start with the transformation law of the gravity g in the form [6.25] which can be written:

$$g' = R^T \left(g - \frac{\partial u}{\partial t} - j(\varpi)\, u - 2\, j(\Omega)\, u \right).$$

Applying [6.26] to g' gives, using [7.42]:

$$\frac{\partial g'}{\partial x'} = \frac{\partial g'}{\partial x} R = R^T M\, R,$$

with:

$$M = \frac{\partial g}{\partial x} - \frac{\partial^2 u}{\partial t \partial x} - j(\varpi)\, \frac{\partial u}{\partial x} - 2\, j(\Omega)\, \frac{\partial u}{\partial x} + 2\, j(u)\, \frac{\partial \Omega}{\partial x}.$$

Hence, we have:

$$div_{x'}\, g' = Tr \left(\frac{\partial g'}{\partial x'} \right) = Tr(R^T M\, R) = Tr(R\, R^T M) = Tr(M),$$

$$div_{x'}\, g' = div\, g - \frac{\partial}{\partial t}\, (div\, u) - Tr \left(j(\varpi)\, \frac{\partial u}{\partial x} \right)$$
$$- 2\, Tr \left(j(\Omega)\, \frac{\partial u}{\partial x} \right) + 2\, Tr \left(j(u)\, \frac{\partial \Omega}{\partial x} \right)$$

Taking into account [6.27], the velocity of transport is divergence free:

$$div\, u = Tr\, (j(\varpi)) = 0.$$

Moreover, the latter three terms have the same form and can be transformed as the following generic expression, using the definition [7.43] of curl and [7.13]:

$$Tr \left(j(v)\, \frac{\partial w}{\partial x} \right) = Tr \left(j(v)\, \frac{1}{2} \left(\frac{\partial w}{\partial x} - \left(\frac{\partial w}{\partial x} \right)^T \right) \right)$$
$$= \frac{1}{2}\, Tr\, (j(v)\, j(curl\, w)) = -v \cdot curl\, w.$$

Hence:

$$div_{x'}\, g' = div\, g + (\varpi + 2\,\Omega) \cdot curl\, u - 2\, u \cdot curl\, \Omega,$$

which becomes, owing to [6.28]:

$$div_{x'}\, g' = div\, g + 2 \parallel \varpi \parallel^2 + 4\, \varpi \cdot \Omega - 2\, u \cdot curl\, \Omega. \qquad [16.55]$$

On the other hand, the transformation law [3.51] of the spinning gives together with [7.20]:

$$\parallel \Omega' \parallel^2 = \parallel R^T (\Omega + \varpi) \parallel^2 = \parallel \Omega + \varpi \parallel^2 = \parallel \Omega \parallel^2 + 2\,\Omega \cdot \varpi + \parallel \varpi \parallel^2. \qquad [16.56]$$

Moreover, using [7.43], applying [6.26] to Ω', owing to [3.51] and [7.23], we have:

$$j(curl_{x'}\, \Omega') = \frac{\partial \Omega'}{\partial x'} - \left(\frac{\partial \Omega'}{\partial x'}\right)^T = R^T \left[\frac{\partial \Omega}{\partial x} - \left(\frac{\partial \Omega}{\partial x}\right)^T\right]$$

$$R = R^T j(curl\, \Omega)\, R = j(R^T curl\, \Omega),$$

As j is a regular map:

$$curl_{x'}\, \Omega' = R^T\, curl\, \Omega.$$

Thus, using the latter result, the transformation law [6.22] of A and [7.20], we have:

$$A' \cdot curl_{x'}\, \Omega' = (A + u) \cdot curl\, \Omega. \qquad [16.57]$$

Taking into account [16.55], [16.56] and [16.57], we show after simplification that:

$$div_{x'}\, g' - 2 \parallel \Omega' \parallel^2 + 2\, A' \cdot curl_{x'}\, \Omega' = div\, g - 2 \parallel \Omega \parallel^2 + 2\, A \cdot curl\, \Omega,$$

which achieves the proof. ■

On the other hand, it is worth noting that we know a Galilean invariant of the stress–mass tensor representing the matter, the mass density ρ, according to [10.35]. This suggests the claim that the two Galilean invariants are proportional:

$$\boxed{div\, g - 2 \parallel \Omega \parallel^2 + 2\, A \cdot curl\, \Omega = -4\,\pi\, k_g \rho,} \qquad [16.58]$$

where k_g is a constant. In particular, if we suppose that $A = 0$, then $g = -grad\,\phi$ and $\Omega = 0$ according to [6.14] and the previous condition reduces to Poisson's equation:

$$\Delta\,\phi = 4\,\pi\,k_g \rho. \qquad [16.59]$$

of which the solution is:

$$\phi(x,t) = -\int \frac{k_g\,\rho(x',t)}{\parallel x - x' \parallel}\, d\mathcal{V}(x'),$$

for an integrable density. For a mass m' concentrated to the position $x'(t)$ at time t, using Dirac's distribution at x', we obtain:

$$\phi(x,t) = -\frac{k_g\,m'}{\parallel x - x'(t) \parallel}$$

which restitutes the Newtonian gravitation (section 3.4), k_g being the gravitational constant.

Nevertheless, it is worth noting that [16.59] has no expected Galilean covariance. The general equation compatible with Galileo's principle of relativity is [16.58].

16.7. Bargmannian coordinates

The set $\mathbb{B}$ of the affine Bargmannian transformations is a Lie subgroup of dimension 11 of $\mathbb{A}ff\,(5)$ called Bargmann's group. The subset $\mathbb{B}_0 \subset \mathbb{B}$ of the linear Bargmannian transformations $\hat{P}$ given by [1.4] is a Lie subgroup of dimension 6 of $\mathbb{GL}(5)$.

In section 13.8, we considered a 5-dimensional space $\hat{\mathcal{U}}$ of which the space–time $\mathcal{U}$ is a submanifold. As $\hat{\mathcal{U}}$ is equipped with a metric, it is a Riemannian manifold. Then, the $\mathbb{B}$-structure is non-integrable. In other words, the orthonormal frames are (non-integrable) moving frames. We would like to find a basis change onto (integrable) natural frames.

Differentiating the expression [13.5] of $\hat{P}$ and accounting for [3.24], an infinitesimal Bargmannian transformation around the identity reads:

$$d\hat{P} = \begin{pmatrix} 0 & 0 & 0 \\ du & j(d\psi) & 0 \\ 0 & du^T & 0 \end{pmatrix}. \qquad [16.60]$$

By theorem 3.1, we know that du and $d\psi$ have the form [3.42]:

$$\hat{\Gamma}(d\hat{X}) = \begin{pmatrix} 0 & 0 & 0 \\ j(\Omega)\,dx - g\,dt & j(\Omega)\,dt & 0 \\ 0 & (j(\Omega)\,dx - g\,dt)^T & 0 \end{pmatrix}, \qquad [16.61]$$

The general expression of the torsion 2-form in a moving frame is:

$$\hat{T}(\delta\hat{X}, d\hat{X}) = \hat{\Gamma}(\delta\hat{X})\,d\hat{X} - \hat{\Gamma}(d\hat{X})\,\delta\hat{X} - [\delta, d]\,\hat{X}, \qquad [16.62]$$

For the connection being symmetric, the torsion must vanish:

$$[\delta, d]\,\hat{X} = \hat{\Gamma}(\delta\hat{X})\,d\hat{X} - \hat{\Gamma}(d\hat{X})\,\delta\hat{X},$$

which gives, for the connection [16.61]:

$$[\delta, d]\,X = 0, \qquad [\delta, d]\,z = 2\,\Omega \cdot (dr \times \delta r) - g \cdot (\delta r\,dt - dr\,\delta t). \qquad [16.63]$$

Now, we are able to find a coordinate chart $\hat{\psi}$ and the corresponding natural frame $\hat{S}_{\hat{\psi}}$. Applying formula [16.40] on the manifold $\hat{\mathcal{U}}$ of dimension 5:

$$[\delta, d]\,\hat{X}' = \delta(\hat{P}^{-1})d\hat{X} - d(\hat{P}^{-1})\delta\hat{X} + \hat{P}^{-1}\,[\delta, d]\,\hat{X},$$

using [14.33] and enforcing this Lie bracket to vanish, we obtain:

$$[\delta, d]\,\hat{X} = \delta\hat{P}\,\hat{P}^{-1}d\hat{X} - d\hat{P}\,\hat{P}^{-1}\delta\hat{X},$$

where $\hat{P}$ is the unknown of the equation. We try a solution of the form:

$$d\hat{P} = \begin{pmatrix} 1_{\mathbb{R}^4} & 0 \\ \Phi(X) & 1 \end{pmatrix},$$

where the 4-row Φ is:

$$\Phi(X) = (-\phi, A^T),$$

with $\phi \in \mathbb{R}$ and $A \in \mathbb{R}^3$. Thus, the equation gives:

$$[\delta, d]\,z = (curl\,A) \cdot (dx \times \delta x) + \left(grad\,\phi + \frac{\partial A}{\partial t}\right) \cdot (\delta x\,dt - dx\,\delta t).$$

Comparing with [16.63] leads to:

$$g = -grad\,\phi - \frac{\partial A}{\partial t}, \qquad \Omega = \frac{1}{2}\,curl\,A.$$

that is just the definition [6.14] of the Galilean gravitation potentials. We recovered the basis change of section 13.8 with the transformation matrix [13.72] previously obtained in a heuristic way and X' being what we called *Bargmannian coordinates*.

DEFINITION 16.3.– We call a *Bargmannian basis* (respectively, *frame*) a $\mathbb{B}$-basis $(\vec{e}_\alpha)$ (respectively, $\mathbb{B}$-frame) in the meaning of section 14.2.3.

Bargmannian bases are in fact orthonormal bases then (non-integrable) moving frames. They do not correspond to Bargmannian coordinate charts. Now, let us consider a Bargmanian basis $\hat{S}$ and two Bargmannian coordinate charts $\hat{\psi}, \hat{\psi}^*$ with associated natural frames:

$$S_{\hat{\psi}} = \hat{S}\,\hat{P}, \qquad S_{\hat{\psi}^*} = \hat{S}\,\hat{P}^*.$$

What is the form of the corresponding Bargmannian coordinate change $\hat{X} \mapsto \hat{X}^*$? Necessarily, we have:

$$\frac{\partial \hat{X}^*}{\partial \hat{X}} = \hat{P}^*\hat{P}^{-1} = \begin{pmatrix} 1_{\mathbb{R}^4} & 0 \\ \Delta\Phi & 1 \end{pmatrix},$$

where $\Delta\Phi = \Phi^* - \Phi$.

According to the Frobenius method, the coordinate change is a solution of this system provided that the following compatibility conditions hold:

$$\frac{\partial \Delta\Phi_\alpha}{\partial X^\beta} = \frac{\partial \Delta\Phi_\beta}{\partial X^\alpha},$$

hence, there exists locally a function $X \mapsto f(X) \in \mathbb{R}$ such that:

$$\Delta\Phi = \frac{\partial f}{\partial X},$$

which is just the gauge transformation [6.16]:

$$\phi^* = \phi - \frac{\partial f}{\partial t}, \qquad A^* = A + grad\,f.$$

16.8. Bargmannian torsors

DEFINITION 16.4.– The *Bargmannian torsor* $\hat{\tau}$ of a particle is a 2-contravariant skew-symmetric affine tensor:

$$\hat{\tau} : A^*T_{\hat{X}}\hat{\mathcal{U}} \times A^*T_{\hat{X}}\hat{\mathcal{U}} \to \mathbb{R} : (\hat{\Psi}_1, \hat{\Psi}_2) \mapsto \hat{\tau}(\hat{\Psi}_1, \hat{\Psi}_2),$$

represented in a Bargmannian frame by a skew-symmetric 6×6 matrix:

$$\boxed{\tilde{\tau} = \begin{pmatrix} 0 & \hat{T}^T \\ -\hat{T} & \hat{J} \end{pmatrix},} \qquad [16.64]$$

where $\hat{T} \in \mathbb{R}^5$, $\hat{J} \in \mathbb{M}_{55}^{skew}$ and of which the components, under the Bargmannian transformation [13.7], are modified according to the transformation law:

$$\tilde{\tau} = \tilde{P}\tilde{\tau}'\tilde{P}^T. \qquad [16.65]$$

Taking into account the structure of the space–time, $\hat{T}$ and $\hat{J}$ are decomposed by blocks:

$$\hat{T} = \begin{pmatrix} m \\ p \\ e \end{pmatrix}, \quad \hat{J} = \begin{pmatrix} 0 & -q^T & -\beta \\ q & -j\,(l) & w \\ \beta & -w^T & 0 \end{pmatrix}, \qquad [16.66]$$

where $m, e, \beta \in \mathbb{R}$ are scalar and $p, q, l, w \in \mathbb{R}^3$. As Galileo's group is a subgroup of Bargmann's one, m is the mass, p is the linear momentum, q is the passage and l is the angular momentum. Now, we would like to reveal the physical meaning of the remaining components e, β and w by applying the transformation law of the torsor [16.65] or, equivalently, its inverse one:

$$\tilde{\tau}' = \tilde{P}^{-1}\tilde{\tau}\tilde{P}^{-T}. \qquad [16.67]$$

Accounting for [13.7], it itemizes in the transformation laws [3.5], [3.6] , [3.7], [3.8] of the components m, p, q, l of the Galilean torsor and:

$$\boxed{e' = e - u \cdot p + \frac{m}{2} \parallel u \parallel^2,} \qquad [16.68]$$

$$\boxed{\beta' = \beta - u \cdot q - \tau_0' \, e' + m \, \eta',} \qquad [16.69]$$

$$\boxed{w' = R^T \left[w + l \times u + \frac{1}{2} \parallel u \parallel^2 q - \eta' \, p + (\beta' + \tau_0' \, e') \, u \right] + e' \, k',} \qquad [16.70]$$

Owing to [3.6], it is easy to see that [16.68] is nothing other than the inverse of the transformation law [13.3], then e is interpreted as the kinetic energy. The last two components are more puzzling. In the spirit of the boost method initiated in section 3.1.2, we consider a Bargmannian frame $\hat{f}$ in which the torsor field $\hat{\tau}$ has a *reduced form*:

$$\tilde{\tau} = \begin{pmatrix} 0 & m & 0 & 0 \\ -m & 0 & 0 & 0 \\ 0 & 0 & -j\,(l_0) & 0 \\ 0 & 0 & 0 & 0 \end{pmatrix}, \qquad [16.71]$$

and we claim that it represents a particle at rest. Now, we perform two steps:

– *Step 1:* we consider another Bargmannian frame $\hat{f}\,'$ obtained from the previous one through a Galilean boost $u = -v$ combined with translations $k' = x$ and $\eta' = z$. Applying the inverse transformation law of [16.67] for a spinless particle and removing the primes, we have:

$$p = m \, v, \qquad q = m \, x, \qquad l = q \times v,$$

$$e = \frac{m}{2} \parallel v \parallel^2, \qquad \beta = m \, z, \qquad w = \frac{m}{2} \parallel v \parallel^2 x - m \, z \, v.$$

We recover the definition of the kinetic energy. Taking into account [13.4], we have:

$$d\beta = m \, dz = e \, dt,$$

where e is the Lagrangian of the free particle (see section 6.2.1), which reveals the meaning of the component β as the action. With this interpretation, the last component reads:

$$w = e \, x - \beta \, v,$$

or a lot better, replacing the kinetic energy by the Lagrangian:

$$w = \mathcal{L} \, x - \beta \, v.$$

We call it the *Lagrangian-action momentum.*

– *Step 2:* for the new observer, the particle is moving in USM as it was gravitation free. To embed it into the gravitation field, we consider the linear transformation:

$$\tilde{Q}^{-1} = \begin{pmatrix} 1 & 0 \\ 0 & \hat{Q}^{-1} \end{pmatrix},$$

where $\hat{Q}^{-1}$ is given by [13.72] in order to work with a Bargmannian coordinate system in which the new components are given by $\tilde{\tau}' = \tilde{P}^{-1}\tilde{\tau}\tilde{P}^{-T}$. After calculating and removing the primes, we see that the components m, p, q, l of the Galilean torsor remain unchanged while:

$$\mathcal{L} = \frac{m}{2} \parallel v \parallel^2 - m\,\phi + m\,A \cdot v, \qquad [16.72]$$

$$\beta = m\,(z + A \cdot x), \qquad w = \frac{m}{2} \parallel v \parallel^2 x - m\,z\,v - \phi\,q + A \times l. \qquad [16.73]$$

We recover the general form [6.13] of the Lagrangian in the gravitation field. Because of the additional quantity in β, we call it *augmented action.* Besides, observing that $l = m\,x \times v$ and using the vector triple product [7.15] leads to:

$$w = \mathcal{L}\,x - \beta\,v,$$

and we recover the expression of the Lagrangian-action momentum.

DEFINITION 16.5.– The Bargmannian torsor is structured into two components:

- the *linear* 4-*momentum* $\hat{T}$, itself substructured into:

  1) the *mass* m;

  2) the *linear momentum* p;

  3) the *Lagrangian* $\mathcal{L}$.

- and the *angular* 4-*momentum* $\hat{J}$, itself substructured into:

  1) the *passage* q;

  2) the *angular momentum* l;

  3) the *augmented action* β;

  4) the *Lagrangian-action momentum* w.

In matrix form, the Bargmannian torsor reads:

$$\boxed{\tilde{\tau} = \begin{pmatrix} 0 & m & p^T & \mathcal{L} \\ -m & 0 & -q^T & -\beta \\ -p & q & -j(l) & w \\ -\mathcal{L} & \beta & -w^T & 0 \end{pmatrix}}. \quad [16.74]$$

16.9. Bargmannian momenta

In section 16.4, we showed that the general action of Galileo's group on the momentum tensors useful for physical applications is affine. For Bargmann's group, it is not so. In fact, as we will see in section 17.6, Bargmann's group is constructed just for the translation part θ in [16.10] to be null. Therefore, the considered action of Bargmann's group on the momentum tensors is just the coadjoint one:

$$\hat{F}' = \hat{F}\,\hat{P}, \qquad \hat{L}' = (\hat{P}^{-1}\,\hat{L} + \hat{C}'\,\hat{F})\,\hat{P}. \quad [16.75]$$

The *Bargmannian momentum tensor* $\hat{\boldsymbol{\mu}}$ of a particle is a mixed 1-covariant and 1-contravariant affine tensor:

$$\hat{\boldsymbol{\mu}} : T_{\hat{\boldsymbol{X}}}\hat{\mathcal{U}} \times A^* T_{\hat{\boldsymbol{X}}}\hat{\mathcal{U}} \rightarrow \mathbb{R} : (\hat{\vec{\boldsymbol{V}}}, \hat{\boldsymbol{\Psi}}) \mapsto \hat{\boldsymbol{\mu}}(\hat{\vec{\boldsymbol{V}}}, \hat{\boldsymbol{\Psi}}),$$

Every Bargmannian linear transformation $\hat{P}$ preserves the metrics [13.8] of the 5-dimensional space, then taking into account [13.9]:

$$\hat{P}^*\,\hat{P} = 1_{\mathbb{R}^5},$$

and differentiating, we have:

$$d\hat{P}^*\,\hat{P} = -\hat{P}^* d\hat{P}.$$

Considering the limit when $\hat{P}$ approaches the identity, we see that any infinitesimal generator $d\hat{P}$ of the Lie algebra $\mathfrak{b}_0$ of $\mathbb{B}_0$ is anti-self-adjoint:

$$d\hat{P}^* = -d\hat{P}.$$

Taking into account [16.5], the dual pairing reads:

$$\hat{\mu}\,\hat{Z} = \hat{\mu}\,d\hat{a} = (\hat{F}, \hat{L})\,(d\hat{C}, d\hat{P}) = \hat{F}\,d\hat{C} + Tr(\hat{L}\,d\hat{P}). \quad [16.76]$$

For any $d\hat{P} \in \mathfrak{b}_0$, the last term is transformed as follows:

$$Tr(\hat{L}\, d\hat{P}) = -Tr(\hat{L}\, d\hat{P}^*) = -Tr(d\hat{P}^*\hat{L}) = -Tr((\hat{L}^* d\hat{P})^*) = -Tr(\hat{L}^* d\hat{P}),$$

then $\hat{L}$ is anti-self-adjoint:

$$\hat{L}^* = -\hat{L}.$$

The structure of Euclidean space (and of Riemannian manifold) induces strong properties. In particular, it allows us to generate new tensors by lowering or raising the indices. This suggests to generate the Bargmannian momenta from the torsors, according to the rule:

$$\hat{F}_\beta = \hat{G}_{\beta\mu}\hat{T}^\mu, \qquad \hat{L}^\alpha_\beta = \frac{1}{2}\,\hat{G}_{\beta\mu}\hat{J}^{\alpha\mu},$$

or in matrix form:

$$\hat{F} = \hat{T}^T\hat{G}, \qquad \hat{L} = \frac{1}{2}\,\hat{J}\,\hat{G}. \qquad [16.77]$$

Conversely, we have:

$$\hat{T} = \hat{G}^{-1}\hat{F}^T, \qquad \hat{J} = 2\,\hat{L}\,\hat{G}^{-1}. \qquad [16.78]$$

Let us remark that if $\hat{J}$ is skew-symmetric, $\hat{L}$ is anti-self-adjoint. Indeed, owing to [14.12], we have:

$$\hat{L}^* = \hat{G}^{-1}\left(\frac{1}{2}\,\hat{J}\,\hat{G}\right)^T\hat{G} = \frac{1}{2}\,\hat{J}^T\hat{G} = -\frac{1}{2}\,\hat{J}\,\hat{G} = -\hat{L}.$$

Now, we have to verify that the law of transformation of torsors:

$$\hat{T}' = \hat{P}^{-1}\,\hat{T}, \qquad \hat{J}' = \hat{P}^{-1}\hat{J}\,\hat{P}^{-T} + \hat{C}'(\hat{P}^{-1}\hat{T})^T - (\hat{P}^{-1}\hat{T})\,\hat{C}'^T, \qquad [16.79]$$

is consistent with the one [16.75] of momenta.

LEMMA 16.1.– If $\hat{L}$ is anti-self-adjoint, the transformation law [16.75] is recast as:

$$\hat{F}' = \hat{F}\,\hat{P}, \hat{L}' = \left[\hat{P}^{-1}\,\hat{L} + \frac{1}{2}\left(\hat{C}'\,\hat{F} - \hat{P}^{-1}\hat{G}^{-1}\hat{F}^T\hat{C}'^T\hat{P}^T\hat{G}\right)\right]\,\hat{P}. \qquad [16.80]$$

PROOF.– As 2-covariant tensor, the metric tensor is transformed as:

$$\hat{G}' = \hat{P}^T \hat{G}\, \hat{P}. \qquad [16.81]$$

Because $\hat{L}$ is anti-self-adjoint, we have:

$$\hat{L} = \frac{1}{2}\,(\hat{L} + \hat{L}) = \frac{1}{2}\,(\hat{L} - \hat{L}^*), \qquad [16.82]$$

and similarly:

$$\hat{L}' = \frac{1}{2}\,(\hat{L}' - \hat{L}'^*) = \frac{1}{2}\,(\hat{L}' - \hat{G}'^{-1}\hat{L}'^T\hat{G}').$$

Taking into account [16.75] and [16.81], we have:

$$\hat{L}' = \frac{1}{2}\,\left[(\hat{P}^{-1}\,\hat{L} + \hat{C}'\,\hat{F})\,\hat{P} - \hat{P}^{-1}\hat{G}^{-1}\,(\hat{L}^T\,\hat{P}^{-T} + \hat{F}^T\,\hat{C}'^T)\,\hat{P}^T\hat{G}\,\hat{P}\right].$$

Owing to [14.12], it holds that:

$$\hat{L}' = \left[\hat{P}^{-1}\,\frac{1}{2}\,(\hat{L} - \hat{L}^*) + \frac{1}{2}\left(\hat{C}'\,\hat{F} - \hat{P}^{-1}\hat{G}^{-1}\hat{F}^T\,\hat{C}'^T\,\hat{P}^T\,\hat{G}\right)\right]\,\hat{P},$$

which achieves the proof, owing to [16.82]. ■

THEOREM 16.7.– The transformation laws [16.79] and [16.75] of Bargmannian torsors and momenta are equivalent through [16.77] and [16.78].

PROOF.– Taking into account [16.81], the transformation law of F is equivalent to the one of T:

$$\hat{F}' = \hat{T}'^T\hat{G}' = (\hat{P}^{-1}\,\hat{T})^T\,\hat{P}^T\,\hat{G}\,\hat{P} = \hat{T}^T\,\hat{G}\,\hat{P} = \hat{F}\,\hat{P}.$$

Likewise, the transformation law of F derives from the ones of T and J because:

$$\hat{L}' = \frac{1}{2}\,\hat{J}'\,\hat{G}' = \frac{1}{2}\,\left[\hat{P}^{-1}\hat{J}\,\hat{P}^{-T} + \hat{C}'(\hat{P}^{-1}\hat{T})^T - (\hat{P}^{-1}\hat{T})\,\hat{C}'^T\right]\,\hat{P}^T\hat{G}\,\hat{P}.$$

After simplification, it holds that:

$$\hat{L}' = \left[\hat{P}^{-1}\left(\frac{1}{2}\,\hat{J}\,\hat{G}\right) + \frac{1}{2}\left(\hat{C}'\,(\hat{T}^T\hat{G}) - \hat{P}^{-1}\hat{G}^{-1}(\hat{G}\,\hat{T})\,\hat{C}'^T\,\hat{P}^T\,\hat{G}\right)\right]\,\hat{P}.$$

Taking into account [16.77] leads to [16.80] and we conclude with lemma 16.1. ■

It is worth noting that this construction is valid only if there exists a Euclidean structure on the tangent space and thus a Riemannian structure on the manifold. It cannot be, for instance, applied in the Galilean framework where the space–time is not Riemannian.

Taking into account the expression [13.73] of the metric embedded in the gravitation field, we obtain the one of $\hat{F}$ by [16.77]:

$$\hat{F} = \begin{pmatrix} m & p^T & \mathcal{L} \end{pmatrix} \begin{pmatrix} -2\phi & A^T & -1 \\ A & 1_{\mathbb{R}^3} & 0 \\ -1 & 0 & 0 \end{pmatrix} = \begin{pmatrix} -\mathcal{H} & \pi^T & -m \end{pmatrix},$$

with:

$$\mathcal{H} = \mathcal{L} + 2\,m\,\phi - p \cdot A, \qquad \pi = p + m\,A.$$

Taking into account the expression [16.72] of the Lagrangian and $p = m\,v$, we recover the expression [6.18] of the Hamiltonian:

$$\mathcal{H} = \frac{1}{2}\,m \parallel v \parallel^2 + m\,\phi.$$

It is worth to observe that we obtained it not by Legendre's transform but by lowering the index of an Euclidean tensor. Besides, we recognized the expression [6.17] of the generalized linear momentum.

Likewise, we deduce the expression of $\hat{L}$ from the one of $\hat{J}$:

$$\hat{L} = \frac{1}{2} \begin{pmatrix} 0 & -q^T & -\beta \\ q & -j(l) & w \\ \beta & -w^T & 0 \end{pmatrix} \begin{pmatrix} -2\phi & A^T & -1 \\ A & 1_{\mathbb{R}^3} & 0 \\ -1 & 0 & 0 \end{pmatrix} = \frac{1}{2} \begin{pmatrix} \alpha & -q^T & 0 \\ -w_1 & -S & -q \\ -\gamma & -w_2^T & 0 \end{pmatrix},$$

with:

$$\alpha = \beta - q \cdot A, \qquad w_1 = w + l \times A + 2\phi\,q,$$

$$\gamma = 2\,\beta\,\phi + w \cdot A, \qquad S = j(l) - q\,A^T, \qquad w_2 = w - \beta\,A.$$

Taking into account [16.73] and the expression [6.18] of the Hamiltonian, we see that α is the *action* and:

$$w_1 = \mathcal{H}\,x - \alpha\,v,$$

which we call the *Hamiltonian-action momentum* and is expanded as:

$$w_1 = \left(\frac{1}{2} m \parallel v \parallel^2 + m\,\phi\right) x - m\,z\,v,$$

It is worth comparing it to the rather similar but distinct expression [3.75] of Laplace–Runge–Lenz vector:

$$w_L = v \times l + m\,\phi\,x = v \times (x \times m\,v) + m\,\phi\,x = \left(m \parallel v \parallel^2 + m\,\phi\right) x - m\,(x \cdot v)\,v.$$

Moreover, in the absence of gravitation, the Hamiltonian $\mathcal{H}$ is reduced to the kinetic energy e, π to the usual linear momentum p and $S = j(l)$.

$$\hat{F} = \begin{pmatrix} -e & p^T & -m \end{pmatrix}, \qquad \hat{L} = \frac{1}{2}\begin{pmatrix} \alpha & -q^T & 0 \\ -w & -j(l) & -q \\ 0 & -w^T & 0 \end{pmatrix}.$$

Considering an infinitesimal generator of Bargmann's group:

$$d\hat{C} = \begin{pmatrix} d\tau_0 \\ dk \\ d\eta \end{pmatrix}, \qquad d\hat{P} = \begin{pmatrix} 0 & 0 & 0 \\ du & j(d\varpi) & 0 \\ 0 & du^T & 0 \end{pmatrix},$$

and using [7.3] and [7.13], the dual pairing [16.76] reads:

$$\hat{\mu}\,\hat{Z} = l \cdot d\varpi - q \cdot du + p \cdot dk - e\,d\tau_0 - m\,d\eta.$$

Comparing with [16.12], we recover the Galilean momenta l, q, p, e and we discover an extra Bargmannian momentum, the mass m.

As for the torsors, it is possible to extend the definition of the momentum tensors to represent the behavior of continua. We have just to give them a vector value. As the space–time $\mathcal{U}$ is a submanifold of the 5-dimensional space $\hat{\mathcal{U}}$, an event $\boldsymbol{X} \in \mathcal{U}$ also belongs to $\hat{\mathcal{U}}$. We consider a field $\boldsymbol{X} \mapsto \hat{\boldsymbol{\mu}}$ where $\hat{\boldsymbol{\mu}}(\boldsymbol{X})$ is Bargmannian momentum tensor with vector value in the tangent space to the space–time at $\boldsymbol{X}$:

$$\hat{\boldsymbol{\mu}} : T_{\boldsymbol{X}}\hat{\mathcal{U}} \times A^* T_{\boldsymbol{X}}\hat{\mathcal{U}} \to T_{\boldsymbol{X}}\mathcal{U} : (\hat{\vec{\boldsymbol{V}}}, \hat{\boldsymbol{\Psi}}) \mapsto \hat{\boldsymbol{\mu}}(\hat{\vec{\boldsymbol{V}}}, \hat{\boldsymbol{\Psi}}).$$

Fixing the value of the argument $\hat{\boldsymbol{\Psi}}$ to the affine function $\hat{1}$ of constant value equal to 1, let us consider the vector valued 1-covariant tensor on $\hat{\mathcal{U}}$ defined by:

$$\hat{\boldsymbol{T}} : T_{\boldsymbol{X}}\hat{\mathcal{U}} \to T_{\boldsymbol{X}}\mathcal{U} : \hat{\vec{\boldsymbol{V}}} \mapsto \hat{\boldsymbol{T}}(\hat{\vec{\boldsymbol{V}}}) = \hat{\boldsymbol{\mu}}(\hat{\vec{\boldsymbol{V}}}, \hat{1}).$$

We recover the definition 13.5 in the chapter on thermodynamics and its extension [13.79] in the presence of gravitation. For this reason, it is named momentum tensor too.

16.10. Poincarean structures

According to Klein's Erlangen program [KLE 72], a geometry is a transformation group (or symmetry group). All the way along this book, we have seen that a physics such as Galileo's is also a transformation group then the corresponding space is equipped with its underlying geometry. That of the Galilean mechanics is atypical because of the loss of space–time metrics but many geometrical tools remain useful, in particular the covariant derivative at the root of the principle of relativity. On the other hand, the classical thermodynamics is governed by Bargmann's group, a set of affine transformations which preserve the metrics [13.8] of the 5-dimensional space. The corresponding geometry is Riemannian.

By opposition to Galilean relativity, that of Lorentz–Poincaré–Einstein is based on the experimental fact that the speed of the light – even if it is huge – has a finite value c for every observer. Which are the underlying transformation group and geometry? In the absence of gravitation, the light rays are straight lines and the particles of light – the photons – move at the constant velocity c in any coordinate system X where the observer takes measures to identify the events. Hence, we are interested in determining the coordinate changes $X' \mapsto X$ preserving the straight lines in the space–time:

$$dx = v\, dt,$$

with $\| v \| = c$. Equivalently, they preserve the relation:

$$\| dx \|^2 - c^2 dt^2 = 0.$$

thus, *Minkowski's metrics* G defined by Gram's matrix:

$$G = \begin{pmatrix} c^2 & 0 \\ 0 & -1_{\mathbb{R}^3} \end{pmatrix}. \qquad [16.83]$$

DEFINITION 16.6.– The affine transformations $dX = P\, dX' + C$ preserving the metrics:

$$P^* P = 1_{\mathbb{R}^4}, \qquad [16.84]$$

are called *Poincarean transformations*. Their set is a Lie subgroup of $\mathbb{A}ff(4)$ of dimension 10 called *Poincaré's group* and is denoted by $\mathbb{P}$. Its linear part $\mathbb{P}_0$ is called *Lorentz group* and its elements are *Lorentzian transformations*.

By a choice of an orthonormal basis, we see that the metrics signature is $(1, 3)$ then $\mathbb{P} = \mathbb{E}(1, 3)$ and $\mathbb{P}_0 = \mathbb{O}(1, 3)$. From a physical point of view, these transformations can be parametrized due to the following result:

THEOREM 16.8.– The Lorentzian transformations can be unically decomposed as the product:

$$P = P_u P_R,$$

of two Lorentzian transformations, P_u associated with a boost u and P_R associated with an orthogonal transformation $R \in \mathbb{O}(3)$:

$$P_u = \begin{pmatrix} \gamma & \dfrac{1}{c^2}\gamma u^T \\ \gamma u & 1_{\mathbb{R}^3} + \dfrac{1}{c^2}\dfrac{\gamma^2}{\gamma+1} u\, u^T \end{pmatrix}, \qquad P_R = \begin{pmatrix} \epsilon & 0 \\ 0 & \epsilon' R \end{pmatrix}, \qquad [16.85]$$

where $\gamma = \mid 1 - \parallel u \parallel^2 / c^2 \mid^{-1/2}$, $\epsilon, \epsilon' \in \{-1, 1\}$.

For the proof, the readers are referred to [VAL 78]. For simplicity, we put $\epsilon = \epsilon' = 1$ in the sequel. Organizing the calculus as in section 1.3.4 by working in $\mathbb{R}^5$, a Poincarean transformation is represented by the 5×5 matrix:

$$P_u = \begin{pmatrix} 1 & 0 & 0 \\ \tau_0 & \gamma & \dfrac{1}{c^2}\gamma u^T R \\ k & \gamma u & R + \dfrac{1}{c^2}\dfrac{\gamma^2}{\gamma+1} u\, u^T R \end{pmatrix}. \qquad [16.86]$$

The extension of the torsor concept to Poincaré–Einstein relativity is straightforward:

DEFINITION 16.7.– The *Poincarean torsor* $\boldsymbol{\tau}$ of a particle is a skew-symmetric 2-contravariant affine tensor:

$$\boldsymbol{\tau} : A^*T_{\mathbf{X}}\mathcal{M} \times A^*T_{\mathbf{X}}\mathcal{M} \to \mathbb{R} : (\boldsymbol{\Psi}_1, \boldsymbol{\Psi}_2) \mapsto \boldsymbol{\tau}(\boldsymbol{\Psi}_1, \boldsymbol{\Psi}_2),$$

represented in a Poincarean frame by a skew-symmetric 5×5 matrix:

$$\boxed{\tilde{\tau} = \begin{pmatrix} 0 & T^T \\ -T & J \end{pmatrix},} \qquad [16.87]$$

where $T \in \mathbb{R}^4$, $J \in \mathbb{M}_{44}^{skew}$ and of which the components, under the Poincarean transformation, are modified according to the transformation law:

$$\tilde{\tau} = \tilde{P}\tilde{\tau}'\tilde{P}^T. \qquad [16.88]$$

Decomposing T and J by blocks as in [3.3], we claim that in a Poincarean coordinate system X' where a particle of mass m_0 and spin l_0 is at rest at the position $x' = 0$ at time $t' = 0$, the torsor reads:

$$\tilde{\tau}' = \begin{pmatrix} 0 & m_0 & 0 \\ -m_0 & 0 & 0 \\ 0 & 0 & -j(l_0) \end{pmatrix}.$$

In the spirit of the boost method of section 3.1.2, let us consider another coordinate system $X = PX' + C$ with a boost $u = v$ and a translation of the origin at $k = x_0$ (hence, $\tau_0 = 0$ and $R = 1_{\mathbb{R}^3}$). Applying the transformation law [16.88] with the Poincarean transformation [16.86] gives the torsor components in the new coordinate system:

$$m = \gamma\, m_0, \qquad p = \gamma\, m_0 v = m\, v,$$

$$q = \gamma \left(m_0 x_0 + \frac{1}{c^2}\, v \times l_0 \right),$$

$$l = Adj \left(1_{\mathbb{R}^3} + \frac{1}{c^2}\, \frac{\gamma^2}{\gamma + 1} v\, v^T \right) l_0 + x_0 \times m\, v,$$

where $\gamma = \mid 1 - \parallel v \parallel^2 / c^2 \mid^{-1/2}$.

Unlike the Galilean relativity where the mass is invariant by Galilean transformation, the mass m, equal to m_0 when the particle is at rest, increases with the velocity up to the infinity when the velocity approaches the speed of the light. The linear momentum p is equal to the product of the mass by the velocity as in Galilean relativity but now with a mass depending on the velocity. The first term of q is similar to the Galilean passage but there is a correction by the spin. The angular momentum l is decomposed into the spin component (but which is now depending on the velocity) and the orbital one.

It is worth noting that the linear 4-momentum:

$$T = m_0 U, \qquad [16.89]$$

where:

$$U = \begin{pmatrix} \gamma \\ \gamma\, v \end{pmatrix}, \qquad [16.90]$$

represents the *velocity vector* $\vec{U}$, relativistically analogous to the Galilean one [1.12] which it approaches when c approaches the infinity. It verifies the condition:

$$U^*(\vec{U}) = U^T G U = c^2. \qquad [16.91]$$

Due to the Riemannian structure of the space–time, we can build the *Poincarean momentum tensor* likewise [16.77]:

$$F = T^T G = \left(e, -p^T\right), \qquad L = \frac{1}{2} J G = \frac{1}{2} \begin{pmatrix} 0 & q^T \\ c^2 q & -j(l) \end{pmatrix}, \qquad [16.92]$$

where the energy is given by:

$$e = m\, c^2,$$

a formula that has already caused a lot of ink to flow. It appears here as a simple index lowering:

$$F_0 = G_{00} T^0.$$

Let us remark that the proof of lemma 16.1 is valid for any Euclidean structure, independently of the space dimension. Then, Poincarean momentum tensors are governed by the transformation law:

$$F' = F\, P, \qquad L' = \left[P^{-1} L + \frac{1}{2}\left(C' F - P^{-1} G^{-1} F^T C'^T P^T G\right)\right] P.$$

Because the first relation [16.92] reads $F = T^*$, for any Poincarean transformation, $P^{-1} = P^*$ and $P^T G\, P = G$, hence the previous relation becomes:

$$F' = F\, P, \qquad L' = P^* L\, P + \frac{1}{2}\left(C' \left(P^* T\right)^* - \left(P^* T\right) C'^*\right). \qquad [16.93]$$

Let us achieve this quick review of relativity by a smidgen of thermodynamics of continua. The motivation in Galilean relativity to introduce the momentum tensor $\hat{\boldsymbol{T}}$ of definition 13.5 was to obtain, according to theorem 13.2, the balance of both the mass and energy as expression of the first principle of thermodynamics 13.1. In Lorentz–Poincaré–Einstein relativity, we just saw that mass and energy are identical, the factor c^2 aside. It is not necessary to use the artifact of the fifth dimension and we work in the space–time only with the *temperature 4-vector* $\vec{\boldsymbol{W}}$ represented by a column:

$$W' = \begin{pmatrix} \beta \\ 0 \end{pmatrix},$$

in a coordinate system X' where the elementary volume is at rest. In another coordinate system X obtained from X' by a boost v, the temperature vector reads:

$$W = P\,W' = \begin{pmatrix} \gamma & \dfrac{1}{c^2}\gamma\, v^T \\ \gamma\, v\; 1_{\mathbb{R}^3} + & \dfrac{1}{c^2}\dfrac{\gamma^2}{\gamma+1} v\, v^T \end{pmatrix} \begin{pmatrix} \beta \\ 0 \end{pmatrix} = \begin{pmatrix} \gamma\,\beta \\ \gamma\,\beta\, v \end{pmatrix}.$$

Then, the absolute temperature is transformed as:

$$\theta' = \frac{\theta}{\gamma} = \theta \mid 1- \parallel v \parallel^2 / c^2 \mid^{1/2}.$$

According to Planck's theory, when the velocity of the elementary volume increases with respect to the observer, the temperature measured by her or him decreases.

In general relativity, we use Levi–Civita connection associated with the Riemannian structure. The *Poincarean friction tensor* is the self-adjoint 1-covariant and 1-contravariant tensor:

$$\mathrm{f} = \frac{1}{2}\left(\nabla \vec{\boldsymbol{W}} + \left(\nabla \vec{\boldsymbol{W}}\right)^*\right).$$

The *Poincarean momentum tensor* is a self-adjoint 1-covariant and 1-contravariant tensor $\boldsymbol{T}$ on the space–time. The relativistic version of the *first principle of the thermodynamics* claims that it is covariant divergence free:

$$\boldsymbol{Div}\,\boldsymbol{T} = \boldsymbol{0}.$$

Inspiring from the *second principle of the thermodynamics* 13.2, the relativistic version claims that the *local production of entropy*:

$$\Phi = \boldsymbol{Div}\left(\boldsymbol{T}\,\vec{\boldsymbol{W}} + \zeta\,\vec{\boldsymbol{N}}\right) - \frac{1}{c^2}\left(\boldsymbol{U}^*(\boldsymbol{f}(\vec{\boldsymbol{U}}))\right)\frac{1}{c^2}\left(\boldsymbol{U}^*(\boldsymbol{T}_I(\vec{\boldsymbol{U}}))\right) \geq 0,$$

of a continuous medium characterized by fields of Planck's potential ζ, velocity vector $\vec{\boldsymbol{U}}$, mass flux vector $\vec{\boldsymbol{N}}$, temperature vector $\vec{\boldsymbol{W}}$ and momentum tensor $\boldsymbol{T}$ is non-negative and vanishes if and only if the process is reversible. The reason is that when the speed of the light approaches the infinite, the linear form $\boldsymbol{U}^*$ represented by the 4-row:

$$U^T G = (\gamma, \gamma\, v^T)\begin{pmatrix} c^2 & 0 \\ 0 & -1_{\mathbb{R}^3} \end{pmatrix} = c^2\left(\gamma, -\frac{1}{c^2}\gamma\, v^T\right),$$

approaches $c^2\boldsymbol{e}^0$ where $\boldsymbol{e}^0$ is the time arrow. Hence we recover the expression [13.56] of the Galilean local production of the entropy. The readers interested in the relativistic thermodynamics of continua are referred to [VAL 78, VAL 81].

16.11. Lie group statistical mechanics

In order to discover the underlying geometric structure of the statistical mechanics, we are interested in the affine maps $\boldsymbol{\Theta}$ on the affine space of momentum tensors, represented by an affine function Θ from $\mathfrak{g}^*$ into $\mathbb{R}$:

$$\boldsymbol{\Theta}(\boldsymbol{\mu}) = \Theta(\mu) = z + \mu\, Z,$$

where $z = \Theta(0) = \boldsymbol{\Theta}(\boldsymbol{\mu}_0)$ and $Z = lin(\Theta) \in \mathfrak{g}$ are the affine components of $\boldsymbol{\Theta}$. If the components of the momentum tensors are modified according to [16.7], the change of affine components of $\boldsymbol{\Theta}$ is given by the induced action:

$$z = z' - \theta(a)\, Ad(a)\, Z', \qquad Z = Ad(a)\, Z'. \tag{16.94}$$

Then, $\boldsymbol{\Theta}$ is a G-tensor. In [SOU 70, SOU 97b], Souriau proposed a statistical mechanics model using geometric tools of Lie group theory. Let $d\lambda$ be a measure on $\boldsymbol{\mu} = orb\,(\mu)$ and a Gibbs probability measure $p\, d\lambda$ with:

$$p = e^{-\Theta(\mu)} = e^{-(z+\mu\, Z)}.$$

The normalization condition $\int_{orb(\mu)} p\, d\lambda = 1$ links the components of $\boldsymbol{\Theta}$, allowing us to express z in terms of Z:

$$z(Z) = \ln \int_{orb(\mu)} e^{-\mu\, Z}\, d\lambda. \tag{16.95}$$

The corresponding entropy and mean momenta are:

$$s = -\int_{orb(\mu)} p \ln p\, d\lambda = z + M\, Z, \quad M = \int_{orb(\mu)} \mu\, p\, d\lambda = -\frac{\partial z}{\partial Z}, \tag{16.96}$$

M satisfying the same transformation law as the one [16.7] of μ. Hence, M are the components of a momentum tensor $\boldsymbol{M}$ which can be identified to the orbit $orb(M)$, which defines a map $\boldsymbol{\mu} \mapsto \boldsymbol{M}$, i.e. a correspondence between two orbits. This construction is formal and, for reasons of integrability, the integrals will be performed only on a subset of the orbit according to a heuristic way explained later on.

People generally consider that the definition of the entropy is relevant for applications insofar the number of particles in the system is very huge. For instance, the number of atoms contained in one mole is Avogadro's number equal to 6×10^{23}. It is worth noting that Vallée and Lerintiu proposed a generalization of the ideal gas law based on convex analysis and a definition of entropy which does not require the classical approximations (Stirling's formula) [VAL 05].

Now, let us reveal the link between Lie group statistical mechanics in the classical Galilean context and the thermodynamics of continua of Chapter 13. In other words, how do we deduce the momentum tensor $\boldsymbol{T}$ from $\boldsymbol{M}$ and Planck's potential ζ from z? We work in five steps:

1) *Step 1: modeling the deformation.* Let us consider N identical particles contained in a box of finite volume V, large with respect to the particles but representing the volume element of the continuum thermodynamics. For the coordinate change [10.6], [10.7]:

$$t = t' + \tau_0, \qquad x = \varphi(t', s'),$$

the Jacobean matrix is given by [10.3]:

$$\frac{\partial X}{\partial X'} = P = \begin{pmatrix} 1 & 0 \\ v & F \end{pmatrix},$$

Moreover, we suppose that the box of initial volume V_0 is at rest in the considered coordinate system ($v = 0$) and the deformation gradient F is uniform in the box, then $dx = F\, ds'$. According to [16.27] and the transformation law [16.26] of the Galilean momentum tensor without boost ($u = 0$), the linear momentum is transformed through $p = F^{-1} p'$. The measure becomes:

$$d\lambda = d^3x\, d^3p\, d^2n = d^3s'\, d^3p'\, d^2n.$$

Replacing the orbit by the subset $V_0 \times \mathbb{R}^3 \times \mathbb{S}^2$ and integrating [16.95] gives for a particle after leaving out the primes:

$$z = \frac{1}{2}\ln(\det(C)) - \frac{3}{2}\ln\beta + C^{te}, \qquad [16.97]$$

where the value of the constant is not relevant in the following.

2) *Step 2: identification.* It is based on the following result.

THEOREM 16.9.– The transformation law of the temperature vector $\vec{\hat{W}}$ is the same as the one of affine maps $\boldsymbol{\Theta}$ on the affine space of momentum tensors through the identification:

$$Z = (-W, 0), \qquad z = m\,\zeta,$$

PROOF.– First, let us verify that the form $Z = (-W, 0)$ does not depend on the choice of the affine frame. Indeed, starting from $Z' = (-W', 0)$ and applying the adjoint representation [16.6] with $dC' = -W'$ and $dP' = 0$, we find that $dC = -W$ and $dP = 0$ with:

$$W = P\,W'.$$

Moreover, using the notations of [16.26] and 16.94] gives:

$$z = z' - \theta(a)\,Ad(a)\,Z' = z' + F_m P\,W'.$$

On the other hand, let $\hat{W}$ be the 5-column [13.10] representing the temperature vector:

$$\hat{W} = \begin{pmatrix} W \\ \zeta \end{pmatrix} = \begin{pmatrix} \beta \\ w \\ \zeta \end{pmatrix}.$$

Taking into account [1.4] and [16.27], it is easy to verify that its transformation law $\hat{W} = \hat{P}\,\hat{W}'$ with the linear Bargmannian transformation [13.5] can be recast as:

$$\begin{pmatrix} W \\ \zeta \end{pmatrix} = \begin{pmatrix} P & 0 \\ F_1 P & 1 \end{pmatrix} \begin{pmatrix} W' \\ \zeta' \end{pmatrix},$$

which is the transformation law of the affine map $\boldsymbol{\Theta}$ provided that $z = m\,\zeta$, which achieves the proof. ■

For the box at rest in the coordinate system X, we put:

$$W = \begin{pmatrix} \beta \\ 0 \end{pmatrix}.$$

– *Step 3: boost method.* A new coordinate system $\bar{X}$ in which the box has the velocity v can be deduced from $X = P\,\bar{X} + C$ by applying a boost $u = -v$ (hence, $k = 0$, $\tau_0 = 0$ and $R = 1_{\mathbb{R}^3}$). The transformation law of vectors gives the new components $\bar{W}^T = (\beta, \beta\,v^T)$ and [16.94] leads to:

$$\bar{z} = z + \frac{m\,\beta}{2} \parallel v \parallel^2 = z + \frac{m}{2\,\beta} \parallel w \parallel^2 .$$

Taking into account [16.97] and leaving out the bars:

$$z = \frac{1}{2}\,\ln(\det(C)) - \frac{3}{2}\,\ln\beta + \frac{m}{2\,\beta} \parallel w \parallel^2 + C^{te}.$$

It is clear from [16.96] that s is Legendre conjugate of $-z$, then, introducing the internal energy:

$$e_{int} = e - \frac{1}{2\,m} \parallel p \parallel^2,$$

the entropy is:

$$s = \frac{3}{2} \ln e_{int} + \frac{1}{2} \ln(\det(C)) + C^{te},$$

and, by $Z = \partial s \,/\, \partial M$, we derive the corresponding momenta:

$$\beta = \frac{\partial s}{\partial e} = \frac{3}{2\,e_{int}}, \qquad w = -grad_p\, s = \frac{3}{2\,e_{int}}\frac{p}{m}.$$

3) *Step 4: link between z and ζ.* Planck's potential ζ is given by theorem 16.9:

$$\zeta = \frac{z}{m} = \frac{1}{2\,m} \ln(\det(C)) - \frac{3}{2\,m} \ln \beta + \frac{1}{2\,\beta} \parallel w \parallel^2 + C^{te}.$$

From [13.42] and [13.36], we obtain the linear 4-momentum $\Pi = (\mathcal{H}_R, -p^T)$ and Cauchy's stresses:

$$\mathcal{H}_R = \rho \left(\frac{3}{2}\frac{k_B T}{m} + \frac{1}{2} \parallel v \parallel^2 \right), \qquad p = \rho v, \qquad \sigma = -q\, 1_{\mathbb{R}^3},$$

where, from the expression of the pressure, we recover the *ideal gas law*:

$$q = \frac{\rho}{m} k_B T = \frac{N}{V} k_B T.$$

17

Symplectic Structure on a Manifold

17.1. Symplectic form

If you need to brush up on symplectic techniques, this would be a good time to consult, for instance, [GUI 84b], [LIB 87], [SOU 70] and [SOU 70].

DEFINITION 17.1.– Let $\eta \mapsto \omega$ be a smooth field of 2-form on a manifold $\mathcal{N}$ such that:

– $Ker\,\omega = \{d\eta \in T_\eta \mathcal{N} \quad \text{s.t.} \quad \iota(d\eta)\,\omega = 0\}$ has a constant dimension;

– ω is closed: $d\omega = 0$.

Then, we say that $\mathcal{N}$ is a *presymplectic manifold* and ω is a *symplectic form*. It can be proved that the rank of ω is even. If rank $\omega = dim\,\mathcal{N}$, $\mathcal{N}$ is called a *symplectic manifold*.

To motivate this definition, let us illustrate it by a simple example. The motion of a spinless particle of mass m, position x at time t and velocity v can be described in the "phase space" by the point:

$$\eta = \begin{pmatrix} t \\ x \\ v \end{pmatrix} \in \mathcal{N} = \mathbb{R}^7.$$

If it is subjected to a Galilean gravitation field, it can be described in Galilean coordinate systems by the skew-symmetric 2-covariant tensor or 2-form:

$$\omega(d\eta, \delta\eta) = m\,\left[(dv - g\,dt) \cdot (\delta x - v\,\delta t) - (\delta v - \,g\,\delta t)\right.$$
$$\left.\cdot(dx - v\,dt) - 2\,\Omega \cdot (\delta x \times dx)\right]. \qquad [17.1]$$

It is represented by the 7×7 matrix:

$$\tilde{\omega} = m \begin{pmatrix} 0 & g^T & -v^T \\ -g & 2\,j(\Omega) & 1_{\mathbb{R}^3} \\ v & -1_{\mathbb{R}^3} & 0 \end{pmatrix},$$

such that:

$$\omega(d\eta, \delta\eta) = \delta\eta^T \tilde{\omega}\, d\eta.$$

In short notations of exterior calculus, the 2-form reads:

$$\omega = m\left[(dv_i - g_i\, dt) \wedge (dx_i - v_i\, dt) - 2\Omega_{ij}\, dx_i \wedge dx_j\right], \qquad [17.2]$$

where – for simplicity's sake – all the indices are lowered, the standard convention of summation on the repeated indices is used and Ω_{ij} is the element at the intersection of the i-th row and the j-th column of the matrix $j(\Omega)$:

– From a mechanical viewpoint, the equation $\iota(d\eta)\,\omega = 0$ is interpreted as the *equation of motion*. Indeed, it reads:

$$\forall \delta\eta, \quad (\iota(d\eta)\,\omega)(\delta\eta) = \omega(d\eta, \delta\eta) = 0,$$

that, taking into account [17.1], leads to:

$$\frac{dx}{dt} = v, \qquad \frac{dv}{dt} = g - 2\,\Omega \times \frac{dx}{dt}, \qquad [17.3]$$

$$v \cdot \frac{dv}{dt} = \frac{d}{dt}\left(\frac{1}{2} \parallel v \parallel^2\right) = g \cdot \frac{dx}{dt}, \qquad [17.4]$$

which, by elimination of the velocity, provides the equation of motion [3.46]. Moreover, the last condition is the balance of power. As it is a consequence of the previous one, rank $\omega = 7 - 1 = 6$. Hence, the manifold $\mathcal{N}$ is presymplectic and the leaves are the trajectories of particles.

– Let us now examine the consequences of the closure condition $d\omega = 0$. Expanding [17.2] and owing to $dt \wedge dt = 0$, we have:

$$\omega = m\left[dv_i \wedge dx_i - v_i\, dv_i \wedge dx_i - g_i\, dt \wedge dx_i - 2\,\Omega_{ij}\, dx_i \wedge dx_j\right], \qquad [17.5]$$

and differentiating:

$$d\omega = m\Big[-dv_i \wedge dv_i \wedge dx_i - \left(\frac{\partial g_i}{\partial t}\, dt + \frac{\partial g_i}{\partial x_j}\, dx_j\right) \wedge dt \wedge dx_i$$
$$-2\left(\frac{\partial \Omega_{ij}}{\partial t}\, dt + \frac{\partial \Omega_{ij}}{\partial x_k}\, dx_k\right) \wedge dx_i \wedge dx_j\Big]. \qquad [17.6]$$

The first terms vanish because same factors are repeated and it remains:

$$d\omega = m\left[-\left(\frac{\partial g_i}{\partial x_j} + 2\,\frac{\partial \Omega_{ij}}{\partial t}\right) dx_i \wedge dx_j \wedge dt - 2\,\frac{\partial \Omega_{ij}}{\partial x_k}\, dx_k \wedge dx_i \wedge dx_j\right],$$

or, assembling the similar terms:

$$\begin{aligned} d\omega = m\,[&\left(\frac{\partial g_3}{\partial x_2} - \frac{\partial g_2}{\partial x_3} + 2\,\frac{\partial \Omega_1}{\partial t}\right) dx_2 \wedge dx_3 \wedge dt \\ &+ \left(\frac{\partial g_1}{\partial x_3} - \frac{\partial g_3}{\partial x_1} + 2\,\frac{\partial \Omega_2}{\partial t}\right) dx_3 \wedge dx_1 \wedge dt \\ &+ \left(\frac{\partial g_2}{\partial x_1} - \frac{\partial g_1}{\partial x_2} + 2\,\frac{\partial \Omega_3}{\partial t}\right) dx_1 \wedge dx_2 \wedge dt \\ &-2\left(\frac{\partial \Omega_1}{\partial x_1} + \frac{\partial \Omega_2}{\partial x_2} + \frac{\partial \Omega_3}{\partial x_3}\right) dx_1 \wedge dx_2 \wedge dx_3]. \end{aligned} \qquad [17.7]$$

Hence, ω is closed provided that:

$$curl\, g + 2\,\frac{\partial \Omega}{\partial t} = 0, \qquad div\, \Omega = 0,$$

and we recover the conditions [6.15] of existence of the potential of the Galilean gravitation.

It is worth noting that to define the Bargmannian coordinate systems, we claim that the Bargmannian connection is symmetric. Now, we prove a result which states a link between this fact and the closure of the Galilean symplectic form [17.1].

THEOREM 17.1.– If the Bargmannian connection is symmetric, the Galilean 2-form [17.1] is closed.

PROOF.– Combining [16.61] and [16.63], we see that the the four former components of the vector valued torsion 2-form, namely $\hat{T}^0$ corresponding to the time and $\hat{T}^i$ corresponding to the space coordinates, are automatically null while:

$$[\delta, d]\, z + \hat{T}^4(\delta\hat{X}, d\hat{X}) = -g \cdot (\delta x\, dt - dx\, \delta t) + 2\,\Omega \cdot (dx \times \delta x).$$

On the other hand, expanding [17.1] gives:

$$\begin{aligned} \omega(d\eta, \delta\eta) = m\,[&(dv \cdot \delta x - \delta v \cdot dx) - v \cdot (\delta t\, dv - dt\, \delta v) \\ &-g \cdot (dt\, \delta x - \delta t\, dx) + 2\,\Omega \cdot (dx \times \delta x)], \end{aligned} \qquad [17.8]$$

then:

$$\omega(d\eta, \delta\eta) = m\left((dv\cdot\delta x - \delta v\cdot dx) - v\cdot(\delta t\, dv - dt\,\delta v) + [\delta, d]z + \hat{T}^4(\delta\hat{X}, d\hat{X})\right).$$

By exterior differentiation, we have:

$$d\omega(\Delta\eta, d\eta, \delta\eta) = m\left(([\Delta,[\delta,d]] + [\delta,[d,\Delta]] + [d,[\Delta,\delta]])\ z + d\hat{T}^4(\Delta\hat{X}, \delta\hat{X}, d\hat{X})\right),$$

in which the former term vanishes because of Jacobi's identity, hence:

$$d\omega = m\, d\hat{T}^4,$$

that achieves the proof. ■

17.2. Symplectic group

DEFINITION 17.2.– A Lie group G smoothly left acting on $\mathcal{N}$ and preserving the symplectic form:

$$\forall a \in G, \qquad L_a^*\omega = \omega,$$

is called a *symplectic group*. We say that the action is a *symplectic action*.

For instance, Galileo's group acts on $\mathcal{N}$ according to [1.8] and [1.13]:

$$t = t' + \tau_0, \qquad x = R\,x' + u\,t' + k, \qquad v = R\,v' + u, \qquad [17.9]$$

and, differentiating with respect to η':

$$dt = dt', \quad dx = R\,dx' + u\,dt', \quad dv = R\,dv',$$
$$\delta t = \delta t', \quad \delta x = R\,\delta x' + u\,\delta t', \quad \delta v = R\,\delta v'.$$

Let us prove that Galileo's group is symplectic by substituting these expressions into the 2-form [17.8], which leads to:

$$\begin{aligned}\omega(d\eta', \delta\eta') = m\,[&(dv'\cdot\delta x' - \delta v'\cdot dx') - v'\cdot(\delta t'\,dv' - dt'\,\delta v')\\ &-(R^T(g - 2\,\Omega\times u))\cdot(dt'\,\delta x' - \delta t'\,dx')\\ &+2\,(R^T\Omega)\cdot(dx'\times\delta x')],\end{aligned} \qquad [17.10]$$

Observing that Galilean transformations are particular cases of Galilean coordinate changes [3.27]:

$$x = R(t')\,x' + x_0(t'),$$

where $\tau_0 = 0$, R is time-independent (then $\varpi = 0$) and $x_0(t') = u\,t' + k$, the transformation laws of the Galilean gravitation components [3.51] and [3.62] give:

$$g' = R^T(g - 2\,\Omega \times u), \qquad \Omega' = R^T\Omega,$$

that proves Galileo's group is symplectic.

17.3. Momentum map

We denote by $Z \cdot \eta$ the infinitesimal action of $Z \in \mathfrak{g}$ at η. For instance, if G is Galileo's group, differentiating [17.9] at constant η' and considering the limit when a approaches the identity, η' approaches η and we obtain the expression of $\delta\eta = Z \cdot \eta$:

$$\delta t = \delta\tau_0, \qquad \delta x = \delta\varpi \times x + \delta u\,t + \delta k, \qquad \delta v = \delta\varpi \times v + \delta u. \qquad [17.11]$$

All the way along this book, we have seen that integrals of the motion are important tools to integrate the equations. We would now like to show how they are linked to symmetry groups.

DEFINITION 17.3.– A map $\psi : \mathcal{N} \to \mathfrak{g}^*$ such that:

$$\forall \eta \in \mathcal{N}, \quad \forall Z \in \mathfrak{g}, \qquad \iota(Z \cdot \eta)\,\omega = -d(\psi(\eta)Z).$$

is called a *momentum map* of G.

The momentum map is the quantity involved in *Noether's theorem*:

THEOREM 17.2.– Let $\mathcal{N}$ be a presymplectic manifold with symplectic group G and ψ be a momentum map. Then, ψ is constant on each leaf of $\mathcal{N}$.

PROOF.– It is sufficient to prove for any $Z \in \mathfrak{g}$ that $\psi(\eta)\,Z$ is an integral of the motion. In other words, $Ker\,\omega \subset Ker\,d(\psi(\eta)\,Z)$ or, according to the definition of a momentum map, $Ker\,\omega \subset Ker\,(\iota(Z \cdot \eta)\,\omega)$, that is:

$$\iota(\delta\eta)\,\omega = 0 \qquad \Rightarrow \qquad \omega(Z \cdot \eta, \delta\eta) = 0,$$

which is true because ω is skew-symmetric, which achieves the proof. ■

Let us illustrate this result by revisiting Kepler's problem solved in section 3.4. In a Newtonian coordinate system X, the spinning Ω vanishes and the gravity g given by [3.65] with $x' = 0$ is collinear to x. After swapping $d\eta$ and $\delta\eta$, the symplectic 2-form [17.8] is reduced to:

$$\omega(\delta\eta, d\eta) = m\,[(\delta v \cdot dx - dv \cdot \delta x) - v \cdot (dt\,\delta v - \delta t\,dv) \\ -g \cdot (\delta t\,dx - dt\,\delta x)]. \qquad [17.12]$$

Now, we are able to find momentum maps:

– Considering only the infinitesimal generator $\delta\tau_0$:

$$\delta\eta = \begin{pmatrix} \delta\tau_0 \\ 0 \\ 0 \end{pmatrix},$$

we find the corresponding momentum map:

$$\eta \mapsto -e = -\left(\frac{1}{2} m \parallel v \parallel^2 + m\phi\right).$$

and we recover the total energy [3.74] as integral of the motion because:

$$\iota(\delta\eta)\,\omega = m\,\delta\tau_0\,(v \cdot dv - grad\,\phi) = -d\,(\delta\tau_0\,(-e))\ .$$

– Considering the generator $\delta\varpi$:

$$\delta\eta = \begin{pmatrix} 0 \\ \delta\varpi \times x \\ \delta\varpi \times v \end{pmatrix},$$

we have, because g is collinear to x:

$$\iota(\delta\eta)\,\omega = m\,\delta\varpi \cdot (dx \times v + x \times dv) = -d\,(l \cdot \delta\varpi)\ .$$

and we recover the angular momentum as momentum map.

For the use of the momentum map, in particular in the problems of the spherical pendulum, Euler's top and Lagrange top's, the readers can consult [CUS 97].

17.4. Symplectic cohomology

THEOREM 17.3.– Let $\mathcal{N}$ be a connected presymplectic manifold with symplectic group G and $\eta \mapsto \mu = \psi(\eta)$ be a momentum map. Then:

$\diamondsuit$ there exists a smooth map θ of G into $\mathfrak{g}^*$ defined by:

$$\theta(a) = \psi(a \cdot \eta) - Ad^*(a)\,\psi(\eta)\,;$$

$\heartsuit$ θ verifies: $\theta(a'a) = \theta(a') + Ad^*(a')\,\theta(a)$; $\spadesuit$ the derivative $f = D\theta(e)$ is a 2-form of the Lie algebra $\mathfrak{g}$ such that:

$$f(Z,[Z',Z'']) + f(Z',[Z'',Z]) + f(Z'',[Z,Z']) = 0,;$$

$\clubsuit$ we have:

$$(D\psi(\eta))(Z \cdot \eta) = \psi(\eta) \circ ad(Z) + \iota(Z)f,$$
$$\omega(Z \cdot \eta, Z' \cdot \eta) = \mu[Z,Z'] + f(Z,Z').$$

For the proof, the readers are referred, for instance, to Souriau [SOU 70, p. 109] (theorem 11.17), or [SOU 97] for its English translation. Replacing η by $a^{-1} \cdot \eta$ in condition $\diamondsuit$, this formula reads:

$$\psi(\eta) = Ad^*(a)\,\psi'(\eta) + \theta(a),$$

where $\psi \mapsto \psi' = a \cdot \psi$ is the induced action of the one of G on $\mathcal{N}$. It is worth observing that it is nothing else [16.10] with $\mu = \psi(\eta)$ and $\mu' = \psi'(\eta)$.

In this sense, *the values of the momentum map are just the momentum G-tensor components* of definition 16.2!

Hence, the condition $\diamondsuit$ can be recast saying that ψ is equivariant:

$$L_a^* \psi = a \cdot \psi.$$

The previous theorem suggests the following notions.

DEFINITION 17.4.– We call *symplectic cocycle* of G a smooth map $cocs$ from G into $\mathfrak{g}^*$ such that:

$\diamondsuit$ $cocs(a'a) = cocs(a') + Ad^*(a')\,cocs(a)$;

$\heartsuit$ $dcocs = D\,cocs\,(e)$ is skew-symmetric.

Formula $\diamondsuit$ is called the *symplectic cocycle identity*.

Let $\mu_0 \in \mathfrak{g}^*$. We call *symplectic coboundary* of G a smooth map $cobs_{\mu_0}$ of G into $\mathfrak{g}^*$ such that:

$$cobs_{\mu_0}(a) = Ad^*(a)\,\mu_0 - \mu_0.$$

The symplectic coboundaries are the obvious solutions of equation [16.8] encountered in section 16.3, hence solutions of the symplectic cocycle identity $\diamondsuit$.

As maps valued in the linear space $\mathfrak{g}^*$, the symplectic cocycles form a linear space and the symplectic coboundaries form a linear subspace thereof. The relation "$cocs_1$ and $cocs_2$ differs by a coboundary" is an equivalence relation and the set of equivalence classes is a linear space called the space of *classes of symplectic cohomology*. The class of symplectic cohomology does not depend on the choice of the momentum map but only on the structure of the Lie group G. Let us determine the one of Galileo's group $\mathbb{GAL}$.

THEOREM 17.4.– The most general symplectic cohomology of Galileo's group is defined by:

$$cocs(a)\,Z' = l(a)\cdot d\varpi' - q(a)\cdot du' + p(a)\cdot dk' - e(a)\,d\tau_0', \qquad [17.13]$$

where the components are:

$$p(a) = m\,u, \qquad e(a) = \frac{1}{2}\,m\,\|\,u\,\|^2, \qquad [17.14]$$

$$q(a) = m\,(k - \tau_0 u), \qquad l(a) = m\,k \times u. \qquad [17.15]$$

The space of symplectic cohomology of Galileo's group is of dimension 1.

PROOF.– We know by theorem 16.3 that, *modulo* a symplectic coboundary, the most general solution of equation $\diamondsuit$ of definition 17.4 is [17.13] of components:

$$p(a) = m\,u, \qquad e(a) = \frac{1}{2}\,m\,\|\,u\,\|^2 + e_1\,\tau_0,$$

$$q(a) = m\,(k - \tau_0 u), \qquad l(a) = m\,k \times u + s\,u.$$

By differentiation, the components of $D\,coc\,(e)$ are:

$$dp = m\,du, \qquad de = \frac{1}{2}\,m\,u \cdot du + e_1\,d\tau_0,$$

$$dq = m\,(dk - \tau_0\,du - d\tau_0\,u), \qquad dl = m\,(dk \times u + k \times du) + s\,du.$$

that leads to:

$$dcocs\,(Z, Z') = m\,(du \cdot dk' - dk \cdot du') + s\,du \cdot d\varpi' - e_1 d\tau_0\, d\tau_0', \qquad [17.16]$$

As the 2-form is skew-symmetric, $s = e_1 = 0$, which achieves the proof. ■

With [17.14] and [17.15], we recover the action on the Galilean momentum tensors defined by [16.23], [16.24] and [16.25] according to physical arguments.

In the following, we also need the following result:

THEOREM 17.5.– Let $\mathcal{N}$ be a presymplectic manifold with symplectic group G and $\eta \mapsto \alpha(\eta)$ be a field of 1-forms such that $\omega = d\alpha$ and:

$$\forall a \in G, \qquad L_a^* \alpha = \alpha.$$

Then, G is a symplectic group and there exists a momentum map ψ of G defined by:

$$\forall Z \in \mathfrak{g}, \qquad \psi(\eta)\, Z = \omega(Z \cdot \eta).$$

Its class of symplectic cohomology is null.

For the proof, the readers are referred, for instance, to Souriau [SOU 70, p. 107] (theorem [11.10]), or [SOU 97] for its English translation.

17.5. Central extension of a group

Bargmann's group was introduced in section 13.2 by heuristic arguments. Now, we would like to construct it in link with the symplectic cohomology. Let us start recalling elements of group extension theory.

DEFINITION 17.5.– Let G be a group and N be an abelian group with the operation denoted additively. A map from $G \times G$ into N is called a *N-cocycle* of G if:

$$\forall a, a', a'' \in G, \quad coc\,(a, a') + coc\,(aa', a'') = coc\,(a, a'a'') + coc\,(a', a''). \quad [17.17]$$

This condition is called the *N-cocycle identity.*

DEFINITION 17.6.– Let θ be a map from a group G into an abelian group N. The map cob_θ from $G \times G$ into N defined by:

$$cob_\theta\,(a, a') = \theta\,(a) + \theta\,(a') - \theta\,(aa'),$$

is called an *N-coboundary* of G.

As maps valued in the abelian group N, the N-cocycles form an abelian group and the coboundaries form a normal subgroup thereof. The relation "coc_1 and coc_2 differs by a coboundary" is an equivalence relation and the set of equivalence classes is a group, called the N-cohomology group of G. Now, *Schreier's theorem* is presented in a simplified version, strictly useful for our needs.

THEOREM 17.6.– Let G be a group, N be an abelian group and a given N-cocycle coc of G. Then, the product $\hat{G} = G \times N$ with the operation:

$$\forall a, a' \in G, \quad \forall \theta, \theta' \in N, \quad (a, \theta)(a', \theta') = (aa', \theta + \theta' + coc(a, a')), \quad [17.18]$$

is a group with normal subgroup N. $\hat{G}$ is called an extension of G by N. Two equivalent N-cocycles of G endow underlying group structures which are isomorphic.

For the demonstration and a more general formulation, the readers are referred to [HAL 53], sections 15.1 and 15.2. The cocycle identity is motivated by the need for the operation of $\hat{G}$ being associative. It is easy to verify that the identity of $\hat{G}$ is $(e, 0)$ and the inversion is given by:

$$(a, \theta)^{-1} = \left(a^{-1}, -\theta - coc\left(a, a^{-1}\right)\right). \quad [17.19]$$

If G and N are Lie groups and the map coc is smooth, the extension $\hat{G}$ with the operation [17.18] is a Lie group. Its dimension is the sum of the ones of G and N. The Lie algebra of $\hat{G}$ (respectively, G, N) is denoted by $\hat{\mathfrak{g}}$ (respectively, $\mathfrak{g}$,$\mathfrak{n}$). As N is abelian, its adjoint representation is obviously the identity of $\mathfrak{n}$. The adjoint representation of G being known, we now want to determine the one of its extension $\hat{G}$. Let us consider two arbitrary elements of $\hat{G}$, namely $\hat{a} = (a, \theta)$, $\hat{a}' = (a', \theta')$. Owing to [17.18] and [17.19], we have:

$$\begin{aligned}\hat{a}^{-1}\hat{a}'\hat{a} &= \left(a^{-1}, -\theta - coc\left(a, a^{-1}\right)\right)\left(a'a, \theta' + \theta + coc\left(a', a\right)\right)\\ &= \left(a^{-1}a'a, \theta' + coc\left(a', a\right) + coc\left(a^{-1}, a'a\right) - coc\left(a, a^{-1}\right)\right).\end{aligned}$$

If $Z \in \mathfrak{g}$, $Y \in \mathfrak{n}$, we can write $\hat{Z} = (Z, Y) \in \hat{\mathfrak{g}}$. Differentiating both members of the previous relation with respect to $\hat{a}' = (a', \theta')$ at $\hat{a}' = e$ and taking into account [18.1], we obtain:

$$Ad\left(\hat{a}^{-1}\right)\hat{Z} = \left(Ad\left(a^{-1}\right)Z, Y + \left(D\, coc\left(e, a\right)\right)(Z, 0) + \left(D\, coc\left(a^{-1}, a\right)\right)(0, Za)\right).$$

Introducing the linear map:

$$B : \mathfrak{g} \to \mathfrak{n} : Z \mapsto Y = \left(D\, coc\left(e, a\right)\right)(Z, 0) + \left(D\, coc\left(a^{-1}, a\right)\right)(0, Za) \quad [17.20]$$

the adjoint representation of the extension $\hat{G}$ is given by:

$$Ad\left(\hat{a}^{-1}\right)\hat{Z} = \left(Ad\left(a^{-1}\right)Z, Y + B\left(a\right)Z\right). \qquad [17.21]$$

Now, we want to determine the co-adjoint representation of the extension. If $\mu \in \mathfrak{g}^*$, $\eta \in \mathfrak{n}^*$, we can write $\hat{\mu} = (\mu, \eta) \in \hat{\mathfrak{g}^*}$. Let $C\left(a\right) : \mathfrak{n}^* \to \mathfrak{g}^*$ be the transpose map ${}^t\left(B\left(a\right)\right)$ such that:

$$\forall Z \in \mathfrak{g}, \forall \eta \in \mathfrak{n}^*, \quad \eta\left(B\left(a\right)Z\right) = \left(C\left(a\right)\eta\right)Z.$$

Taking into account [17.21], we have:

$$\hat{\mu}\left(Ad\left(\hat{a}^{-1}\right)\hat{Z}\right) = \left(Ad^*\left(a\right)\mu + C\left(a\right)\eta\right)Z + \eta Y.$$

Owing to definition [18.2], the co-adjoint representation of the extension $\hat{G}$ has the form:

$$Ad^*\left(\hat{a}\right)\hat{\mu} = \left(Ad^*\left(a\right)\mu + C\left(a\right)\eta, \eta\right). \qquad [17.22]$$

17.6. Construction of a central extension from the symplectic cocycle

Our aim is now to understand the link between the group extension cocycles and symplectic cocycles. Before presenting the main result, we need two results.

LEMMA 17.1.– If *coc* is an N-cocycle of G, C is the map from $\mathfrak{n}^*$ into $\mathfrak{g}^*$ defined in the previous section and $\eta \in \mathfrak{g}^*$, then $C\left(\bullet\right)\eta$ verifies the symplectic cocycle identity.

PROOF.– The starting idea is introducing the map g from $G \times G$ into N defined by $g\left(a, b\right) = coc\left(b, a\right) + coc\left(a^{-1}, ba\right)$, so that the map B defined in the previous section satisfies:

$$\forall Z \in \mathfrak{g}, \quad B\left(a\right)Z = \left(Dg\left(e, a\right)\right)\left(Z, 0\right). \qquad [17.23]$$

On the other hand, using the N-cocycle identity, we obtain the following three relations:

$$coc\left(b, aa'\right) = coc\left(b, a\right) + coc\left(ba, a'\right) - coc\left(a, a'\right), \qquad [17.24]$$

$$coc\left(a'^{-1}a^{-1}, baa'\right) = coc\left(a'^{-1}, a^{-1}baa'\right) + coc\left(a^{-1}, baa'\right) - coc\left(a'^{-1}, a^{-1}\right), \qquad [17.25]$$

$$coc\left(a^{-1}, baa'\right) = coc\left(a^{-1}ba, a'\right) + coc\left(a^{-1}, ba\right) - coc\left(ba, a'\right). \qquad [17.26]$$

Introducing [17.26] into [17.26], next adding member to member the resulting relation to [17.24], we find:

$$g(b, aa') = coc(b, a) - coc(a, a') + coc\left(a'^{-1}, a^{-1}baa'\right) + coc\left(a^{-1}ba, a'\right)$$
$$+ coc\left(a^{-1}, ba\right) - coc\left(a'^{-1}, a^{-1}\right).$$

Differentiating both members and taking into account [17.23] leads to:

$$B(aa')Z = (Dcoc(e, a))(Z, 0) + \left(Dcoc\left(a'^{-1}, a'\right)\right)\left(0, \left(Ad\left(a^{-1}\right)Z\right)a'\right)$$
$$+ (Dcoc(e, a'))\left(\left(Ad\left(a^{-1}\right)Z\right), 0\right) + \left(Dcoc\left(a^{-1}, a\right)\right)(0, Za).$$

Owing to the definition [17.20] of the map B, we have:

$$B(aa')Z = \left(B(a) + B(a') \circ Ad\left(a^{-1}\right)\right)Z.$$

As Z is arbitrary, we have:

$$B(aa') = B(a) + B(a')\,Ad\left(a^{-1}\right).$$

By transposition, we obtain:

$$C(a'a) = C(a) + Ad^*(a)\,C(a').$$

Taking the value of each member for $\eta \in \mathfrak{g}^*$ shows the symplectic cocycle identity (definition 17.4, formula $\diamondsuit$) is verified by the map $cocs(\bullet) = C(\bullet)\,\eta$, which achieves the proof. ■

LEMMA 17.2.– If *cob* is an N-coboundary of G, then $C(\bullet)\,\eta$ is a symplectic coboundary.

PROOF.– Let *cob* be associated with the map θ from G into N:

$$cob(a, a') = \theta(a) + \theta(a') - \theta(aa').$$

Differentiating both members gives:

$$(Dcob(a', a))(Z', Z) = (D\theta(a'a))(Z'a + a'Z) - (D\theta(a'))(Z') - (D\theta(a))(Z).$$

Owing to the definition [17.20] of the map B, we have:

$$B(a)Z = (Dcob(e, a))(Z, 0) + \left(Dcob\left(a^{-1}, a\right)\right)(0, Za),$$
$$B(a)Z = (D\theta(e))\left(Ad\left(a^{-1}\right)Z\right) - (D\theta(e))(Z).$$

Transposing and taking the value for $\eta \in \mathfrak{g}^*$ of each member shows that $cob\,(\bullet) = C\,(\bullet)\,\eta$ is the symplectic coboundary associated with $\mu = {}^t\,(D\theta\,(e))\,\eta$. ■

Now, we can state and prove the main result.

THEOREM 17.7.– Let $\mathcal{M}$ be a connected symplectic manifold with 2-form ω, G be a Lie group with a symplectic action $x \mapsto x' = a \cdot x$ on $\mathcal{M}$, $\psi : \mathcal{M} \to \mathfrak{g}^*$ be a momentum map and $cocs$ be an element of its class of symplectic cohomology. Let be given a principal N-bundle $\pi : \hat{\mathcal{M}} \to \mathcal{M}$, a 1-form $\hat{\alpha}$ on $\hat{\mathcal{M}}$, invariant by the action of $\hat{G}$ and such that $\pi^*\omega = d\hat{\alpha}$. If:

a) the Lie group $\hat{G}$ is a central extension of G by N with the N-cocycle coc,

b) $\hat{G}$ acts on $\hat{\mathcal{M}}$ in such way that:

$$\hat{a} \cdot \pi\,(\hat{x}) = \pi\,(\hat{a} \cdot \hat{x})\,,; \qquad [17.27]$$

c) $\hat{G}$ acts on $\mathcal{M}$ by:

$$\forall \hat{a} = (a, \theta) \in \hat{G} = G \times N, \quad \forall x \in \mathcal{M}, \quad \hat{a} \cdot x = a \cdot x, \qquad [17.28]$$

then:

1) ♢ for the symplectic form $\hat{\omega} = d\hat{\alpha}$, for any $\mu \in \mathfrak{g}^*$ and $\eta \in \mathfrak{n}^*$, the map:

$$\hat{\psi} : \hat{\mathcal{M}} \to \hat{\mathfrak{g}}^* : \hat{x} \mapsto \hat{\psi}\,(\hat{x}) = (\psi\,(\pi\,(\hat{x})) + \mu, \eta)\,, \qquad [17.29]$$

is a momentum map;

2) ♡ modulo a symplectic coboundary, we have:

$$C\,(a)\,\eta = cocs\,(a)\,. \qquad [17.30]$$

PROOF.– As $d\hat{\alpha}$ is the pull-back of ω by π, we have for any tangent vectors $\delta\hat{x}$ and $d\hat{x}$ to $\hat{M}$ at $\hat{x}$:

$$(d\hat{\alpha})\,(\delta\hat{x}, d\hat{x}) = \omega\,(D\pi\,(\delta\hat{x})\,, D\pi\,(d\hat{x})) = (\iota\,(D\pi\,(\delta\hat{x}))\,\omega)\,(D\pi\,(d\hat{x}))\,.$$

In particular, for any $\hat{Z} \in \hat{\mathfrak{g}}$:

$$(d\hat{\alpha})\,\left(\hat{Z} \cdot \hat{x}, d\hat{x}\right) = (\iota(D\pi\,(\hat{Z} \cdot \hat{x}))\,\omega)\,(D\pi\,(d\hat{x})). \qquad [17.31]$$

Let us transform the left-hand member. As $\hat{\alpha}$ is conserved by the symplectic action, its Lie derivative with respect to the tangent vector $\hat{Z} \cdot \hat{x}$ vanishes:

$$\mathcal{L}_{\hat{Z} \cdot \hat{x}}\,\hat{\alpha} = 0.$$

Using Cartan's formula gives:

$$d(\iota(\hat{Z} \cdot \hat{x})\, \hat{\alpha}) = -\iota(\hat{Z} \cdot \hat{x})\, d\hat{\alpha}.$$

Hence:

$$(d\hat{\alpha})\, (\hat{Z} \cdot \hat{x}, d\hat{x}) = -d\, (\hat{\alpha}(\hat{Z} \cdot \hat{x})).$$

As the action of $\hat{G}$ on $\hat{\mathcal{M}}$ is symplectic, it has a momentum map $\hat{\psi} : \hat{\mathcal{M}} \to \hat{\mathfrak{g}}^*$ such that:

$$\hat{\psi}\,(\hat{x})\, \hat{Z} = \hat{\alpha}\,(\hat{Z} \cdot \hat{x}),$$

with a null class of symplectic cohomology because of theorem 17.5. Owing to the decomposition $\hat{Z} = (Z, Y) \in \hat{\mathfrak{g}} = \mathfrak{g} \times \mathfrak{n}$, we put:

$$\hat{\psi}\,(\hat{x})\, \hat{Z} = \phi\,(\hat{x})\, Z + \chi\,(\hat{x})\, Y.$$

Finally, the left-hand member of [17.31] becomes:

$$(d\hat{\alpha})\, (\hat{Z} \cdot \hat{x}, d\hat{x}) = -d\left[\phi\,(\hat{x})\, Z + \chi\,(\hat{x})\, Y\right]. \qquad [17.32]$$

Next, let us transform the right-hand member of [17.31]. Combining [17.27] and [17.28] gives:

$$\pi\,(\hat{a} \cdot \hat{x}) = \hat{a} \cdot \pi\,(\hat{x}) = a \cdot \pi\,(\hat{x})\,. \qquad [17.33]$$

Differentiating with respect to $\hat{a}$ at $\hat{a} = e$ in the direction $d\hat{a} = \hat{Z} = (Z, Y)$ gives:

$$D\pi\,(\hat{Z} \cdot \hat{x}) = Z \cdot \pi(\hat{x}).$$

Thus:

$$(\iota(D\pi\,(\hat{Z} \cdot \hat{x}))\omega)\, (D\pi\,(d\hat{x})) = (\iota(Z \cdot \pi(\hat{x}))\, \omega)\, (D\pi\,(d\hat{x})).$$

As $\psi : \mathcal{M} \to \mathfrak{g}^*$ is a momentum map, the right-hand member of [17.31] becomes:

$$(\iota(D\pi\,(\hat{Z} \cdot \hat{x}))\, \omega)\, (D\pi\,(d\hat{x})) = -D(\psi(\pi(\hat{x}))\, Z)\, (D\pi(d\hat{x})),$$

or:

$$(\iota(D\pi\,(\hat{Z} \cdot \hat{x}))\, \omega)\, (D\pi(d\hat{x})) = -d(\psi(\pi(\hat{x}))\, Z). \qquad [17.34]$$

Introducing [17.32] and [17.34] into [17.31] gives:

$$d\left(\phi\left(\hat{x}\right)-\psi\left(\pi\left(\hat{x}\right)\right)\right)Z.+d\left(\chi\left(\hat{x}\right)\right)Y=0.$$

As $\hat{Z}=(Z,Y)$ is arbitrary, we obtain:

$$\phi\left(\hat{x}\right)=\psi\left(\pi\left(\hat{x}\right)\right)+\mu,\qquad \chi\left(\hat{x}\right)=\eta,$$

where $\mu\in\mathfrak{g}^*$ and $\eta\in\mathfrak{n}^*$ are arbitrary constants of integration, which proves the structure of the momentum map $\hat{\psi}$ is given by [17.29].

Finally, we know the class of symplectic cohomology of $\hat{G}$ is null. Owing to [17.22], [17.33] and [17.29], the symplectic cocycle of $\hat{\psi}$ is for $\hat{a}=(a,\theta)$:

$$\begin{aligned}&cocs\left(\hat{a}\right)=\hat{\psi}\left(\hat{a}\cdot\hat{x}\right)-Ad^*\left(\hat{a}\right)\hat{\psi}\left(\hat{x}\right)\\&=\left(\psi\left(a\cdot\pi\left(\hat{x}\right)\right)+\mu,\eta\right)-\left(Ad^*\left(a\right)\left(\psi\left(\pi\left(\hat{x}\right)\right)+\mu\right)+C\left(a\right)\eta,\eta\right).\end{aligned}$$

Putting $x=\pi\left(\hat{x}\right)$, we have:

$$cocs\left(\hat{a}\right)=\left(\psi\left(a\cdot x\right)-Ad^*\left(a\right)\psi\left(x\right)-C\left(a\right)\eta-Ad^*\left(a\right)\mu+\mu,0\right).$$

Owing to ♢ of theorem 17.3 and definition [17.4] of the symplectic coboundary, we obtain:

$$cocs\left(\hat{a}\right)=\left(cocs\left(a\right)-C\left(a\right)\eta-cobs_\mu\left(a\right),0\right).$$

As the class of symplectic cohomology vanishes and the choice of the symplectic coboundary is arbitrary, we can impose that the symplectic cocycle of $\hat{\psi}$ is null, hence we have:

$$cocs\left(a\right)=C\left(a\right)\eta+cobs_\mu\left(a\right).$$

The left-hand member is a symplectic cocycle and, because of lemma 17.1, so is the first term of the right-hand member. Hence, we prove that, modulo a symplectic coboundary, we have:

$$C\left(a\right)\eta=cocs\left(a\right),$$

which achieves the proof. ■

The main interest of theorem 17.7 is when the manifold $\mathcal{M}$ is given with a symmetry group G of $\mathcal{M}$ of which the class of symplectic cohomology $cocs$ does not vanish. To obtain a suitable cocycle, we have to satisfy condition [17.30]. Let us apply this method to find a central extension of Galileo's group by $\mathbb{R}$ which has a null class of symplectic cohomology. Let $(a^n)_{n=1,\ldots,10}$ the system of local coordinates on $\mathbb{GAL}$ composed of τ_0, k^i, u^i, R^i_j. Hence, condition [17.30] leads to a system of 10 partial derivative equations with the right-hand member composed of second-order polynomial functions of the local coordinates. We try a polynomial function *coc* which is linear with respect to the second argument:

$$coc\,(a, a') = Tr\,(\Lambda\,(a)\,R') - \chi\,(a)^T\,u' + \pi\,(a)^T\,k' - \epsilon\,(a)\,\tau_0'. \qquad [17.35]$$

Now, we have to determine $\Lambda\,(a) \in \mathbb{M}_{33}$, $\chi\,(a) \in \mathbb{R}^3$, $\pi\,(a) \in \mathbb{R}^3$ and $\epsilon\,(a) \in \mathbb{R}$ by identification. Introducing the general expression of Galileo's symplectic cocycle given by theorem 17.4 and [17.35] into [17.30] gives the system of equations:

$$j^{-1}\left((R\Lambda\,(a))^a\right) - u \times \chi\left(a^{-1}\right) + k \times \pi\left(a^{-1}\right) = -k \times u, \qquad [17.36]$$

$$-\,k + \chi\left(a^{-1}\right) - \tau_0\pi\left(a^{-1}\right) = -\left(k - \tau_0 u\right), \qquad [17.37]$$

$$\pi\left(a^{-1}\right) = -u, \qquad [17.38]$$

$$\epsilon\left(a^{-1}\right) = \frac{1}{2}\,\parallel u \parallel^2, \qquad [17.39]$$

where $(A)^a$ is the skew-symmetric part of the matrix A. Condition [17.39] gives $\epsilon\left(a^{-1}\right)$ and, taking into account [1.15]:

$$\epsilon\,(a) = \frac{1}{2}\,\parallel u \parallel^2 .$$

Also, condition [17.38] gives $\pi\left(a^{-1}\right)$ and, taking into account [1.15]:

$$\pi\,(a) = R^T u.$$

Finally, condition [17.37] entails that χ vanishes and so is Λ because of condition [17.36]. Hence, the extension cocycle [17.35] becomes:

$$coc\,(a, a') = u^T R\,k' + \frac{1}{2}\,\parallel u \parallel^2\,\tau_0'. \qquad [17.40]$$

Owing to the linearity of *coc* with respect to the second argument, we let the readers verify that the elements of the central extension $\hat{G}$ are affine transformations which can be represented by the matrix $\hat{P}$ given by [13.7].

The central extension of Galileo's group $\mathbb{GAL}$ obtained by theorem 17.7 is nothing else but the Bargmann's group $\mathbb{B}$ and its class of symplectic cohomology is null.

The original extension cocycle proposed by Bargmann [BAR 54]:

$$coc_B(a,a') = \frac{1}{2}\left(u^T R\, k' - k^T R\, u' + u^T R\, u'\, \tau_0'\right),$$

can be deduced from [17.40] by adding the coboundary associated with the map:

$$\theta(a) = \frac{1}{2}\, k^T u.$$

Although the result is already known since Bargmann's pioneering work [BAR 54], the present method seems to be simpler to handle. Moreover, studying the momentum map of the extended group, we emphasized the link between the extension cocycles and the symplectic ones.

Reformulated in cohomological terms, Simms result [SIM 68] says that if the second cohomology space $H^2(\mathfrak{g};\mathfrak{n})$ is null, then the second cohomology group $H^2(G;N)$ so is, with the trivial action of G into N. Without such restrictive condition on the cohomology space, the present method is based on a family of homomorphisms from $H^2(G;N)$ into the first cohomology group $H^1(G;\mathfrak{g}^*)$ with the coadjoint representation of G. These homomorphisms are linearly depending on $\eta \in \mathfrak{n}^*$.

17.7. Coadjoint orbit method

There is another way to introduce the symplectic form. Inspiring from the Hamiltonian formalism of sections 6.1 and 6.2.2, we introduce *Cartan 1-form*:

$$\alpha = \pi_i\, dx_i - \mathcal{H}\, dt,$$

of which the components are the generalized linear momentum [6.17] and the Hamiltonian [6.18] of a particle:

$$\pi_i = m\,(v_i + A_i), \qquad \mathcal{H} = \frac{m}{2}\, v_i v_i + m\,\phi.$$

Its external differential:

$$d\alpha = d\pi_i \wedge dx_i - d\mathcal{H} \wedge dt, \tag{17.41}$$

is expanded as:

$$d\alpha = m\left[\left(dv_i + \frac{\partial A_i}{\partial x_j}\, dx_j + \frac{\partial A_i}{\partial t}\, dt\right)\right.$$
$$\left.\wedge dx_i - v_i\, dv_i \wedge dt - \frac{\partial \phi}{\partial x_i}\, dx_i \wedge dt - \frac{\partial \phi}{\partial t}\, dt \wedge dt\right],$$

where the last term vanishes, then:

$$d\alpha = m\left[dv_i \wedge dx_i - v_i\, dv_i \wedge dt + \left(\frac{\partial \phi}{\partial x_i} + \frac{\partial A_i}{\partial t}\right) dt \wedge dx_i + \frac{\partial A_j}{\partial x_i}\, dx_i \wedge dx_j\right],$$

that, according to [6.14], leads to [17.5], hence the symplectic form ω is the exterior differential of Cartan 1-form:

$$\omega = d\alpha.$$

In addition, it is worth noting that [17.41] is expanded as:

$$d\alpha = d\pi_i \wedge dx_i - \left(\frac{\partial \mathcal{H}}{\partial x_i}\, dx_i + \frac{\partial \mathcal{H}}{\partial \pi_i}\, d\pi_i + \frac{\partial \mathcal{H}}{\partial t}\, dt\right) \wedge dt,$$

where the last term vanishes, which can be recast as:

$$d\alpha = d\pi_i \wedge dx_i + \frac{\partial \mathcal{H}}{\partial x_i}\, dt \wedge dx_i - \frac{\partial \mathcal{H}}{\partial \pi_i}\, d\pi_i \wedge dt - \frac{\partial \mathcal{H}}{\partial x_i}\frac{\partial \mathcal{H}}{\partial \pi_i}\, dt \wedge dt,$$

that leads to the new expression of the symplectic form:

$$\omega = d\alpha = \left(d\pi_i + \frac{\partial \mathcal{H}}{\partial x_i}\, dt\right) \wedge \left(dx_i - \frac{\partial \mathcal{H}}{\partial \pi_i}\, dt\right). \qquad [17.42]$$

as *factorization of symplectic form into the canonical equations* [6.19] and [6.21].

It is this kind of construction that we aim to extend in a more general context for the classical dynamics, as in the previous example, and the relativist one. We wish the formulation of the equation of motion to be covariant, which is natural in relativity but less obvious in classical mechanics. To develop some intuition of a suitable generalization, we consider the usual case where $A = 0$ for which the factorized symplectic form reduces to:

$$\omega = (dp_i - g_i dt) \wedge \left(dx_i - \frac{p_i}{m} dt\right).$$

Our starting point is to spot the linear momentum p and the gravity g as basic ingredients. Hence, the construction lays in three steps:

– the linear momentum is generalized as the momentum affine tensor;

– in relativity's spirit, the gravity is generalized as an affine connection;

– the momentum and the connection are factorized to give a symplectic form.

Now, we need an important result, *Kirillov–Kostant–Souriau theorem*, revealing the *orbit symplectic structure*:

THEOREM 17.8.– Let G be a Lie group and an orbit of the coadjoint representation $orb\,(\mu) \subset \mathfrak{g}^*$. Then:

♢ The inclusion map $orb\,(\mu) \to \mathfrak{g}^*$ is a regular imbedding. A vector $d\mu \in T_\mu \mathfrak{g}^*$ is tangent to the orbit if there exists $Z_d \in \mathfrak{g}$ such that:

$$d\,\mu = \mu \circ ad\,(Z_d) + dcocs\,(Z_d) = -ad^*(Z_d)\mu + dcocs\,(Z_d).$$

♡ The orbit $orb\,(\mu)$ is a symplectic manifold of which the symplectic form is defined by:

$$\omega_{KKS}(d\mu, \delta\mu) = \mu\,[Z_d, Z_\delta] + dcocs\,(Z_d, Z_\delta),$$

The dimension of the orbit is even.

♠ G is a symplectic group and any $\mu \in \mathfrak{g}^*$ is its own momentum.

The readers can find a demonstration in [SOU 70, pp. 116–118] (theorem 11.34).

DEFINITION 17.7.– Let $\eta \mapsto \omega(\eta)$ be a smooth field of 2-forms on a bundle $\pi : \mathcal{N} \to \mathcal{B}$. If each fiber is a presymplectic (respectively, symplectic) manifold, we say that $\mathcal{N}$ is a *presymplectic bundle* (respectively, a *symplectic bundle*).

For instance, $\mathfrak{g}^*$ carrying the orbit symplectic structure is a symplectic bundle.

Closer to the present formulation, we can cite two groups of papers. The first ones use the concept of "mechanical connection", implicitly in [SMA 70a], [SMA 70b], [ABR 78], and explicitly in [KUM 81], [GUI 84a], [SHA 89], [SIM 91], [MON 90] and [MAR 92]. The second group is composed of works in relation with the theory of Yang-Mills fields: [STE 78a, STE 78b, GUI 84b, CUR 85, DUV 80, DUV 82a, DUV 82b].

17.8. Connections

Let $\pi : \mathcal{F} \to \mathcal{M}$ be a G-principal bundle of affine frames with the free action $(a, f) \mapsto f' = a \cdot f$ on each fiber. In section 16.3, we built the associated G-principal bundle:

$$\hat{\pi} : \mathfrak{g}^* \times \mathcal{F} \to (\mathfrak{g}^* \times \mathcal{F})/G : (\mu, f) \mapsto \boldsymbol{\mu} = orb(\mu, f),$$

for the free action:

$$(a, (\mu, f)) \mapsto (\mu', f') = a \cdot (\mu, f) = (a \cdot \mu, a \cdot f),$$

where the action on $\mathfrak{g}^*$ is [16.7]. Clearly, the orbit $\boldsymbol{\mu} = orb(\mu, f)$ can be identified to the momentum G-tensor $\boldsymbol{\mu}$ of components μ in the G-frame f. The orbit space $(\mathfrak{g}^* \times \mathcal{F})/G$ is sometimes denoted by $\mathfrak{g}^* \times_G \mathcal{F}$. The manifold $\mathfrak{g}^* \times \mathcal{F}$ plays the role of the "phase space" considered in the elementary example of the introduction.

Let $ver_f = Ker\,(D\pi)$ the vertical space at f. An *Ehresmann connection* on the G-principle bundle $\mathcal{F}$ is a field of supplementary subspaces hor_f in $T_f\mathcal{F}$:

$$T_f\mathcal{F} = ver_f \oplus hor_f.$$

The decomposition $df = df_v + df_h$ is unique and the map $hor : T_f\mathcal{F} \to hor_f : df \mapsto df_h$ is called the horizontal projection.

Alternatively, a connection can be defined by a field of $\mathfrak{g}$-valued 1-forms $\tilde{\Gamma}$ on $\mathcal{F}$ such that $hor_f = Ker\,\tilde{\Gamma}$ and:

$\diamondsuit$ $\tilde{\Gamma}$ is *vertical*: $\forall df_h \in hor_f, \qquad \tilde{\Gamma}(df_h) = 0;$

$\heartsuit$ $\tilde{\Gamma}(Z \cdot f) = Z;$

$\spadesuit$ $\tilde{\Gamma}$ is *Ad-equivariant*: $L_a\tilde{\Gamma} = Ad(a)\,\tilde{\Gamma}$ where $Ad(a)$ is the adjoint representation.

The covariant derivative $\tilde{\nabla}_{\overrightarrow{d\boldsymbol{X}}}\boldsymbol{\mu}$ of a momentum field $\boldsymbol{X} \mapsto \boldsymbol{\mu}(\boldsymbol{X})$ in a moving frame $\boldsymbol{X} \mapsto f(\boldsymbol{X})$ is defined by:

$$\tilde{\nabla}_{dX}\mu = d\mu - (\tilde{\Gamma}(df)) \cdot \mu, \qquad [17.43]$$

that can read:

$$\tilde{\nabla}_{\overrightarrow{d\boldsymbol{X}}}\boldsymbol{\mu} = orb(\tilde{\nabla}_{dX}\mu, f) = orb(d\mu - (\tilde{\Gamma}(df)) \cdot \mu, f). \qquad [17.44]$$

As G is a subgroup of $\mathbb{A}ff(n)$, the connection is decomposed as $\tilde{\Gamma} = (\Gamma_A, \Gamma)$ where:

– The $\mathfrak{gl}(n)$-valued 1-form Γ is a classical linear connection describing the infinitesimal motion of the basis $S = lin(f)$. In general relativity, it represents the gravitation.

– The $\mathbb{R}^n$-valued 1-form Γ_A is the affine part of the connection describing the infinitesimal motion of the origin of the affine frame f. Its physical meaning is not so strong as the gravitation but it represents the observer, as discussed in section 5.3.2.

It is worth noting that usual connections are defined for a right action of G on the components: $\mu \cdot a = a^{-1} \cdot \mu$ and, differentiating around the identity, there is a sign change in the infinitesimal action of $\mathfrak{g}$: $\mu \cdot Z = -Z \cdot \mu$. Hence, the rule to swap the usual connections $\tilde{\Gamma}'$ for the corresponding ones $\tilde{\Gamma}$ considered here is:

$$\tilde{\Gamma}' = -\tilde{\Gamma}. \qquad [17.45]$$

For more details on connections and in particular on affine connections, the readers are referred, for instance, to [KOB 63].

17.9. Factorized symplectic form

According to the dual pairing [16.5], we can put momenta and connections into duality:

$$\mu \tilde{\Gamma} = F \Gamma_A + Tr\,(L\Gamma).$$

This suggests to introduce the factorized 2-form:

$$\boxed{\omega = \frac{1}{2}\, d\mu \wedge \tilde{\Gamma},} \qquad [17.46]$$

exterior product of the $\mathfrak{g}^*$-valued 1-form $d\mu$ and the $\mathfrak{g}$-valued 1-form $\tilde{\Gamma}$, hence a scalar valued 1-form.

We begin with a preliminary result.

LEMMA 17.3.– Let $f \mapsto \tilde{\Gamma}$ be a field of connection 1-form on a G-principal bundle $\pi : \mathcal{F} \rightarrow \mathcal{M}$. The group G left acts on the dual $\mathfrak{g}^*$ of the Lie algebra of G and on the associated G-principal bundle: $\hat{\pi} : \mathfrak{g}^* \times \mathcal{F} \rightarrow (\mathfrak{g}^* \times \mathcal{F})/G$ by: $a \cdot \eta = a \cdot (\mu, f) = (a \cdot \mu, a \cdot f)$. Then:

– The tangent space to $\mathfrak{g}^* \times \mathcal{F}$ at η is a direct sum:

$$T_\eta(\mathfrak{g}^* \times \mathcal{F}) = ver_\eta \oplus hor_\eta,$$

where $ver_\eta = Ker\,(D\hat{\pi})$ is the space of vertical vectors and $hor_\eta = \mathfrak{g}^* \times hor_f$ is the space of horizontal vectors.

– Any tangent vector $d\eta = (d\mu, df)$ can be decomposed in a unique way as $d\eta = d\eta_v + d\eta_h$, where:

$$d\eta_v = (\tilde{\Gamma}(df)) \cdot \eta,$$

$$d\eta_h = (\nabla_{dX}\,\mu, hor\,(df)), \quad \text{with:} \quad D\pi(df) = \overrightarrow{d\boldsymbol{X}}.$$

PROOF.– When Z_d spans $\mathfrak{g}$, the vector $d\eta_v = Z_d \cdot \eta$ generates the whole space ver_η. As the action of G on $\mathcal{F}$ is free, its isotropy group at η is reduced to the unity. Thus:

$$dim\,(ver_\eta) = dim\,(orb\,(\eta)) = dim\,G - dim\,(iso\,(\eta)) = dim\,G.$$

Clearly, the corresponding infinitesimal action: $\tau_\eta\ :\ \mathfrak{g} \to ver_\eta$ is one-to-one. On the other hand, it holds:

$$dim\,(hor_\eta) = dim\,\mathfrak{g}^* + dim\,(hor_f) = dim\,\mathfrak{g}^* + dim\,M.$$

Besides, we have:

$$dim\,(\mathfrak{g}^* \times \mathcal{F}) = dim\,\mathfrak{g}^* + dim\,M + dim\,G = dim\,(ver_\eta) + dim\,(hor_\eta).$$

Hence, let us consider: $d\eta \in ver_\eta \cap hor_\eta$, and: $Z_d = \tau_\eta^{-1}(d\eta) \in \mathfrak{g}$. Since df is vertical and horizontal, then: $Z_d = 0$, and: $d\mu = Z_d \cdot \mu = 0$. Finally, $d\eta$ vanishes. Consequently:

$$T_\eta(\mathfrak{g}^* \times \mathcal{F}) = ver_\eta \oplus hor_\eta.$$

Now, we can give an algorithm to determine the direct factors of any tangent vectors, by calculating:

– the infinitesimal generator: $Z_d = \tilde{\Gamma}(df)$;

– the vertical part: $d\eta_v = (d\mu_v, df_v)$, with: $d\mu_v = Z_d \cdot \mu, \qquad df_v = Z_d \cdot f$;

– the horizontal part: $d\eta_h = (d\mu_h, df_h)$, by:

$$d\mu_h = d\mu - d\mu_v = d\mu - (\tilde{\Gamma}(df)) \cdot \mu = \nabla_{dX}\mu,$$

with: $D\pi(df) = \overrightarrow{d\boldsymbol{X}}$,

$$df_h = df - df_v = df - (\tilde{\Gamma}(df)) \cdot f = hor\,(df).$$

that achieves the proof. ■

We can now state the main result:

THEOREM 17.9.– Let $f \mapsto \tilde{\Gamma}$ be a field of connection 1-form on a G-principal bundle $\pi : \mathcal{F} \to \mathcal{M}$. The group G left acts on $\mathfrak{g}^*$ by the affine representation [16.10]:

$$\mu' = a \cdot \mu = Ad^*(a)\,\mu + cocs(a), \qquad [17.47]$$

and on the G-principal bundle $\hat{\pi} : \mathfrak{g}^* \times \mathcal{F} \to (\mathfrak{g}^* \times \mathcal{F})/G$ by: $a \cdot \eta = a \cdot (\mu, f) = (a \cdot \mu, a \cdot f)$. Let $\eta \to \omega$ be a smooth field of 2-form on $\mathfrak{g}^* \times \mathcal{F}$ defined by:

$$\omega = \frac{1}{2}\, d\mu \wedge \tilde{\Gamma}.$$

Then:

$\diamondsuit$ Let $\boldsymbol{\mu} = orb(\mu, f)$ be an orbit in $\mathfrak{g}^* \times \mathcal{F}$, and $\psi_{\boldsymbol{\mu}} = pr_1 \circ 1_{\boldsymbol{\mu}}$ be the projection on the first component $\eta = (\mu, f) \mapsto \mu$, restricted to the orbit. Then, the smooth map $\psi_{\boldsymbol{\mu}}$ is a submersion.

$\heartsuit$ We have:

$$\omega(\delta\eta, \delta\eta) = \frac{1}{2}\,(d\,\mu\, Z_\delta - \delta\,\mu\, Z_d) \qquad \text{with:} \quad Z_d = \tilde{\Gamma}(df), \qquad Z_\delta = \tilde{\Gamma}(\delta f).$$

On each orbit, ω is the pull-back of Kirillov–Kostant–Souriau symplectic form ω_{KKS} by $\psi_{\boldsymbol{\mu}}$:

$$\omega = \psi_{\boldsymbol{\mu}}^*\,(\omega_{KKS}),$$

and is invariant by left action:

$$L_a^*\,\omega = \omega.$$

The G-principal bundle $\mathfrak{g}^* \times \mathcal{F}$ is a presymplectic bundle of symplectic form ω.

$\spadesuit$ $\psi_{\boldsymbol{\mu}}$ is a momentum map and:

$$\psi_{\boldsymbol{\mu}} \circ L_a = Ad^*(a)\,\psi_{\boldsymbol{\mu}} + cocs(a).$$

$\clubsuit$ The equation of motion is:

$$d\eta \in Ker(\omega) \quad \Leftrightarrow \quad \tilde{\nabla}_{dX}\mu = d\,\mu + ad^*(\tilde{\Gamma})\mu - dcocs\,(\tilde{\Gamma}) = 0, \qquad [17.48]$$

and the momentum $\boldsymbol{\mu} = orb(\mu, f)$ is parallel-transported:

$$\tilde{\nabla}_{\overrightarrow{d\mathbf{X}}}\boldsymbol{\mu} = \mathbf{0}.$$

PROOF.– The action of G on $\boldsymbol{\mu}$ being free, the corresponding infinitesimal action: $\tau_\eta : \mathfrak{g} \to T_\eta \boldsymbol{\mu}$, is one-to-one. The tangent map to $\psi_{\boldsymbol{\mu}}$ is clearly: $D\psi_{\boldsymbol{\mu}} : d\eta_v \mapsto d\mu_v = Z_d \cdot \mu$, with: $Z_d = \tau_\eta^{-1}(d\eta_v)$, and: $\mu = pr_1(\eta)$. When Z_d spans $\mathfrak{g}$, the vector $d\mu_v = Z_d \cdot \mu$ generates the whole space ver_μ. Thus, $D\psi_{\boldsymbol{\mu}}$ is surjective and $\psi_{\boldsymbol{\mu}}$ is a submersion at η, which proves $\diamondsuit$.

Lemma 17.3 claims that each tangent vector to $\mathfrak{g}^* \times \mathcal{F}$ admits a unique decomposition into vertical and horizontal parts:

$$d\eta = d\eta_v + d\eta_h, \quad \delta\eta = \delta\eta_v + \delta\eta_h, \quad d\eta_v, \delta\eta_v \in ver_\eta, \quad d\eta_h, \delta\eta_h \in hor_\eta.$$

By linearity, it holds:

$$\omega(d\eta, \delta\eta) = \omega(d\eta_v, \delta\eta_v) + \omega(d\eta_v, \delta\eta_h) + \omega(d\eta_h, \delta\eta_v) + \omega(d\eta_h, \delta\eta_h).$$

Taking into account the properties $\diamondsuit$ and $\heartsuit$ of the connection 1-form, we have:

$$\omega(d\eta_v, \delta\eta_v) = \frac{1}{2}\,(d\,\mu_v\,\tilde{\Gamma}(Z_\delta \cdot f) - \delta\,\mu_v\,\tilde{\Gamma}(Z_d \cdot f)) = \frac{1}{2}\,(d\,\mu_v\,Z_\delta - \delta\,\mu_v\,Z_d), \tag{17.49}$$

$$\omega(d\eta_v, \delta\eta_h) = \frac{1}{2}\,(d\mu_v\,\tilde{\Gamma}(\delta f_h) - \delta\mu_h\,\tilde{\Gamma}(Z_d \cdot f)) = -\frac{1}{2}\,\delta\mu_h\,Z_d, \tag{17.50}$$

$$\omega(d\eta_h, \delta\eta_v) = \frac{1}{2}\,(d\,\mu_h\,\tilde{\Gamma}(Z_\delta \cdot f) - \delta\mu_v\,\tilde{\Gamma}(df_h)) = \frac{1}{2}\,d\mu_h\,Z_\delta, \tag{17.51}$$

$$\omega(d\eta_h, \delta\eta_h) = \frac{1}{2}\,(d\mu_h\,\tilde{\Gamma}(\delta f_h) - \delta\mu_h\,\tilde{\Gamma}(df_h)) = 0. \tag{17.52}$$

It follows that:

$$\omega(d\eta, \delta\eta) = \frac{1}{2}\,(d\mu_v\,Z_\delta - \delta\mu_v\,Z_d + d\mu_h\,Z_\delta - \delta\mu_h\,Z_d).$$

Recombining vertical and horizontal parts, we obtain:

$$\omega(d\eta, \delta\eta) = \frac{1}{2}\,(d\,\mu\,Z_\delta - \delta\,\mu\,Z_d). \tag{17.53}$$

Now, let us suppose that $\delta\eta$ and $d\eta$ are both vertical vectors, i.e. tangent to the orbit. Since G acts on $\mathfrak{g}^*$ by the affine representation [17.47], it can be written that:

$$d\,\mu = \mu \circ ad\,(Z_d) + (D\,cocs(e))\,Z_d, \quad \delta\,\mu = \mu \circ ad\,(Z_\delta) + (D\,cocs(e))\,Z_\delta.$$

Consequently, we have:

$$\omega(\delta\eta, d\eta) = \mu\,[Z_d, Z_\delta] + dcocs\,(Z_d, Z_\delta) = \omega_{KKS}(d\mu, \delta\mu),$$

which allows us to identify ω and ω_{KKS}. In other words, ω is the pull-back of Kirillov–Kostant–Souriau symplectic form ω_{KKS} by ψ_μ on each orbit. By the way, it is worth to notice that the two terms of expression [17.53] are equal:

$$\omega(d\eta, \delta\eta) = d\,\mu\,Z_\delta = -\delta\,\mu\,Z_d. \quad [17.54]$$

Let us now consider:

$$L_a^*\,\omega = \frac{1}{2}\,L_a^*\,(d\mu \wedge \tilde{\Gamma}) = \frac{1}{2}\,L_a^*\,(d\mu) \wedge L_a^*\,(\tilde{\Gamma}).$$

As a is constant, we have:

$$L_a^*\,(d\,\mu) = DL_a(d\mu) = d\,(a \cdot \mu) = d\,(Ad^*(a)\,\mu + cocs(a)) = Ad^*(a)\,d\,\mu.$$

Because of the property ♠ of the connection 1-form $\tilde{\Gamma}$, it holds:

$$L_a^*\,\omega = \frac{1}{2}\,(Ad^*(a)\,d\mu\,) \wedge (Ad\,(a)\,\tilde{\Gamma}).$$

By the definition of the coadjoint representation, we have:

$$L_a^*\,\omega = \frac{1}{2}\,d\mu \wedge (Ad\,(a^{-1})\,Ad\,(a)\,\tilde{\Gamma}) = \frac{1}{2}\,d\mu \wedge \tilde{\Gamma} = \omega.$$

ω is invariant by left action.

Next, we wish to prove that ω is a symplectic form. Let us consider any $d\eta_v \in Ker\,\omega \cap ver_\eta$. The motion equation of ω, restricted to the orbit μ is: $\forall \delta\eta_v \in ver_\eta, \quad (\iota\,(d\eta_v)\,\omega)\,\delta\eta_v = \omega(d\eta_v, \delta\eta_v) = \omega_{KKS}(d\mu_v, \delta\mu_v) = 0.$

Owing to [17.54], $\forall Z_\delta \in \mathfrak{g}$, $d\mu_v\,Z_\delta = 0$, to be satisfied by: $d\mu_v = Z_d \cdot \mu = 0$. The infinitesimal generator Z_d belongs to the Lie algebra of the isotropy group $iso\,(\mu)$, denoted by $\mathfrak{h}_\mu$. We have:

$$d\eta_v \in Ker\,\omega \cap ver_\eta = Ker\,D\psi_\mu = \{d\eta_v = (0, Z_d \cdot f), \;\; Z_d \in \mathfrak{h}_\mu\}.$$

Consequently, $dim\,(Ker\,\omega) = dim\,\mathfrak{h}_\mu = dim\,(iso\,(\mu)) = dim\,G - dim\,(orb\,(\mu))$ is constant since the orbit of μ is an immersed manifold in $\mathfrak{g}^*$. Besides, as the orbit of μ is symplectic, ω_{KKS} is closed and: $d\,\omega = d\,(\psi_\mu^*\,\omega_{KKS}) = \psi_\mu^*\,(d\,\omega_{KKS}) = 0.$

Each orbit in $\mathfrak{g}^* \times \mathcal{F}$, which is a fiber of the bundle, is presymplectic. Hence, $\mathfrak{g}^* \times \mathcal{F}$ is a presymplectic G-principal bundle, which proves ♡.

Because μ is its own momentum (point ♠ of Theorem 17.8), one has for any $Z_d \in \mathfrak{g}$:

$$\begin{aligned}\iota\,(Z_d \cdot \eta)\,\omega &= \iota\,(D\psi_{\boldsymbol{\mu}}(Z_d \cdot \eta))\,\omega_{KKS} = \iota\,(Z_d \cdot \mu)\,\omega_{KKS}\\ &= -\,d\,(\mu\, Z_d) = -\,d\,(\psi_{\boldsymbol{\mu}}(\eta)\, Z_d).\end{aligned}$$

Then, $\psi_{\boldsymbol{\mu}}$ is a momentum map on $\mathfrak{g}^* \times \mathcal{F}$. As G acts on $\mathfrak{g}^*$ by the coadjoint representation:

$$\begin{aligned}\psi_{\boldsymbol{\mu}}(a \cdot \eta) &= \psi_{\boldsymbol{\mu}}(a \cdot \mu, a \cdot f) = a \cdot \mu = Ad^*(a)\,\mu + cocs(a)\\ &= Ad^*(a)\,\psi_{\boldsymbol{\mu}}(\eta) + cocs(a),\end{aligned}$$

that proves ♠.

Finally, let us deduce the motion equation, starting from:

$$\forall\,\delta\eta \in T_\eta(\mathfrak{g}^* \times \mathcal{F}),\quad (\iota\,(d\eta)\,\omega)\,\delta\eta = \omega(d\eta, \delta\eta) = \frac{1}{2}\,(d\mu\, Z_\delta - \delta\mu\, Z_d) = 0.$$

First, let us choose $\delta\eta$ horizontal. Then: $Z_\delta = 0$, and: $\delta\mu = \delta\mu_h \in \mathfrak{g}^*$. Thus, one has:

$$\forall\,\delta\mu \in \mathfrak{g}^*,\quad (\iota\,(d\eta)\,\omega)\,\delta\eta = -\frac{1}{2}\,\delta\mu\, Z_d = 0. \qquad [17.55]$$

Consequently, $Z_d = \tilde{\Gamma}(df) = 0$ and $d\eta$ is horizontal. Next, by choosing $\delta\eta$ vertical, it holds:

$$\forall\, Z_\delta \in \mathfrak{g},\quad (\iota\,(d\eta)\,\omega)\,\delta\eta = \frac{1}{2}\,d\mu\, Z_\delta = 0. \qquad [17.56]$$

Hence, $d\mu = d\mu_h = 0$, and by lemma 17.3:

$$d\mu_h = \nabla_{dX}\,\mu = 0,$$

and, owing to [17.44], the momentum is parallel-transported. Taking into account [17.43] and ♢ of theorem 17.8, we obtain [17.48]:

$$\nabla_{dX}\,\mu = d\,\mu + ad^*(\tilde{\Gamma})\mu - dcocs\,(\tilde{\Gamma}) = 0,$$

that achieves the proof. ■

Using the swap rule [17.45], Equation [17.48] reads:

$$\nabla_{dX}\,\mu = d\,\mu - ad^*(\tilde{\Gamma}')\mu + dcocs\,(\tilde{\Gamma}') = 0. \qquad [17.57]$$

It is worth to remark that, if $dcocs = 0$, it is nothing else *Euler-Poincaré equation* [POI 01]. In fact, [17.57] generalizes this equation when the class of symplectic cohomology of the group is not null, especially for the important case of Galileo's group.

17.10. Application to classical mechanics

Its symmetry group is Galileo's one, $\mathbb{GAL}$. Owing to [16.26], the action [16.7] reads:

$$F = F'\,P^{-1} + F_m(C,P), \qquad L = (P\,L' + C\,F')\,P^{-1} + L_m(C,P),$$

where the symplectic cocycle coc has components F_m and L_m. Differentiating arround the identity, one has:

$$\begin{aligned}
&Z\cdot\mu = (dC, dP)\cdot(F, L),\\
&Z\cdot\mu = (-F\,dP + (DF_m(e))(dC, dP),\\
&dP\,L - L\,dP + dC\,F + (DF_m(e))(dC, dP)).
\end{aligned}$$

Applying [17.43] and using [17.45] gives $\tilde{\nabla}\mu = (\tilde{\nabla}F, \tilde{\nabla}L)$ with:

$$\begin{aligned}
\tilde{\nabla}F &= dF - F\,\Gamma' + (DF_m(e))(\Gamma'_A, \Gamma') = \nabla F + (DF_m(e))(\Gamma'_A, \Gamma'),\\
\tilde{\nabla}L &= dL + \Gamma' L - L\,\Gamma' + \Gamma'_A F + (DL_m(e))(\Gamma'_A, \Gamma')\\
&= \nabla L + \Gamma'_A F + (DL_m(e))(\Gamma'_A, \Gamma').
\end{aligned}$$

The Galilean connection Γ' is given by [3.38] and Γ'_A by [5.56] that reads with the standard notations of the differential geometry:

$$\Gamma'_A(dX) = dX - \nabla_{dX}\,C,$$

where:

$$C = \begin{pmatrix} 0 \\ x \end{pmatrix}.$$

Let us detail the calculation for the component F. Taking into account the expression [3.38] of the Galilean gravitation, the linear covariant derivative reads:

$$\nabla F = \left(-de, dp^T\right) - \left(-e, p^T\right) \begin{pmatrix} 0 & 0 \\ j(\Omega)\, dx - g\, dt & j(\Omega)\, dt \end{pmatrix},$$

$$\nabla F = \left(-de - p \cdot (\Omega \times dx - g\, dt), (dp + \Omega \times p\, dt)^T\right).$$

Moreover, the most general symplectic cocycle, given by theorem 17.4, is given by [16.27]:

$$F_m(C, P) = m \left(-\frac{1}{2} \parallel u \parallel^2, u^T\right).$$

Owing to [3.42], one has:

$$(DF_m(e))(\Gamma'_A, \Gamma') = \left(0, m\, du^T\right) = \left(0, m\, (\Omega \times dx - g\, dt)^T\right).$$

Owing to theorem 17.4 and $p = m\, v$, the equation of motion $\clubsuit$ of theorem 17.9 reads:

– balance of energy:

$$\tilde{\nabla} e = de + p \cdot (\Omega \times dx - g\, dt) = de - p \cdot g\, dt = 0,$$

– balance of linear momentum:

$$\tilde{\nabla} p = dp + \Omega \times p\, dt + \underline{m\, (\Omega \times dx - g\, dt)} = dp - m\, (g - 2\,\Omega \times v)\, dt = 0.$$

Likewise, owing to [5.60], the explicit calculation of the affine covariant derivative of the component L gives:

– balance of passage:

$$\tilde{\nabla} q = dq + \Omega \times (q - \underline{m\, x})\, dt - p\, dt = 0,$$

– balance of angular momentum:

$$\tilde{\nabla} l = dl + \Omega \times l\, dt - q \times (g\, dt - \Omega \times dx) + p \times (\Omega \times x)\, dt = 0.$$

The terms resulting from the non-null class of symplectic cohomology of Galileo's group are underlined and absolutely necessary to find the expected equation fitting the experiments of classical dynamics. After dividing by dt and some simplifications, we can recast them as:

$$\dot{e} = g \cdot p, \quad \dot{p} = m\,(g - 2\,\Omega \times v),$$
$$\dot{q} = p, \quad \dot{l} + \Omega \times l_0 = x \times \; m\,(g - 2\,\Omega \times v).$$

We recover the equations of motion [5.59], [5.63] and [5.64] in absence of other forces, excepted that the balance of mass is replaced by the one of energy.

It is possible to obtain these equations of motion by using [17.57]. Considering the representation of a Galilean transformation a by a 5×5 matrix [1.18], we have by differentiation around the identity:

$$Z_d = d\tilde{P} = \begin{pmatrix} 0 & 0 & 0 \\ d\tau_0 & 0 & 0 \\ dk & du & j(d\varpi) \end{pmatrix}.$$

Applying [18.4] gives the Lie bracket of two infinitesimal Galilean transformations:

$$[Z_d, Z_\delta] =$$
$$\begin{pmatrix} 0 & 0 & 0 \\ 0 & 0 & 0 \\ d\tau_0\, \delta u - \delta\tau_0\, du + \delta\varpi \times dk - d\varpi \times \delta k & \delta\varpi \times du - d\varpi \times \delta u & j(\delta\varpi \times d\varpi) \end{pmatrix}.$$

Owing to [18.5], we obtain the infinitesimal coadjoint representation of Galileo's group $\mu' = ad^*(Z_d)\mu$ represented by:

$$e' = -du \cdot p, \quad p' = -d\varpi \times p,$$
$$q' = -d\varpi \times q + d\tau_0 p, \quad l' = -d\varpi \times l + du \times q - dk \times p.$$

On the other hand, *dcocs* is given for Galileo's group by [17.16] with $s = e_1 = 0$. Hence, applying [17.57] and combining with the expressions [3.38] and [5.56] of the Galilean connection allows to recover once again the balance of energy, linear momentum, passage and angular momentum.

It is worth verifying that the dimension of the kernel of the symplectic form is equal to 1, then the leaves of the null foliation of ω are curves representing the trajectories. Indeed, the dimension of Galileo's group $\mathbb{GAL}$ is 10. The one of the dual $\mathfrak{gal}^*$ of its Lie algebra so is. The dimension of the $\mathbb{GAL}$-frame bundle $\mathcal{F}$ is the dimension of the space-time $\mathcal{M}$ plus the one of $\mathbb{GAL}$. Thus the dimension of the "phase space" is:

$$dim\,(\mathfrak{gal}^* \times \mathcal{F}) = 2\,dim\,(\mathbb{GAL}) + dim\,\mathcal{M} = 2 \times 10 + 4 = 24.$$

On the other hand the equation of motion ♣ of theorem 17.9 is equivalent to $2 \times dim(\mathbb{GAL}) = 20$ scalar 'canonical equations' [17.55] and [17.56]. Moreover, there are 3 extra equations to satisfy $p = m\,v$ that have to be added to the 20 canonical equations to give 23 equations. The dimension of the Kernel of ω is, as expected, equal to $24 - 23 = 1$.

17.11. Application to relativity

A noteworthy fact is that the class of symplectic cohomology of Poincaré's group is null. Then, the corresponding equation of motion is more simple. For the right action of the Poincaré's group on the momentum tensors, the transformation law is given by [16.94]:

$$F' = F\,P, \qquad L' = P^*\,L\,P + \frac{1}{2}\,(C'\,(P^*T)^* - (P^*T)\,C'^*)\,.$$

Replacing P by $P^{-1} = P^*$, we deduce the corresponding left action:

$$F' = F\,P^*, \qquad L' = P\,L\,P^* + \frac{1}{2}\,(C'\,(P\,T)^* - (P\,T)\,C'^*)\,.$$

The affine covariant derivative of the momentum tensor is given by [17.43]:

$$\tilde{\nabla} F = dF - F\,\Gamma^*, \qquad \tilde{\nabla} L = dL - \Gamma\,L - L\,\Gamma^* - \frac{1}{2}\,(\Gamma_A T^* - T\,\Gamma_A^*)\,.$$

To work with the usual connections, we apply the swap rule [18.45] and next, leaving out the primes, we obtain:

$$\tilde{\nabla} F = dF + F\,\Gamma^*, \qquad \tilde{\nabla} L = dL + \Gamma\,L + L\,\Gamma^* + \frac{1}{2}\,(\Gamma_A T^* - T\,\Gamma_A^*)\,.$$

Moreover, differentiating [16.84], we see that the elements of the Lie algebra $\mathfrak{p}_0$ of Lorentz group are anti-self-adjoint. As the connection 1-form takes its values in $\mathfrak{p}_0$, one has:

$$\Gamma^* = -\Gamma,$$

and ♣ of theorem 17.9 leads to:

$$\tilde{\nabla} F = \nabla F = 0, \qquad \tilde{\nabla} L = \nabla L + \frac{1}{2}\,(\Gamma_A T^* - T\,\Gamma_A^*) = 0, \qquad [17.58]$$

where the usual linear covariant derivatives occur:

$$\nabla F = dF - F\,\Gamma, \qquad \nabla L = dL + \Gamma\,L - L\,\Gamma.$$

For Levi-Civita connection, we have in tensor notations, because of [14.43]:

$$\nabla F_\beta = \nabla(G_{\beta\mu} T^\mu) = G_{\beta\mu} \nabla T^\mu = 0,$$

and because Gram's matrix is regular, we conclude that:

$$\nabla T = 0.$$

Hence the first relation in [17.58] is the classical equation of the motion for "test particles" in relativity while the second one is new and gives a covariant form of the balance of angular momentum.

Once again, we can determine the dimension of the kernel of the symplectic form. Likewise Galileo's group, the dimension of the "phase space" is:

$$dim\,(\mathfrak{p}^* \times \mathcal{F}) = 2\,dim\,\mathbb{P} + dim\,\mathcal{M} = 2 \times 10 + 4 = 24.$$

Moreover, there are 4 extra equations [16.90] to satisfy but under the condition [16.92], then in fact only $4 - 1 = 3$ independant equations that have to be added to the 20 canonical equations to give 23 equations and the dimension of the Kernel of ω is, as expected, equal to $24 - 23 = 1$.

18

Advanced Mathematical Tools

18.1. Vector fields

For the sake of simplicity, all the objects considered in this chapter are assumed to be of class C^∞ and are qualified of *smooth*.

A vector field $\overrightarrow{\boldsymbol{V}}$ can be seen as a derivation operator on the smooth scalar fields f according to:

$$\overrightarrow{\boldsymbol{V}}(f) = df\, \overrightarrow{\boldsymbol{V}}.$$

The *flow* generated by a vector field $\overrightarrow{\boldsymbol{V}}$ is the curve $t \mapsto \boldsymbol{X}' = \varphi_t(\boldsymbol{X})$ solution of the ordinary differential equation:

$$\frac{d}{dt}(\varphi_t(\boldsymbol{X})) = \overrightarrow{\boldsymbol{V}}(\varphi_t(\boldsymbol{X})),$$

with the initial condition $\varphi_0(\boldsymbol{X}) = \boldsymbol{X}$.

The *Lie derivative* of the tensor field $\boldsymbol{T}$ with respect to the vector field $\overrightarrow{\delta \boldsymbol{X}}$ is the vector field:

$$\mathcal{L}_{\overrightarrow{\delta \boldsymbol{X}}} \boldsymbol{T} = \lim_{t \to 0} \frac{1}{t} \left(\varphi_t^* \boldsymbol{T} - \boldsymbol{T}\right),$$

where φ_t is the flow generated by $\overrightarrow{\delta \boldsymbol{X}}$.

The *Lie bracket* of two vector fields $\overrightarrow{\delta \boldsymbol{X}}$ and $\overrightarrow{d\boldsymbol{X}}$ is the vector field $[\overrightarrow{\delta \boldsymbol{X}}, \overrightarrow{d\boldsymbol{X}}]$ such that for any smooth scalar field f:

$$\overrightarrow{[\delta, d]\boldsymbol{X}} = [\overrightarrow{\delta \boldsymbol{X}}, \overrightarrow{d\boldsymbol{X}}] = \overrightarrow{\delta \boldsymbol{X}}(\overrightarrow{d\boldsymbol{X}}(f)) - \overrightarrow{d\boldsymbol{X}}(\overrightarrow{\delta \boldsymbol{X}}(f)).$$

The Lie bracket is skew-symmetric and verifies *Jacobi's identity*:

$$[\Delta, [\delta, d]] + [\delta, [d, \Delta]] + [d, [\Delta, \delta]] = 0.$$

Moreover, we have:

$$\mathcal{L}_{\overrightarrow{\delta X}} \overrightarrow{dX} = [\overrightarrow{\delta X}, \overrightarrow{dX}].$$

18.2. Lie group

A *Lie group* is a differential manifold G which is also a group, such that the group multiplication $(a, a') \mapsto a\,a'$ and inversion $a \mapsto a^{-1}$ are smooth maps.

The tangent space T_eG to a Lie group G at the identity is called its *Lie algebra* and is denoted by $\mathfrak{g}$. It is a linear space of finite dimension.

Any Lie group G left linearly acts on its Lie algebra $\mathfrak{g}$ by the *adjoint representation*:

$$Ad\,(a) : \mathfrak{g} \to \mathfrak{g} : Z \mapsto Z' = aZa^{-1}. \tag{18.1}$$

G left linearly acts on the dual $\mathfrak{g}^*$ of $\mathfrak{g}$ by the *coadjoint representation* Ad^* such that:

$$\forall Z \in \mathfrak{g}, \forall \mu \in \mathfrak{g}^*, \qquad (Ad^*(a)\mu)\,(Z) = \mu\left(Ad\left(a^{-1}\right) Z\right). \tag{18.2}$$

For all $Y, Z \in \mathfrak{g}$, we define the *Lie bracket* $[Y, Z]$ as the *infinitesimal adjoint representation*:

$$ad(Y)\,Z = D_e(Ad(\bullet)Y)\,Z, \tag{18.3}$$

that is an element of $\mathfrak{g}$. The Lie bracket is a bilinear skew-symmetric map verifying *Jacobi's identity*:

$$[Z, [Z', Z'']] + [Z', [Z'', Z]] + [Z'', [Z, Z']] = 0.$$

For a matrix group G, [18.3] gives:

$$ad(Y)\,Z = d(aYa^{-1}),$$

where we differentiate at $Y = C^{te}$, $a = e$ in the direction $da = Z$, which gives:

$$ad(Y)\,Z = [Y, Z],$$

with:

$$[Y, Z] = ZY - YZ. \qquad [18.4]$$

The *infinitesimal coadjoint representation* ad^* is defined by differentiating [18.2] at constant μ and Z, $a = e$ in the direction $da = Y$:

$$(ad^*(Y)\mu)(Z) = -\mu\,(Ad\,(Y)\,Z) = -\mu[Y, Z]. \qquad [18.5]$$

As the formula is true for all Z, it reads in short:

$$ad^*(Y)\mu = -\mu \circ ad(Y).$$

Let G be a Lie group smoothly acting on a manifold $\mathcal{M}$. The *orbit space* $\mathcal{M} \,/\, G$ is a manifold for which the canonical map $\pi : \mathcal{M} \to \mathcal{M} \,/\, G$ is a submersion if and only if the set of the couples $(\boldsymbol{X}, \boldsymbol{Y})$ of points belonging to the same orbit is a closed submanifold of the product manifold $\mathcal{M} \times \mathcal{M}$.

A Lie group (and any of its subgroups) naturally acts onto itself by translation $(a \cdot a' = a\,a')$. For any Lie subgroup H, the orbit manifold $G \,/\, H$ exists, G smoothly acts onto $G \,/\, H$ and:

$$dim(G \,/\, H) = dim(G) - dim(H).$$

If a point $\boldsymbol{X} \in \mathcal{M}$ is such that its orbit $orb(\boldsymbol{X})$ is a locally closed subset of $\mathcal{M}$, then $orb(\boldsymbol{X})$ is a submanifold of $\mathcal{M}$ for which the canonical map $f_{\boldsymbol{X}} : G \,/\, iso(\boldsymbol{X}) \to orb(\boldsymbol{X})$ is a diffeomorphism and:

$$dim(orb(\boldsymbol{X})) = dim(G) - dim(iso(\boldsymbol{X})). \qquad [18.6]$$

An *invariant* function (or *invariant*) by the group G is a constant function on the orbits of $\mathcal{M}$. If the functions $f_1(\boldsymbol{X}), \ldots, f_p(\boldsymbol{X})$ are invariant, for any function $h(y_1, \ldots, y_p)$, the function $h'(\boldsymbol{X}) = h(f_1(\boldsymbol{X}), \ldots, f_p(\boldsymbol{X}))$ is invariant. We say that $f_1, \ldots, f_p$ generate h' and that h' is *functionally dependent* on $f_1, \ldots, f_p$. A *functional basis* of the orbit is a minimal set of independent invariant functions $f_1, \ldots, f_p$ generating all the invariant functions and the maximum number of independent invariants is p.

If the dimension of the orbit of $\boldsymbol{X}$ is m, it is locally defined in $\mathcal{M}$ of dimension n by $(n - m)$ independent equations $f_1(\boldsymbol{X}) = 0, \ldots, f_{n-m}(\boldsymbol{X}) = 0$, then the set $\{f_1, \ldots, f_{n-m}\}$ is a functional basis of the orbit and the maximum number of independent invariants is:

$$n_{inv} = dim(\mathcal{M}) - dim(orb(\boldsymbol{X})). \qquad [18.7]$$

Generally, the dimension of the orbits is varying when $\boldsymbol{X}$ is running in $\mathcal{M}$ and the orbits of maximum dimension are called *generic orbits* while the other ones are called *singular orbits*.

As example, let us consider the group $\mathbb{SO}(3)$ acting on $\mathbb{R}^3$ by $(R, x) \mapsto R \cdot x = Rx$.

– *Generic orbits.* If $x \neq 0$, the orbit is the sphere of radius $\| x \|$. The isotropy group of x is composed of the rotations of axis x. The rule [18.6] gives $dim(orb(x)) = 3 - 1 = 2$ and allows us to recover the fact that the orbit is a surface. According to [18.7], the functional basis contains $3 - 2 = 1$ invariant, for instance $\| x \|$. All the other invariant functions have the form $h\left(\| x \|\right)$.

– *Singular orbits.* If $x = 0$, the orbit is $\{0\}$. The isotropy group is composed of all the rotations. The rule [18.6] gives $dim(orb(x)) = 3 - 3 = 0$ and allows us to recover the fact that the orbit is a point. According to [18.7], the functional basis contains $3 - 0 = 3$ invariant functions. The invariant functions of the basis are, for instance, $x^1 = x^2 = x^3 = 0$.

18.3. Foliation

Let $\mathcal{M}$ be a manifold and $\boldsymbol{X} \mapsto \mathcal{T}(\boldsymbol{X}) \subset T_{\boldsymbol{X}}\mathcal{M}$ be a field of vector spaces. An *integral manifold* of this field is a manifold $\mathcal{N}$ embedded into $\mathcal{M}$ such that at any $\boldsymbol{X} \in \mathcal{N}$ the tangent space to $\mathcal{N}$ is $\mathcal{T}(\boldsymbol{X})$. The field $\boldsymbol{X} \mapsto \mathcal{T}(\boldsymbol{X})$ is a *foliation* of $\mathcal{M}$ if every $\boldsymbol{X} \in \mathcal{M}$ belongs to an integrable manifold. A field $\boldsymbol{X} \mapsto \mathcal{T}(\boldsymbol{X})$ is a foliation if and only if for any tangent vector fields $\boldsymbol{X} \mapsto \overrightarrow{d\boldsymbol{X}}$ and $\boldsymbol{X} \mapsto \overrightarrow{\delta \boldsymbol{X}}$:

$$\overrightarrow{d\boldsymbol{X}}, \overrightarrow{\delta \boldsymbol{X}} \in \mathcal{T}(\boldsymbol{X}) \qquad \Leftrightarrow \qquad \left[\overrightarrow{d\boldsymbol{X}}, \overrightarrow{\delta \boldsymbol{X}}\right] \in \mathcal{T}(\boldsymbol{X}).$$

Then, $\mathcal{M}$ has a partition in connected integral manifolds called *leaves*.

18.4. Exterior algebra

Let $(i_1, \ldots, i_p)$ and $(j_1, \ldots, j_p)$ be two p-uples of integers $1 \leq i_1, \ldots, i_p, j_1, \ldots, j_p \leq n$. The *Kronecker symbol* is defined by:

$$\varepsilon^{i_1, \ldots, i_p}_{j_1, \ldots, j_p} = \begin{cases} 1 \text{ if } (i_1, \ldots, i_p) \text{ is an even permutation of } (j_1, \ldots, j_p) \\ -1 \text{ if } (i_1, \ldots, i_p) \text{ is an odd permutation of } (j_1, \ldots, j_p) \\ 0 \qquad \text{otherwise} \end{cases}.$$

Let us now consider tensors over the tangent space $\mathcal{T} = T_{\boldsymbol{X}}\mathcal{M}$ to a manifold of dimension n. The *anti-symmetrization* of a p-covariant tensor $\boldsymbol{T}$ is the tensor $\boldsymbol{T}^a = Alt(\boldsymbol{T})$ such that:

$$\boldsymbol{T}^a(\overrightarrow{d_1\boldsymbol{X}},\ldots,\overrightarrow{d_p\boldsymbol{X}}) = \varepsilon_{1,\ldots,p}^{i_1,\ldots,i_p}\,\boldsymbol{T}(\overrightarrow{d_{i_1}\boldsymbol{X}},\ldots,\overrightarrow{d_{i_p}\boldsymbol{X}}),$$

where the sum over the multi-index $(i_1,\ldots,i_p)$ is taken over the group of permutations of $(1,\ldots,p)$.

A *p-form* (or *exterior form* of *degree* p) is an anti-symmetric p-covariant tensor of which the sign changes by swapping two arguments. It is preserved by Alt. The degree of ω is denoted by $deg(\omega)$. The 0-forms are scalars and the 1-forms are linear forms. The set of p-forms is a linear space denoted by $\wedge^p\mathcal{T}^*$. If $(\boldsymbol{e}^i)$ is a cobasis, the set of p-forms:

$$\boldsymbol{e}^{i_1}\wedge\ldots\wedge\boldsymbol{e}^{i_p} = Alt(\boldsymbol{e}^{i_1}\otimes\ldots\otimes\boldsymbol{e}^{i_p}),$$

is a basis of $\wedge^p\mathcal{T}^*$, then:

$$dim(\wedge^p\mathcal{T}^*) = \binom{n}{p} = \frac{n!}{p!\,(n-p)!}.$$

The *exterior product* of a p-form ω and a q-form θ is the $(p+q)$-form:

$$\omega\wedge\theta = Alt(\omega\otimes\theta).$$

It is:

– associative: $(\omega\wedge\theta)\wedge\varphi = \omega\wedge(\theta\wedge\varphi)$;

– distributive: $\omega\wedge(\theta+\varphi) = \omega\wedge\theta+\omega\wedge\varphi$;

– anticommutative: $\omega\wedge\theta = (-1)^{deg(\omega)\,deg(\theta)}\,\theta\wedge\omega$.

Equipped with the sum and operation $\wedge$, the set $\wedge\mathcal{T}^* = \wedge^0\mathcal{T}^*\oplus\ldots\oplus\wedge^n\mathcal{T}^*$ is an algebra called the *exterior algebra*.

The *interior product* of a vector $\overrightarrow{d\boldsymbol{X}}$ and a p-form ω is the $(p-1)$-form $\iota(\overrightarrow{d\boldsymbol{X}})\omega$ defined by:

$$(\iota(\overrightarrow{d\boldsymbol{X}})\omega)\,(\overrightarrow{d_1\boldsymbol{X}},\ldots,\overrightarrow{d_{p-1}\boldsymbol{X}}) = \omega\,(\overrightarrow{d\boldsymbol{X}},\overrightarrow{d_1\boldsymbol{X}},\ldots,\overrightarrow{d_{p-1}\boldsymbol{X}}),$$

with the convention that if $p=0$, $\iota(\overrightarrow{d\boldsymbol{X}})\omega = 0$. If $p=1$, $\iota(\overrightarrow{d\boldsymbol{X}})\omega$ is the value of the linear form ω for the vector $\overrightarrow{d\boldsymbol{X}}$. The properties of the interior product are:

- $\iota(\overrightarrow{d\boldsymbol{X}})(\omega + \theta) = \iota(\overrightarrow{d\boldsymbol{X}})\omega + \iota(\overrightarrow{d\boldsymbol{X}})\theta$;
- $\iota^2 = 0$;
- $\iota(\overrightarrow{d\boldsymbol{X}})(\omega \wedge \theta) = (\iota(\overrightarrow{d\boldsymbol{X}})\omega) \wedge \theta + (-1)^{deg(\omega)}\, \omega \wedge (\iota(\overrightarrow{d\boldsymbol{X}})\theta)$.

The *exterior derivative* of a smooth field of p-forms $\boldsymbol{X} \mapsto \omega(\boldsymbol{X})$ is a smooth field of $(p+1)$-forms $\boldsymbol{X} \mapsto d\omega(\boldsymbol{X})$ such that:

$$d\omega\,(\overrightarrow{d_0\boldsymbol{X}}, \overrightarrow{d_1\boldsymbol{X}}, \ldots, \overrightarrow{d_p\boldsymbol{X}}) = Alt(((D\omega)\,\overrightarrow{d_0\boldsymbol{X}})\,(\overrightarrow{d_1\boldsymbol{X}}, \ldots, \overrightarrow{d_p\boldsymbol{X}})).$$

If f is a field of 0-forms (i.e. a scalar field), df is its usual derivative. The properties of the exterior derivative are:

- $d(\omega + \theta) = d\omega + d\theta$;
- $d^2 = 0$;
- $d(\omega \wedge \theta) = (d\omega) \wedge \theta + (-1)^{deg(\omega)}\, \omega \wedge (d\theta)$.

If $\boldsymbol{F} : \mathcal{M} \to \mathcal{M}$ is a smooth map, the properties of the pull-back of exterior forms are:

- $\boldsymbol{F}^*(\omega \wedge \theta) = (\boldsymbol{F}^*\omega) \wedge (\boldsymbol{F}^*\theta)$;
- $\boldsymbol{F}^*(d\omega) = d(\boldsymbol{F}^*\omega)$.

The fundamental relation between the interior product, the exterior derivative and the Lie derivative is given by *Cartan's formula*:

$$\mathcal{L}_{\overrightarrow{\delta\boldsymbol{X}}}\,\omega = d\,(\iota(\overrightarrow{\delta\boldsymbol{X}})\omega) + \iota(\overrightarrow{\delta\boldsymbol{X}})d\omega.$$

Many properties of exterior forms can be extended to vector valued forms. Let $\mathcal{R}$ be a linear space of finite dimension. For instance, the exterior product of an $\mathcal{R}^*$-valued 1-form ω and an $\mathcal{R}$-valued 1-form θ is the scalar valued 2-form:

$$\omega \wedge \theta = Alt(\omega \otimes \theta).$$

with:

$$\forall \overrightarrow{d_1\boldsymbol{X}}, \overrightarrow{d_2\boldsymbol{X}} \in \mathcal{T}, \qquad (\omega \otimes \theta)(\overrightarrow{d_1\boldsymbol{X}}, \overrightarrow{d_2\boldsymbol{X}}) = (\omega(\overrightarrow{d_1\boldsymbol{X}}))\,\theta(\overrightarrow{d_2\boldsymbol{X}}).$$

18.5. Curvature tensor

Linear spaces are intuitively perceived as flat. In the framework of differential geometry, this fact can be expressed by noting that a linear space is naturally equipped with a covariant differential Γ vanishing in the coordinate systems X associated with basis. However, it is worth remarking that in any other coordinate system X' deduced from X by a non-affine coordinate change ("curvilinear coordinates"), the transformation law [14.35] of the covariant differential shows that Γ' is not null.

A manifold is said to be *flat* if there exist coordinate systems covering the manifold and in which the covariant differential vanishes. Conversely, let $\mathcal{M}$ be a manifold of dimension n equipped with a (not necessarily null) covariant differential Γ given in some coordinate systems X. Is this manifold flat? If the answer is yes, there exist new coordinate systems X' for which $\Gamma' = 0$. Owing to [14.35], the Jacobian matrix $P = \partial X / \partial X'$ is solution of the differential system:

$$dP = -\Gamma(dX)\, P. \qquad [18.8]$$

As the covariant differential is symmetric, there are $n^2(n+1)/2$ independent scalar equations in n^2 unknowns P^α_β. In general, this system is overdetermined and has no solution, unless the compatibility condition:

$$\delta\,(dP) = d\,(\delta P),$$

is fulfilled, that is taking into account [18.8] and differentiating:

$$\delta\,(dP) - d\,(\delta P) = (d\Gamma(\delta X) - \delta\Gamma(dX))\, P + \Gamma(\delta X)\, dP - \Gamma(dX)\, \delta P = 0.$$

Using [18.8] once again, it holds:

$$\delta\,(dP) - d\,(\delta P) = R\,(dX, \delta X) P = 0,$$

with:

$$R\,(dX, \delta X) = d\Gamma(\delta X) - \delta\Gamma(dX) + \Gamma(dX)\,\Gamma(\delta X) - \Gamma(\delta X)\,\Gamma(dX). \qquad [18.9]$$

The matrix P being regular, the manifold is flat if the compatibility condition:

$$R\,(dX, \delta X) = 0.$$

is satisfied. Conversely, if this quantity does not vanish, the manifold is *curved* and [18.9] characterizes the manifold curvature. The map R is bilinear, skew-symmetric and is valued in the linear space $\mathbb{M}_{nn}$ of the $n \times n$ matrices. It represents a vector-valued linear tensor $\mathbf{R}$, called *Riemann–Christoffel curvature tensor*. By the choice of a basis of the tangent space, we have the decomposition:

$$R^{\delta}_{\gamma}(dX, \delta X) = R^{\delta}_{\alpha\beta\gamma} dX^{\alpha} \delta X^{\beta},$$

where the components of the curvature tensor read:

$$R^{\delta}_{\alpha\beta\gamma} = \Gamma^{\delta}_{\alpha\mu}\Gamma^{\mu}_{\beta\gamma} - \Gamma^{\delta}_{\beta\mu}\Gamma^{\mu}_{\alpha\gamma} + \frac{\partial \Gamma^{\delta}_{\beta\gamma}}{\partial X^{\alpha}} - \frac{\partial \Gamma^{\delta}_{\alpha\gamma}}{X^{\beta}}, \qquad [18.10]$$

and are skew-symmetric in α and β.

Bibliography

[ABR 78] ABRAHAM R., MARDEN J.E., *Foundation of Mechanics*, 2nd ed., Addison-Wesley Pub. Co., Reading, MA, 1978.

[BAR 54] BARGMANN V., "On unitary representation of continuous groups", *Annals of Mathematics*, vol. 59, pp. 1–46, 1954.

[BOU 70] BOURBAKI N., *Éléments de Mathématique*, Hermann, Paris, 1970.

[CAR 23] CARTAN É., "Sur les variétés à connexion affine et la théorie de la relativité généralisée (première partie)", *Annales de l'École Normale Supérieure*, vol. 40, pp. 325–412, 1923.

[CAR 24] CARTAN É., "Sur les variétés à connexion affine et la théorie de la relativité généralisée (suite)", *Annales de l'École Normale Supérieure*, vol. 41, pp. 1–25, 1924.

[CAU 23] CAUCHY A.-L., "Recherches sur l'équilibre et le mouvement intérieur des corps solides ou fluides, élastiques ou non élastiques", *Bulletin des Sciences par la Société Philomathique*, available at: gallica.bnf.fr, pp. 9–13, 1823.

[CAU 27] CAUCHY A.-L., "De la pression ou tension dans un corps solide", *Exercices de mathématiques*, vol. 2, pp. 42–56, 1827.

[CHE 06] CHEVALLIER D., *Introduction à la théorie des groupes de Lie réels*, Coll. Mathematiques à l'Université, Ellipses, Paris, 2006.

[COS 07] COSSERAT E., COSSERAT F., "Sur la mécanique générale", *Comptes Rendus de l'Académie des Sciences*, vol. 145, pp. 1139–1142, 1907.

[CUR 85] CURTIS W.D., MILLER R., *Differential Manifolds and Theoretical Physics*, Academy Press, New York, 1985.

[CUS 97] CUSHMAN R.H., BATES L.M., *Global Aspects of Classical Integrable Systems*, Birkhäuser Verlag, Basel, 1997.

[DES 03] DE SAXCÉ G., VALLÉE C., "Affine tensors in shell theory", *Journal of Theoretical and Applied Mechanics*, vol. 41, no. 3, pp. 593–621, 2003.

[DES 11] DE SAXCÉ G., VALLÉE C., "Affine tensors in mechanics of freely falling particles and rigid bodies", *Mathematics and Mechanics of Solid Journal*, vol. 17, no. 4, pp. 413–430, 2011.

[DUV 80] DUVAL C., "On the prequantum description of spinning particles in an external gauge field", *Differential geometrical methods in mathematical physics: Proceedings of the conferences held at Aix-en-Provence, September 3–7, 1979 and Salamanca, September 10–14, 1979*, GARCÍA P.L., PÉREZ-RENDÓN A., SOURIAU J.M. (eds), Springer-Verlag, Berlin, 1980.

[DUV 82a] DUVAL C., "Einstein-Yang-Mills-Higgs general covariance and mass formulas", *Proc. of the Meeting 'Geometry and Physics'*, pp. 239-251, Florence, 12–15 October 1982.

[DUV 82b] DUVAL C., HORVÁTHY P., "Particles with internal structure: the geometry of classical motions and conservation laws", *Annals of Physics*, vol. 142, no. 1, pp. 10–32, 1982.

[DUV 85] DUVAL C., BURDET G., KÜNTZLE H.P., "Perrin, Bargmann structures and Newton-Cartan theory", *Physical Review D*, vol. 31, pp. 1841–1853, 1985.

[DUV 91] DUVAL C., GIBBONS G., HORVÁTHY P., "Celestial mechanics, conformal structures and gravitational waves", *Physical Review D*, vol. 43, pp. 3907–3922, 1991.

[GRA 04] GRABOWSKA K., GRABOWSKI J., URBAŃSKI P., "AV-differential geometry: Poisson and Jacobi structures", *Journal of Geometry and Physics*, vol. 52, pp. 398–446, 2004.

[GUI 84a] GUICHARDET A., "On rotation and vibration motions of molecules" *Annales de l'Institut Henri Poincaré*, vol. 40, pp. 329–342, 1984.

[GUI 84b] GUILLEMIN V., STERNBERG S., *Symplectic Techniques in Physics*, University Press, Cambridge, 1984.

[HAL 53] HALL M., *The Theory of Groups*, The Macmillan Co., New York, 1953.

[KLE 72] KLEIN F., *Vergleichende Betrachtungen über neuere geometrische Forschungen*, Erlangen, 1872 (Reprinted with additional notes in Mathematische Annalen 43, pp. 460–497, 1893).

[KOB 63] KOBAYASHI S., NOMIZU K., *Foundations of Differential Geometry*, John Wiley & Sons, New York, vol. 1, 1963.

[KOB 72] KOBAYASHI S., *Transformation Groups in Differential Geometry*, Springer, 1972.

[KUM 81] KUMMER M., "On the construction of the reduced phase space of a Hamiltonian system", *Indiana University Mathematics Journal*, vol. 30, pp. 281–291, 1981.

[KUN 72] KÜNZLE H.P., "Galilei and Lorentz structures on space-time: comparison of the corresponding geometry and physics", *Annales de l'Institut Henri Poincaré, section A*, vol. 17, no. 4, pp. 337–362, 1972.

[LIB 87] LIBERMANN P., MARLE C.-M., *Symplectic Geometry and Analytical Mechanics*, D. Reidel, Dordrecht, 1987.

[MAR 92] MARSDEN J.E., *Lectures on Mechanics*, Lecture Note Series vol. 174, Cambridge University Press, 1992.

[MIS 73] MISNER W.M., THORNE K.S., WHEELER J.A., *Gravitation*, W. H. Freeman and co., San Francisco, 1973.

[MON 90] MONTGOMERY R., "Isoholonomic problems and some applications", *Communications in Mathematical Physics*, vol. 128, pp. 565–592, 1990.

[NOL 73] NOLL W., "Lectures on the foundations of continuum mechanics and thermodynamics", *Archive for Rational Mechanics and Analysis*, vol. 52, pp. 62–92, 1973.

[PER 53] PÉRÈS J., *Mécanique générale*, Masson, Paris, 1953.

[POI 01] POINCARÉ H., "Sur une forme nouvelle des équations de la Mécanique", *Comptes Rendus de l'Académie des Sciences*, vol. 132, no. 7, pp. 369–371, 1901.

[RIC 01] RICCI-CURBASTO G., LEVI-CIVITA T., "Méthodes de calcul différentiel absolu et leurs applications", *Mathematische Annalen*, vol. 54, pp. 125–137, 1901.

[RYC 84] RYCHLEWSKI J., "On Hooke's law", *Prikladnaya Matematika i Mekhanika*, vol. 48, no. 3, pp. 303–314, 1984.

[SAL 00] SALENÇON J., *Mécanique des milieux continus*, Éditions de l'École polytechnique, Palaiseau, 2000.

[SHA 89] SHAPERE A., WILCZECK F., "Geometry of self-propulsion at low Reynolds number", *Journal of Fluid Mechanics*, vol. 198, pp. 557–585, 1989.

[SIM 68] SIMMS D.J., *Lie Groups and Quantum Mechanics*, Lectures Notes in Mathematics vol. 52, Springer-Verlag, Berlin, 1968.

[SIM 91] SIMO J.C., LEWIS D.R., MARSDEN J.E., "Stability of relative equilibria I: the reduced energy method", *Archive for Rational Mechanics and Analysis*, vol. 115, pp. 15–59, 1991.

[SMA 70a] SMALE S., "Topology and mechanics", *Inventiones mathematicae*, vol. 10, pp. 305–331, 1970.

[SMA 70b] SMALE S., "Topology and mechanics", *Inventiones mathematicae*, vol. 11, pp. 45–64, 1970.

[SOU 70] SOURIAU J.-M., *Structure des systèmes dynamiques*, Dunod [out of print], Paris, 1970.

[SOU 92] SOURIAU J.-M., *Calcul linéaire*, Presses Universitaires de France [1964, out of print] and Jacques Gabay [republishing], Paris, 1992.

[SOU 97a] SOURIAU J.-M., "Milieux continus de dimension 1, 2 ou 3: statique et dynamique", *Actes du 13ème Congrès Français de Mécanique, Poitiers-Futuroscope*, pp. 41–53, 1997.

[SOU 97b] SOURIAU J.-M., *Structure of Dynamical Systems, a Symplectic View of Physics*, Birkhäuser Verlag, New York, 1997.

[SOU 07] SOURIAU J.-M., Grammaire de la Nature, available at: http://www.jmsouriau.com/Grammaire_de_la_nature.htm, 2007.

[SOU 08] SOURIAU J.-M., *Géométrie et relativité*, Hermann [1964, out of print] and Jacques Gabay [republishing], Paris, 2008.

[STE 78a] STERNBERG S., "On the influence of field theories in our physical conception of geometry", *Lecture Notes in Mathematics*, Springer-Verlag, vol. 676, pp. 399–407, New York, 1978.

[STE 78b] STERNBERG S., UNGAR T., "Classical and prequantized mechanics without Lagrangians and Hamiltonians", *Hadronic Journal*, vol. 1, pp. 33-76, 1978.

[THO 56] THOMSON W. (LORD KELVIN), "Elements of a mathematical theory of elasticity", *Philosophical Transactions of the Royal Society*, vol. 156, pp. 481–498, 1856.

[THO 90] THOMSON W. (LORD KELVIN), *Mathematical and Physical Papers. Elasticity, Heat, Electromagnetism*, Cambridge University Press, Cambridge, vol. 3, 1890.

[TOU 58] TOUPIN R., "World invariant kinematics", *Archive for Rational Mechanics and Analysis*, vol. 1, pp. 181–211, 1958.

[TRU 52] TRUESDELL C., "The mechanical foundation of elasticity and fluid dynamics", *Archive for Rational Mechanics and Analysis*, vol. 1, pp. 125–171, 1952.

[TRU 60] TRUESDELL C., TOUPIN R., *The Classical Field Theories, Encyclopedia of Physics, S. Flügge, Vol III/1, Principles of Classical Mechanics and Field Theory*, Springer-Verlag, Berlin, 1960.

[TUL 88] TULCZYJEW W., URBAŃSKI P., GRABOWSKI J., "A pseudocategory of principal bundles", *Atti della Reale Accademia delle Scienze di Torino*, vol. 122, pp. 66–72, 1988.

[VAL 78] VALLÉE C., Lois de comportement des milieux continus dissipatifs compatibles avec la physique relativiste, PhD Thesis, University of Poitiers, 1978.

[VAL 81] VALLÉE C., "Relativistic thermodynamics of continua", *International Journal of Engineering Science*, vol. 19, no. 5, pp. 589–601, 1981.

[VAL 95] VALID R., *The Nonlinear Theory of Shells through Variational Principles, from Elementary Algebra to Differential Geometry*, Wiley, Chichester, 1995.

[VAL 05] VALLÉE C., LERINTIU C., "Convex analysis and entropy calculation in statistical mechanics", *Proceedings of A. Razmadze Mathematical Institute*, vol. 137, pp. 111–129, 2005.

[WAL 84] WALPOLE L.J., "Fourth-rank tensors of the thirty-two crystal classes: multiplication tables", *Proceedings of the Royal Society of London A*, vol. 391, pp. 149–179, 1984.

Index

G, H, K

L, M

N, P, R

S, T, V

Other titles from

iSTE

in

Mechanical Engineering and Solid Mechanics

2015

KARLIČIĆ Danilo, MURMU Tony, ADHIKARI Sondipon, MCCARTHY Michael
Non-local Structural Mechanics

SAB Karam, LEBÉE Arthur
Homogenization of Heterogeneous Thin and Thick Plates

2014

ATANACKOVIC M. Teodor, PILIPOVIC Stevan, STANKOVIC Bogoljub, ZORICA Dusan
Fractional Calculus with Applications in Mechanics: Vibrations and Diffusion Processes

ATANACKOVIC M. Teodor, PILIPOVIC Stevan, STANKOVIC Bogoljub, ZORICA Dusan
Fractional Calculus with Applications in Mechanics: Wave Propagation, Impact and Variational Principles

CIBLAC Thierry, MOREL Jean-Claude
Sustainable Masonry: Stability and Behavior of Structures

ILANKO Sinniah, MONTERRUBIO Luis E., MOCHIDA Yusuke
The Rayleigh–Ritz Method for Structural Analysis

LALANNE Christian
Mechanical Vibration and Shock Analysis – 5-volume series – 3rd edition
Sinusoidal Vibration – volume 1
Mechanical Shock – volume 2
Random Vibration – volume 3
Fatigue Damage – volume 4
Specification Development – volume 5

LEMAIRE Maurice
Uncertainty and Mechanics

2013

ADHIKARI Sondipon
Structural Dynamic Analysis with Generalized Damping Models: Identification

ADHIKARI Sondipon
Structural Dynamic Analysis with Generalized Damping Models: Analysis

BAILLY Patrice
Materials and Structures under Shock and Impact

BASTIEN Jérôme, BERNARDIN Frédéric, LAMARQUE Claude-Henri
Non-smooth Deterministic or Stochastic Discrete Dynamical Systems: Applications to Models with Friction or Impact

EL HAMI Abdelkhalak, BOUCHAIB Radi
Uncertainty and Optimization in Structural Mechanics

KIRILLOV Oleg N., PELINOVSKY Dmitry E.
Nonlinear Physical Systems: Spectral Analysis, Stability and Bifurcations

LUONGO Angelo, ZULLI Daniele
Mathematical Models of Beams and Cables

SALENÇON Jean
Yield Design

2012

DAVIM J. Paulo
Mechanical Engineering Education

DUPEUX Michel, BRACCINI Muriel
Mechanics of Solid Interfaces

ELISHAKOFF Isaac *et al.*
Carbon Nanotubes and Nanosensors: Vibration, Buckling and Ballistic Impact

GRÉDIAC Michel, HILD François
Full-Field Measurements and Identification in Solid Mechanics

GROUS Ammar
Fracture Mechanics – 3-volume series
Analysis of Reliability and Quality Control – volume 1
Applied Reliability – volume 2
Applied Quality Control – volume 3

RECHO Naman
Fracture Mechanics and Crack Growth

2011

KRYSINSKI Tomasz, MALBURET François
Mechanical Instability

SOUSTELLE Michel
An Introduction to Chemical Kinetics

2010

BREITKOPF Piotr, FILOMENO COELHO Rajan
Multidisciplinary Design Optimization in Computational Mechanics

DAVIM J. Paulo
Biotribolgy

PAULTRE Patrick
Dynamics of Structures

SOUSTELLE Michel
Handbook of Heterogenous Kinetics

2009

BERLIOZ Alain, TROMPETTE Philippe
Solid Mechanics using the Finite Element Method

LEMAIRE Maurice
Structural Reliability

2007

GIRARD Alain, ROY Nicolas
Structural Dynamics in Industry

GUINEBRETIÈRE René
X-ray Diffraction by Polycrystalline Materials

KRYSINSKI Tomasz, MALBURET François
Mechanical Vibrations

SIH George C. *et al.*
Particle and Continuum Aspects of Mesomechanics

Printed and bound by CPI Group (UK) Ltd, Croydon, CR0 4YY
13/09/2026
14961079-0005